LABORATORY AUDITING
FOR QUALITY AND
REGULATORY COMPLIANCE

DRUGS AND THE PHARMACEUTICAL SCIENCES

Executive Editor

James Swarbrick

PharmaceuTech, Inc.
Pinehurst, North Carolina

Advisory Board

DRUGS AND THE PHARMACEUTICAL SCIENCES
A Series of Textbooks and Monographs

LABORATORY AUDITING FOR QUALITY AND REGULATORY COMPLIANCE

Donald C. Singer
GlaxoSmithKline, Collegeville, Pennsylvania, U.S.A.

Raluca-Ioana Stefan
University of Pretoria, Pretoria, South Africa

Jacobus F. van Staden
University of Pretoria, Pretoria, South Africa

CRC Press
Taylor & Francis Group
Boca Raton London New York

CRC Press is an imprint of the
Taylor & Francis Group, an **informa** business

CRC Press
Taylor & Francis Group
6000 Broken Sound Parkway NW, Suite 300
Boca Raton, FL 33487-2742

First issued in paperback 2019

© 2010 by Taylor & Francis Group, LLC
CRC Press is an imprint of Taylor & Francis Group, an Informa business

No claim to original U.S. Government works

ISBN-13: 978-1-57444-570-1 (hbk)
ISBN-13: 978-0-367-39246-8 (pbk)

A CIP record for this book is available from the British Library.

Library of Congress Cataloging-in-Publication Data available on application

Visit the Taylor & Francis Web site at
http://www.taylorandfrancis.com

and the CRC Press Web site at
http://www.crcpress.com

PREFACE

There is an inherent similarity between planning and performing an audit of a laboratory and planning and operating a laboratory. Each area of operation that builds integrity and ensures accuracy and consistency in results is an area that is scrutinized by an effective audit. The laboratory is a facility requiring sufficient services (electrical, water, air) to keep it operating continuously and efficiently. The services that support the facility, its equipment, its operations, and its people should be monitored. An audit can determine if quality is built into the services. Laboratory space, the designation of areas for particular operations, flow of materials, and flow of people serve to design an appropriate plan to meet the needs of customers and employees for testing capability and laboratory safety. An audit can identify if the plans do indeed meet the needs of customers. Training and background of scientists who perform the sample handling, testing, and reporting of results must fit the expertise required to successfully carry out the laboratory operations. And an audit can identify strengths and weaknesses of staff expertise, hiring, and retention of scientists.

This book is focused on analytical (biology, biochemistry, chemistry, and microbiology) laboratories that support regulated industry. Our definition of regulated industry is consumer product industry where Good Manufacturing Practices (GMPs) and good laboratory practices are spoken in similar terms. Laboratory operations in pharmaceutical, food, cosmetic, diagnostics, and medical device industries follow some published standards and many best practices (Singer, 2001). This book is not intended to cover all laboratory operations, but to identify certain tools, techniques, approaches, and philosophies that can be used to evaluate the quality of most laboratory operations.

We challenge you to become a diligent student of successful auditing practices. We also challenge you to become an expert in the interpretation of published standards and best practices related to the laboratory operation that you intend to audit. This is a dynamic practice, because it can and will change as your needs and your firm's needs change.

If you are a laboratory manager, it will also be beneficial to know how to scrutinize your own laboratory in preparation for an external auditor or just to help improve your laboratory operation.

We would like also to encourage you to network with the largest organization of quality professionals in the world, the American Society for Quality (ASQ). They can provide you with additional expertise in the areas of auditing, quality improvement, and quality management. They are the one-stop shopping source for quality. Don has been a member of ASQ and their Food Drug and Cosmetic Division for over twenty years, and the professional relationships and networking with colleagues have provided an ongoing resource of information-sharing and expertise about quality, laboratories, practices, and standards.

Donald C. Singer
Raluca-Ioana Stefan
Jacobus F. van Staden

CONTENTS

INTRODUCTION

The first book about laboratory quality auditing (Singer and Upton, 1993) was written over ten years ago. Since that period of time, many changes have occurred that directly affect the survival of laboratories in GMP-regulated industries. Mergers of large corporations on the one hand decreased the number of internal quality control laboratories, but on the other hand influenced the increase in size, number and capabilities of contract laboratories. Much more testing is being outsourced than ever before and thus dependency on the quality of outsourced services is critically increasing. The development of global standards through international collaboratives of scientists (International Standards Organization, ISO; International Conference on Harmonization, ICH) have not only developed manufacturing quality criteria, but also have developed improved laboratory standards. As laboratory accreditation became a familiar term to global manufacturers and regulatory agency support laboratories, available guidance such as ISO Guide 25 (1990) had to evolve, and the result was ISO 17025 (1999). Additional industry-specific interpretation followed the original ISO 17025 with the AOAC version (2001) that was written for food and pharmaceutical quality testing laboratories. Regulatory agencies such as the Food and Drug Administration (U.S.) began to implement a plan to have their laboratories evaluated and meet the criteria of ISO 17025. The increasing influence of computers and information technologies on data generation, data analysis, and data storage has caused an increasing regulatory scrutiny on this type of data handling in our laboratories.

All of these significant changes are the reasons that we developed a more thorough, up-dated book to build around the basic knowledge of the quality audit relevant to laboratories.

We have also developed a more global perspective of this critical area to fit in with the dynamic environment laboratory management finds itself. My fellow authors, Dr. Stefan and Dr. van Staden, add a strong international experience and flavor to quality auditing of laboratories. We are very excited to offer our readers this desk reference and resource of pertinent information that will help design an effective audit and will help improve quality of a laboratory operation.

Donald C. Singer

I

QUALITY AND AUDITING

QUALITY CONTROL AND QUALITY ASSURANCE IN A LABORATORY

A quality assurance (QA) program must be planned and implemented to provide confidence in a laboratory's execution of its business. The most effective means of evaluating a QA program is auditing.

The quality audit has been defined as a "management tool used to evaluate, confirm, or verify activities related to quality." It is a constructive process (Mills, 1989).

Performing a quality audit has become a routine activity of any business that seeks quality improvement. The results of a quality audit are carefully evaluated and used for developing objectives, which will assist a business in improving the quality process that is already in place. Alternatively, the results can also be used to develop a quality improvement plan that will improve the strengths and reduce or remove the weaknesses that could create future problems.

Every product that is manufactured by pharmaceutical, food, cosmetic, medical device, or biotechnology firms has characteristics that need to be quantified or qualified by laboratory testing. Every sample of body tissue or fluid submitted by a physician, veterinarian, clinic, or hospital for diagnosis requires laboratory testing. Diagnostic test kits for use at home by the public or in laboratories are batch sampled and tested by the manufacturers' laboratories prior to their release for sale. Often, some diagnostic test kits are collaboratively tested (Association of Official Analytical Chemists) by a number of laboratories to confirm (or validate) the accuracy of their results. Just about anything that is used to improve, prevent deterioration of, maintain, or diagnose health in humans or animals is subjected to testing in a laboratory. Quality control and quality assurance are the necessary processes that play the role of a check and balance system in a laboratory.

Analytical testing for known characteristics should have corresponding known standards for comparison. Analytical testing for unknown characteristics must have both known standards and controls to ensure that whatever result is obtained is reliable. The reliability of analytical testing is the means for building trust in the customer and credibility of the testing laboratory. Customers demand trustworthy, consistent analytical practices, which result from tight quality control and quality assurance processes.

The laboratory environment consists of people, facility, instrumentation, chemicals, supplies, and samples submitted for analysis. There has to be a logical and scientific manner of organization and management that can drive the laboratory system. Each laboratory works in an individual manner, so it requires a customized approach and attitude. Every laboratory depends on consistency, and the development of a system which will ensure consistency is dependent on the following factors: well-managed and adequately trained people, well-maintained facilities, calibrated instrumentation, high quality chemicals, adequate supplies, and proper handling of samples. Safety must also be woven through the fabric of the laboratory. These factors will help to form a laboratory environment that can meet the requirements of any kind of laboratory quality audit.

The laboratory quality audit can be used as a tool to help to increase credibility and substantiate trust and confidence of the customers for the laboratory's capabilities. The laboratory quality audit will evaluate the strengths and weaknesses of the quality control/quality assurance processes. Then, the audit can be used to improve the processes and build a better system for the benefit of the laboratory owners, employees, and customers.

What is quality control of a laboratory? Quality control is defined as the operational techniques and activities that sustain quality of a product or service that will satisfy given needs (American Society for Quality Control, 1983).

Each instrument requires periodic calibration, by physical or chemical means, where appropriate. Chemicals used in analytical testing should be of the purity required by the procedures. Known standards, where available, are routinely used to check instrument and method variation. When testing unknown materials, known standards can be used for benchmarks. Consistency in the preparation of testing materials (samples, reagents, media, etc.), in the use of instrumentation, in following appropriate methodology, and in documenting results are results of a successful quality control process.

What is quality assurance in a laboratory? Quality assurance is defined as all those planned or systematic actions necessary to provide adequate confidence that a product or service will satisfy given needs (American Society for Quality Control, 1983).

Written procedures and adequate documentation of all quality control practices, training, and analytical results make up one part of a laboratory quality assurance process. The other part of the process is an experienced quality assurance staff who manages and performs internal audits of the quality control and quality assurance processes.

The quality control and quality assurance processes in a laboratory are usually defined by internal and external regulatory requirements. The minimum criteria for quality control and quality assurance processes should not differ significantly from regulatory requirements and guidance, and thus can provide a more defined basis for an audit. Employees of the laboratory directly affect the quality control and quality assurance processes. Hiring and training criteria have become an important part of quality control and quality assurance, and should be a significant measurement in an audit.

Employees, quality control/quality assurance, and customer interactions are the three most important areas in an audit of a laboratory. If these three areas are well developed and documented, and if the laboratory can satisfactorily perform the testing of which it claims to have experience and capability, then having a larger facility and higher levels of instrumentation technology are advantages and not requirements.

THE REGULATED INDUSTRIES

Many consumer product or service industries are regulated. Where laboratory testing is utilized in these industries, predetermined guidelines are followed for evaluating the safety, efficacy, and overall quality of the products or services. These guidelines originate either internally or externally. There are many external organizations which develop guidelines that have become industry standards, such as government agencies, accreditation groups, or industry forums.

The United States Food and Drug Administration (FDA) enforces the 1938 Food, Drug, and Cosmetic Act, which was revised in 1976. The FDA inspectors follow guidelines set forth in an Investigations Operations Manual (2003). The FDA monitors how industry follows guidelines set forth in documents called current Good Manufacturing Practice (GMP) in Manufacturing, Processing, Packing, or Holding of Drugs (1978), and current GMP in Manufacturing, Processing, Packing, or Holding of Foods for Human Consumption (1979), and Good Laboratory Practices for Nonclinical Laboratory Studies (1978). Food and Drug Administration inspections occur either on a periodic basis or with higher frequency based on the following:

1. Customer complaints or reported adverse reactions.
2. Voluntary recalls by a firm.
3. Food and Drug Administration product sampling program finds a deviation from product quality or product claim.
4. Raw material or packaging component manufacturer problems.
5. Pre approval inspection for a new drug application (NDA).
6. Current GMP inspection for manufacturing of new clinical supplies, medical devices, or diagnostic products.
7. Approval of a sterile product manufacturing facility.
8. Good Laboratory Practice inspection for support of nonclinical testing.

The 1997 FDA Modernization Act was an initiative that was developed to help streamline FDA organization and procedures. The initiative was followed by a "21st century" approach to inspections. A document titled "Pharmaceutical cGMPs for the 21st Century: A Risk-Based Approach" was written to explain a new effort to bring together science-based risk management and quality control systems. One intent of the approach was to increase the responsibility of the manufacturer for quality systems and direct FDA resources for inspections to the higher risk (to patient safety) product manufacturing operations, while reducing inspections of low-risk product manufacturing operations. Complementing the GMP approach to FDA inspections, the FDA has published some guidelines for laboratory competence. Two useful documents are, Guide to Inspections of Pharmaceutical Quality Control Laboratories (FDA, 1993), and Guide to Inspections of Microbiological Quality Control Laboratories (FDA, 1993).

The United States Department of Agriculture (USDA) oversees the agricultural industry, which includes meat, poultry, egg, and dairy products. Testing for microbiological quality attributes, defect action levels, nutritional labeling, and pesticide residues are part of product evaluation programs. United States Department of Agriculture and Food Safety Inspection Service (FSIS) regulated laboratories follow standard methodology such as the Official Methods of Analysis (AOAC, 2003). In 2001, the USDA/FSIS laboratories established a program called Accredited Laboratory Program (ALP) and contracted AOAC to assist in performing onsite evaluations of private laboratories, which helped ensure compliance to the regulations under the Federal Meat Inspection Act and Poultry Products Inspection Act.

The United States Environmental Protection Agency (EPA) has the authority to monitor for air, soil, and water pollution and to enforce standards to protect natural habitats. Sampling techniques and methodology for testing samples have been approved by the EPA for public drinking water and wastewater (American Public Health Association, 1999). Maximum allowable levels of contaminants in public drinking water have been set (Environmental Protection Agency, 1988).

The United States Pharmacopeial Convention oversees a forum of professionals who represent the pharmaceutical industry. The forum consists of a committee of experts (COE) and each expert leads a specialty area committee of professionals (e.g. analytical microbiologists, analytical chemists, pharmacists, pharmaceutical scientists, and physicians). The "bible" of this forum is the US Pharmacopeia and National Formulary (USP-NF XXVII, 2004). The USP-NF is a reference for the biological, chemical, and physical attributes that are used to determine the purity of pharmaceutical raw ingredients or compounds. Published annually, the USP-NF also recommends test protocols for laboratories evaluating the attributes of each pharmaceutical material. A growing new addition to the USP is an informational guidance about nutritional and dietary supplements. Other countries have similar references of pharmacopeial standards, such as the European Pharmacopeia, the British Pharmacopeia, the German Pharmacopeia, and the Japanese Pharmacopeia.

The World Health Organization (WHO) is a forum of professionals who represent various nations' interests in health. Expert committees have written documents which include GMP for pharmaceutical products (WHO, 2003) and Good practices for national pharmaceutical control laboratories (WHO, 2002).

Since "there has been an increasing trend towards the development of broad spectrum accreditation programs that apply the same principles of good laboratory practices to laboratories working in any field of science or technology" (Bell, 1989), the International Organization of Standardization (ISO) published a set of documents which were developed by the International Laboratory Accreditation Conference. The ISO Guide 25, considered to be a generic accreditation document of criteria (Bell, 1989) for technical competence of a testing laboratory, was the first relevant document from ISO. The ISO 9000 series, published in 1987, partially revised in 1999 and again in 2001, is becoming the single source of standardized requirements for the design and implementation of a quality system. These requirements were initially interpreted and adapted in an industry-specific manner (Marquardt et al., 1991). To prevent movement away from standardization, the ISO Technical Committee TC176 met in 1990. They agreed on a strategy for developing worldwide acceptance of global standardization. One of their goals was to seek harmonization between ISO guides and European Community standards (EN 45000 series) dealing with, in part, operation assessment and accreditation of laboratories. International Organization of Standardization 17025, general requirements for the competence of testing and calibration laboratories, were written, and approved in 1999. The AOAC Analytical Laboratory Accreditation Criteria Committee (ALACC) improved ISO 17025:1999 by adding text relevant to chemical and microbiological laboratories in the pharmaceutical and food industries (AOAC, 2001).

An organization formed in 1990 to develop harmonized guidelines for the pharmaceutical industry as the need for global standards increased over time. Representatives from regulatory agencies and industry associations from Europe, United States, and Japan met and formed an International Conference on Harmonization (ICH) of Technical Requirements for Registration of Pharmaceuticals for Human Use. International Conference on Harmonization implemented the formation of technical committees to develop such guidelines as Validation of Analytical Methods (ICH, Q2), Stability Testing of New Drugs and Products (ICH, Q1), and Test Procedures and Acceptance Criteria for Biotechnological/Biological Products (ICH, Q6B).

GOALS OF A LABORATORY QUALITY AUDIT

The simple objective of the complex process, if thorough, of auditing a laboratory's quality program is to evaluate the activities and existing documentation and determine if they meet predetermined standards. A total quality program in a laboratory is usually developed to assure that all activities are performed with the objective of meeting certain standards, both internal and external. Some standards are generated internally, e.g. corporate quality improvement process, routine quality assurance/quality control protocols (including Standard Operating Procedure), or annual accreditation. A corporate quality improvement program may, in part, require that certain laboratory operations become routine and that customers are given information when they need it (for example, a customer service agreement is written and a customer service contact person is identified, facsimile capabilities exist and a computerized database is used that can track samples). Quality assurance programs often provide standards for receiving, handling, coding, and testing samples, and for recording and reporting test results. Quality control programs usually provide the standards for validating instrumentation, determining the suitability of reagents and test procedures, and provide requirements for training of analysts.

A laboratory quality program should have an established set of goals, which every effort is made to reach. Objectives are set to provide the means for achieving the established goals. These objectives must be measurable. It is these measurable objectives on which an audit is based.

Competency is the key objective of any laboratory quality program. Laboratory accreditation is a "formal recognition that a testing laboratory is competent to carry out specific tests or specific types of tests" (Schock, 1989). The accreditation of a laboratory can be either an internal or external standard, or both. The audit for accreditation is an on-site examination of the laboratory to determine if it meets the accreditation criteria (Schock, 1989). Some sources of standards or criteria that are generated externally are government agencies, customer requirements for contracted laboratory services, and procurement quality control requirements. The standards generating agencies in the U.S. government that are concerned with foods, drugs, cosmetics, and medical devices are at all levels of government, i.e. federal, state, and local. Some of the federal agencies are the FDA, EPA, USDA, and FSIS. Internationally, as mentioned in an earlier chapter, the standards generating group has developed a basis for laboratory competency, which is now known as ISO 17025. Laboratories in the food, drug, cosmetic, and medical device industries must meet specific criteria developed and enforced by the U.S. government's FDA. The EPA has set standards for air, water, and ground contamination levels, as well as standards for testing for contaminants in these natural resources.

When a contract laboratory provides testing support for a manufacturing facility, the manufacturer usually provides the laboratory with the test methodology, and criteria must be met to ensure consistency in testing and reporting. Routinely, some overlap exists in the criteria that a laboratory must meet originating from a food, drug, cosmetic, or medical device manufacturer and the criteria generated by government agencies related to the same products. If products must be registered or licensed prior to marketing approval by an agency such as the FDA, all documentation relevant to finished product testing could be evaluated first. The supporting infrastructure and quality programs of a laboratory can lend credence to the integrity of the testing results.

Procurement quality programs usually require sampling and testing of purchased materials, ingredients, and finished products. Testing and documentation criteria are commonly set by the clients, including requirements for calibration of instruments, use of reference standards, and other laboratory control measures.

AUDIT TOOLS: OBSERVATION, WORKING KNOWLEDGE, AND AUDIT DOCUMENTS

A properly conducted audit of a laboratory should be documented thoroughly in a format that will make information simple to find and evaluate. Documenting data and recording comments are the most effective permanent record of pertinent information. An auditor should rely on memory only for a short time (minutes or hours) and only if carrying a pen and notepaper or a microcassette recorder is not possible, for example, in a sterile testing area.

Prior to conducting an audit, the laboratory is contacted to set a date for the on-site visit. An agenda must be prepared, listing the areas to be evaluated and specific testing to be observed. Results of previous audits should be reviewed to determine if follow-up to previous concerns is necessary. Results of recent government agency inspections should be requested, although proprietary information may need to be hidden before the evaluations are shared. Any documentation that can be shared prior to the on-site audit should be requested and reviewed ahead of time. The laboratory is sent the agenda ahead of time. They are then asked to confirm that the time and personnel will be available when the actual on-site audit takes place.

There are three areas in a laboratory from which information should be derived during the on-site audit: personnel, documentation, and observation of testing procedures. Even under the best circumstances and relationships, there is usually a limit to an auditor's contact with the personnel who actually perform the testing. Almost every analyst, when correctly performing a test procedure, does it in a unique way. These unique differences provide a challenge for the auditor. An auditor must observe the performance of the procedure.

First, the auditor reads and interprets the procedure. Then the auditor observes the actual conditions and manipulations of the test procedure by the individual who is routinely assigned to perform that procedure.

The preparation of supplies, materials, test site, and instruments that will be used are all significant parts of a properly performed test procedure. As the analyst performs the test, reads the results, and records the results, the auditor should note any aspects of the test that could lead to error. There are a few test procedures, which are repeated on consecutive days or require more than 1 day to complete (e.g. microbiological testing). Since most laboratories carry out routine testing daily, enough overlap of tests occur that allow an auditor can observe the different steps of a single procedure in 1 or 2 days, even if they are performed on different samples.

If a reference standard calibration is a preparatory step in a procedure, the auditor observes where the standard was stored and how it was handled before and during the calibration. It is also important to observe the type of documentation kept to assure the stability and confidence of the standard, as well as consistency in the calibration.

The laboratory director or supervisor of the specialty area (toxicology, microbiology, chemistry, etc.) usually responds to questions from the auditor. It is not only logical, but imperative that audit questions be directed to the most knowledgeable person in a laboratory area. Those key persons are the only sources of complete, up-to-date information about the test procedures that are being followed. Those individuals know the background of the changes that have been made to procedures in their fields of expertise. They are also familiar with the precision and sensitivity of each procedure performed in their areas. Nowhere else could an auditor expect to find a resource more familiar with the specialized synergy of the laboratory, procedures, and personnel in their area of scientific testing.

Observation permits the auditor to see the actions required to perform a test. The analyst should carry out the test as it is written. If an analyst deviates from the written procedure, the deviation(s) can be a source of error and invalidate the test. A deviation must be documented and reviewed before any further testing is performed. A procedure should be revised, reviewed, and revalidated before changes are implemented.

If an auditor is to observe and accurately evaluate an experienced analyst performing a test, the auditor should understand the technical intricacies of the test. In other words, the most accurate evaluation is made by a qualified auditor who has performed the test enough times to become intimately familiar with it.

The auditor who is experienced in the test methodology will be the best person to make a fair, accurate evaluation of the client laboratory's test conditions. Thus, it is strongly suggested that a chemist audit chemistry testing, a microbiologist audit microbiology testing, a toxicologist audit toxicology testing, and so on. Since there are many laboratory audits that involve a variety of testing areas, it is not uncommon to have more than one specialist perform an audit. In fact, a team approach to an audit is common and very productive when a variety of specialty areas are involved. A team composed of qualified individuals with different fields of expertise (e.g. chemistry, microbiology, metrology, and toxicology) can audit many areas in a laboratory at the same time. The team concept is effective, efficient, and professional. The preparation for a team audit is very important. A plan must be well thought out and agreed upon in order to accomplish daily objectives.

II

NONTECHNICAL PARAMETERS FOR EVALUATION

QUALITY PROGRAMS, DOCUMENTATION, AND ORGANIZATIONAL STRUCTURE

Quality Programs

There should be an established program that monitors and evaluates all laboratory procedures and the competence of the laboratory staff. The program should be well documented and provide the auditor with evidence that the laboratory is effective in the following areas:

1. Accuracy of data
2. Timely reporting of test data
3. Action plans to correct problems
4. Adequate training of analysts
5. Proficiency testing.

Accuracy of Data

All raw data and calculations are written into a bound laboratory notebook, or on prenumbered forms, along with a signature of the analyst, and are checked and signed by a second person for accuracy and completeness. Where instrumentation produces raw data (e.g. chromatographs, spectra, and other integrated printouts), either the raw data are attached or cross-referenced in the written record. Any calculations performed should be explained either in the written method or an explanation should be detailed in the laboratory notebook. If electronic notebook systems are used, then those systems must have validated electronic signature, audit trail, and calculations. Any systems using barcode reading should be validated.

An instrumentation calibration and maintenance program must be well documented. There should be evidence that reference standards have been checked and used where appropriate and necessary. Identity of test and measuring instruments used in a test procedure should be included in the test documentation, to ensure ease of tracing at a later date.

A procedure of particular importance in government regulated laboratories is the chain of custody of specimens. The procedure should include guidelines for the documentation of names and dates for each person-to-person transfer of a specimen, and the tracking of use quantities for each sample removed from the original specimen. This type of protocol is a legal requirement for the tracking of controlled (drug) substances, but can be applied to any specimen.

Timely Reporting of Data

Determining what routine time allotments for analyses of materials are appropriate requires good judgment. Depending on the use of a manual, semiautomated or fully automated procedure, testing time can be rapid (results in seconds) or lengthy (results in hours or days). The chosen method must be appropriate for the precise results, and in many cases for the speed which results are available to make crucial decisions, e.g., medical decisions.

There should be a record of the starting and completion date and time of analyses. A review is performed on completed reports and procedures, which define the test period and amount of time required for calculations and reporting results. This can provide adequate information to evaluate the timeliness of reporting test data.

Action Plans to Correct Problems

When a problem is identified in the quality program, there must be a documentation of the assessment of the problem, an investigation, corrective action taken, and a follow-up review assessing the effect of the corrective action. A written procedure should be available that outlines the steps to follow when unexpected or out-of-specification results occur.

An auditor should request to review an example of a recent unexpected or out-of-specification result investigation, and perform an evaluation on the documentation and action(s) taken to see if it follows the written procedure(s).

Competency of Analysts

Education and experience of analysts do not provide confidence in the capability of analysts to perform their laboratory activities, but are a critical parameter for determining their competency. In addition, adequate training and supervision significantly affect the performance of analysts.

An appropriate education or prior experience is critical to the nature of the activity that an analyst performs. A microbiologist must have a strong understanding of microbial growth parameters and physiology as a foundation for subsequent training in aseptic practices and investigations of the source of microbial contaminants. A chemist must have an understanding of the chemical breakdown of certain molecular structures as a foundation for training in instrumentation such as gas liquid chromatography, high pressure liquid chromatography, and mass spectrometry. An appropriate level of education should match the level of decision-making and interpretation of results for any identified analyst or supervisor.

Many of the regulated industries require unique training on the handling and testing of specific materials, e.g. pharmaceuticals, medical devices, and nutritional supplements. Much of the unique training occurs on-the-job.

A training file should exist for each analyst. The training file should contain a current job description, a curriculum vitae that identifies the educational background and prior training to the current job, and all records that indicate the type and frequency of training that the analyst has completed. Training includes self-training exercises such as reading relevant standard operating procedures and policies, and supervised training for particular analytical procedures. Where appropriate, training should include criteria for meeting acceptable standards in performing the trained activities or procedures. A training program can be internal or external to the laboratory. Training must occur before a new or revised technique/procedure is implemented.

It is important that competency of a laboratory analyst is assessed and documented in their training file. A combination of assessments is considered acceptable. Where registration or certification in particular fields of industry includes examination of scientists (e.g., American College of Microbiology's Registered Microbiologist) (PDA, 2001), both documentation of the certification and any subsequent recertification or requalification are acceptable. Periodic proficiency testing can be used to validate each analyst's performance on particular tests with particular materials. Also, any observations made by a competent person of an analyst performing a test, on a periodic basis, can be another record of their performance.

Although it is difficult to determine the quality of supervision in 1–3 days audit, observation of an analyst performing a procedure can provide an accurate assessment of the success of a training program and type of supervision that exists.

Proficiency Testing

Proficiency testing will be described in more detail later in this book. The primary consideration is that some type of periodic interlaboratory or intralaboratory "check sample" testing program exists and is followed. The nature of the testing performed will dictate what types of samples are appropriate, and availability of certain types of proficiency test samples may restrict the frequency of testing. Without this type of program, though, a laboratory cannot measure itself relative to the status quo. Confidence in the accuracy and precision of test procedures in a laboratory should be partly based on the comparison of results from testing reference specimens in two or more laboratories.

Documentation

How critical is documentation? It is extremely critical in a laboratory. From the recording of test results to the explanations written in a preventative maintenance log for an instrument, documenting information carries a lot of weight.

Documentation is the accepted method for recording information for future reference and investigation. Some types of information that are routinely recorded are:

- Procedures for:

  Quality assurance/quality control program
  Receipt and storage of samples
  Sampling
  Analytical testing
  Validation
  Calibration
  Data recording
  Operation of instruments
  Reagent or test material preparation
  Maintenance of standard operating procedures

- Training records
- Organizational charts
- Sampling schedules
- Analytical testing assignments
- Supplies inventory
- Expiration dating of reagents
- Instrument calibration data
- Instrument maintenance
- Methods validation data
- Analytical test results
- Test conditions: time, temperature, etc.
- Facility maintenance and cleaning

The above list is only a sample. If data can be recorded for future analysis, tracking, trending, or evidence of any laboratory activity, it is appropriate and necessary that it be documented.

An auditor should check to see that records are written in indelible ink. It is preferable that most records be kept in permanently bound notebooks or electronic systems. In general, original data (raw data) is recorded in permanently bound laboratory notebooks, which have been permanently coded with some numerical system. When in use, notebooks should be protected from damage. When in permanent storage, conditions must exist which prevent loss of records by damage or tampering (such as by microfilm archiving). Electronic systems must have routine back-up capabilities and long-term storage (archive) protection. The auditor should observe the entries in a laboratory notebook. Each entry must be accompanied by the signature or initials of the author and the date that the data was entered. It is of utmost importance that the data be traceable to the time and person who made the entry. In many regulated industries it is mandatory that a second signature also accompany data entries as proof that a review of the data has taken place.

There must be approved written procedures for handling, storage, preparation, and testing of a sample. Adequate records should exist to allow anyone to trace a sample from receipt to testing, according to the written procedures.

There should be a technical review of all procedures at least every 2 years. A change control system is necessary to acknowledge procedural revisions and allow for corrections to be made where problems or errors are found. Revisions to procedures should be reviewed and approved by the appropriate competent analyst in the laboratory. Whenever a procedure is changed, a document describing the change and reason must accompany the procedure and become part of the permanent record. The permanent record (or file) contains the history of all procedure revisions.

Computers are increasingly being used as alternative means of keeping permanent records. Original data can be entered automatically (from an instrument) or manually into a database. The software program can provide data storage, retrieval, calculations, and much more. Users of a computerized database must follow regulatory guidelines when raw data is generated (FDA, 21 CFR 11). It is important that if computer systems store original laboratory data, the auditor should ask if a built-in security system exists to prevent tampering with the data and how it prevents loss of data. Evaluate how corrections to original data are controlled. For example, some programs have been written to either prevent changes to original data or to allow changes to be made accompanied by an automatically entered permanent marking adjacent to the changed data (e.g. # or *symbol). Another technological advance in use is a legal alternative to the handwritten signature. An electronic signature, retinal optical scanning, and fingerprinting tools are options for security. Part of an effective audit of a laboratory computer system is to observe original data entry and ask the operator to attempt to make a correction to the data after it has been entered.

It is recommended that a computer system be validated (PMA, 1986, 1990) before it is used to handle critical data. An auditor should review the validation protocol and results.

Documentation is mentioned again in various other sections of this book. A successful quality assurance program in a laboratory is proven, in part, by the thoroughness of the laboratory's documentation. The requirement for documentation is absolute.

Organizational Structure

The importance of an effective organizational structure should not be overlooked. It has been stated that a primary objective of an analytical laboratory is to produce "high quality analytical data through the use of analytical measurements that are accurate, reliable, and adequate for the intended purpose" (Garfield, 1984). The type of working environment and reliability of data is a direct result of the effectiveness of the organizational structure.

The auditor should request an up-to-date organizational chart of the laboratory staff. The chart should be accompanied by descriptions of the communication that occurs for all laboratory procedures. Effective lines of communication are the source of timely actions and reporting. Any procedures that specify information reporting responsibilities should coincide with the present organizational structure. Laboratory management should review procedures after any organizational change that may affect reporting responsibilities. At least two procedures that state action response guidelines should be compared to the organizational chart. If discrepancies exist, a determination must be made as to why they exist and what is necessary to correct them.

There are an unlimited number of possibilities for a laboratory organizational structure. Regardless of the titles chosen (group leader, supervisor, section head, manager, and director) the key elements for strong communication in the laboratory are the reporting relationships between the analysts and their immediate supervisor, and between the supervisor and his or her manager. All raw data and calculations from analysts should be reviewed by the laboratory supervisor, for accuracy or completeness or scientific conclusions, or all three. When problems occur, e.g. instrumentation error or loss of integrity of a client's specimen, all actions taken by an analyst to resolve problems must first be reviewed by the laboratory supervisor. The supervisor's boss should be notified before or concurrent with the action plan, and it is his or her responsibility to notify the client at the appropriate time. Written procedures must exist that list the steps taken, the internal communication that must occur, and the documentation required to describe problems that develop, and also the action taken to resolve the problems. Notification of the client should be part of the procedure. Timely and reliable problem resolution depends on all the appropriate personnel following the written procedures as described above. Problems that appear in the areas of the laboratory that are audited (calculations, instrument calibration, training, procedures, etc.) can usually be related to poor communications.

PERSONNEL

The single most influential factor for a successful laboratory is the hiring and management of qualified people. Laboratory accreditation programs and federal regulations for Good Laboratory Practices specify the educational requirements for a laboratory director, laboratory supervisor, and laboratory analyst.

Written position descriptions should be available for evaluation. The requirements for education and specialist training, where necessary, should at least meet laboratory accreditation requirements. Individual personnel files should contain a curriculum vitae which can be reviewed by an auditor and compared to the description of the position currently held by the individual. An auditor should be aware of individuals without adequate scientific background performing complex tasks that require more technical experience than they possess. A list of assigned duties should be requested to determine which tests certain individuals are performing, and which individuals are assigned quality control responsibilities. This list should be used with personnel files to compare with the experience, education, and training of the individuals who perform the tests. Complex tests should not be performed under any circumstances by individuals without adequate scientific background. Individuals who make scientific judgments and decisions about technical issues should have an appropriate educational background for those decisions.

It is possible that an analyst's experience as documented in their curriculum vitae is sufficient to allow them to perform certain tests. But, it may be more appropriate for a lab to require additional qualification exercises to support their competency.

Training records should also be part of each individual's personnel file. Documentation should exist that can show the ability of an individual to perform specified tests and meet accuracy criteria. This documentation should provide evidence of training, experience, or skills of an individual. Training records include internal checklists of procedures or instrumentation reviewed by the supervisor and approved by the lab director. Copies of school course records with grades and workshop completion certificates are other acceptable records of an individual's training.

Certification by an external scientific-based organization is an additional means of proving the scientific knowledge of an analyst or supervisor. Some certifications include a compilation of education and experience submitted for review. Other certifications require mandatory coursework (college level) and taking a scientific exam to qualify (e.g. American Society for Microbiology Registered Microbiologist certification). After certification has been acquired, often continuing education is necessary to maintain or requalify for the certification. The value of certification is providing a proof of competency for the laboratory analyst to their employer and auditors.

The size of a staff is a product of laboratory space and budget. A laboratory may have a small number of people performing a small number of routine tests; or a laboratory may have a small number of people performing a large variety of tests. The latter situation could be cost effective and highly competitive if the personnel have appropriate skills and education. If not, the situation is stressful and can take a toll on accuracy of results. An auditor should ascertain whether the staffing and experience of the personnel are congruent with the intended output of the laboratory.

FACILITY AND ENVIRONMENT

Maintenance and housekeeping of the building enclosing the laboratory and upkeep of the grounds surrounding the building are a reflection of laboratory management.

There must be adequate space and utilities (water, electric, lighting, compressed gases, heating, and ventilation) to assure that laboratory workers can perform all procedures as safely and efficiently as possible. The quality and quantity of available water and electricity directly affect the efficiency and quality of work performed.

As a general rule, at least 15 linear feet of bench space should be available for each laboratory analyst (Environmental Protection Agency, 1978). Storage space for specimens, reagents, and documents should be sufficient and accessible. An adequate number of electrical outlets, designed with safety in mind (adequately marked for amperage or voltage maximums, grounded, and easily accessible) should be available to ensure continuous use of instrumentation. Back-up battery power packs are commonly used to prevent loss of instrument analysis time during power surges or outages. Sufficient lighting is very important in areas where visual measurements are performed. A source of laboratory water, the purity of which is determined by its intended use (e.g. USP Purified Water, NCCLS type 1 reagent grade water, and distilled water), must be easily accessible and monitored routinely for microbiological/analytical quality and for any other required specifications.

The building should have an appropriate solid and liquid waste management system, including proper drainage of nonhazardous liquid waste and a routine trash pick-up program. Designated receptacles for hazardous waste, broken glass, and other materials with potential health and safety risks should be well labeled and part of the waste management program.

The design of the laboratory includes two critical features that affect sanitation and housekeeping: (1) floors, walls, and ceilings materials, and (2) layout of furniture and equipment. The results of a routine schedule for floor cleaning (and waxing where appropriate) are usually obvious to an auditor and any visitors. Walls and ceilings should be made of inert building materials and kept in good condition, i.e. without signs of chipping or peeling paint, water leaks, or any type of debris.

A well-designed laboratory can provide impetus for efficiency and success. There should be a physical separation between areas where critical measurements are performed and airborne debris is expected (e.g. handling of dehydrated microbiological media). Also, there should be a physical separation of sample receipt, testing, and storage. Any testing that requires very strict environmental conditions should be performed in a separate, environmentally controlled enclosure (e.g. sterility testing or particulate testing).

Just as an HVAC (heating, ventilation, and air cooling) system is specified and maintained by a building's maintenance staff (Belsky, 1991), the heating and other environmental control needs of many tests should be monitored and maintained by the laboratory staff. The laboratory should have adequate heating and refrigeration space for all of its testing requirements. The size and number of ovens, environmental chambers, and refrigerated storage units will be a function of the testing requirements and available laboratory space. During an audit, it is important to assess the oven, environmental or refrigeration space available and determine if the specified conditions are being met. Adequate space helps assure that environmental conditions will be met. All environmentally controlled units (ovens, incubators, water baths, walk-in chambers, etc.) must be monitored on a routine basis. Procedures and documentation should be available demonstrating the monitoring process, i.e. comparing routine measurements to the appropriate predetermined specifications (e.g. temperature, humidity, and air flow). Also, it is advisable to keep permanent records of monitored measurements by using recorders and charts, or computers, on a daily basis. Many laboratories manually record measurements once or twice each day, Monday through Friday, but there is no assurance that the specifications are continuously met over a 24 hr period or on Saturday or Sunday.

III

TECHNICAL PARAMETERS FOR EVALUATION

SAMPLE CONTROL

An audit should cover the procedures that laboratory personnel perform that detail the receipt, documentation, storage, and handling of samples for analysis. Representative sampling procedures will not be discussed here, although if laboratory personnel perform any initial samples collection, written procedures should exist, and the procedures should be reviewed in an audit. Procedures should include size of samples collected, how they should be collected, frequency of collection, and how the sample is subsequently preserved and transported to the laboratory. Conditions of preservation for each sample type should be specified, i.e., refrigeration (for heat sensitivity or stability), use of sterile bags/containers (for aseptic samples), or amber bottles (for light sensitivity). In appropriate circumstances, such as laboratory performing its own sampling, a reference or justification should be written to describe how and why samples are taken and what they represent. This can be an important part of generating meaningful test results.

The sample receiving procedures must be documented by a recordkeeping system. Whether a computerized system or a manually written log is used, the records must be current. Information necessary for identifying each sample should be required before a sample is accepted for testing. Every sample must be given a distinct identification number and labeled with that number and any other information for proper storage and handling. Also look for labeling subsamples during test preparation with the same sample identification number. This is another good practice. Chain-of-custody type documentation is a legal requirement in many government regulated laboratories, but is also advantageous in private laboratories as a preventative measure to assist in criminal court actions. Chain-of-custody type documentation is routinely used for controlled drug substance and forensic samples.

Procedures that state the method of handling and storage of different types of samples are based on the initial identification provided by the originators of the sample and logical preservation techniques. It is obvious that the facility must have adequate space and environmental condition flexibility to store samples under specified conditions such as refrigeration or absence of light.

A detailed description of subsampling of samples for analysis should exist for each type of sample. Reserve samples for additional testing should be a consideration if an adequate original quantity is received. In certain situations, a tracking system might be used to detail quantities of each subsample used in a test, by recording in a book, or by recording quantities used or remaining on the label of the sample storage container. Presently, it is common practice to record quantities of controlled drug substances as they are used in testing. It would be prudent to track rare materials.

INSTRUMENTATION

Instrumentation plays a crucial role in the quality assurance. Therefore, the first step in an audit is to obtain an inventory of instruments along with all SOPs controlling the calibration and preventive maintenance program. "The choice of analysis instrument is imposed by the sample and its matrix" (Baiulescu et al., 1991). The sample acts as "glue" between method and instrument (Aboul-Enein et al., 2000). Therefore, the method should be chosen accordingly with the complexity of the matrix of sample and sample history, and the operational parameters of the instrument should be correlated with the sensitivity and selectivity of the method. In this regard, it is imposed to have high-qualified operators that can perform the analysis professionally, as the selection of operational parameters of the instrument is very important for quality and reliability of the analytical information. Accordingly, the audit should determine if the instrumentation used in the laboratory is interconnected with the methods utilized for analysis, type of samples (history and complexity), sampling procedures as well as if the personnel are qualified to operate the instruments.

A question is imposed: is it good to utilise dedicated instruments for routine analysis in such a laboratory? And will the audit always approve such instruments? The problem of the dedicated instruments is that they are most of the time provided with two buttons: "Start" and "Stop"—very easy to operate (even an unqualified person can do it), but one cannot change the operational parameters accordingly with the sensitivity and selectivity related to the method and sample. They usually can be used for the analysis of a small range of samples with a certain matrix complexity (e.g., glucose analyzers). The first problem that occurs is the loss of selectivity—one cannot find two blood samples or two water samples containing the same compounds in the same amounts; accordingly interferences may take place and the accuracy of the analytical information, which is critical for such analysis is lost. The sensitivity of the dedicated instrument may also not always be enough for the analysis of the analyte(s) in all types of samples. Therefore, it is a very difficult task for the audit to determine if the existing dedicated instruments are correctly utilized for the analyte(s) determination in certain matrices.

Another task of the audit is to determine the moral ageing of the instruments as the quality of the apparatus is crucial. Old instruments should be replaced with new instruments that can assure a better sensitivity as well as an increased signal/noise (S/N) ratio. The evolution of instrumentation is correlated with the evolution of materials, science, and technology.

Calibration and maintenance of instrumentation are essential. The general procedure for laboratory calibration outlines the systematic approach: qualified personnel, relevant, detailed procedures, and recordkeeping (Singer and Upton, 1993). Elements that must be checked includes: identity of instrument, identity of personnel (check for qualifications), frequency of calibration, source (traceability) and appropriateness of calibration standard, and action taken as a result of calibration (Singer and Upton, 1993). The instrument calibration sticker should contain the following information: asset number, date calibration expires, name of person calibrating the unit, and a reference to the notebook containing the calibration data.

Accordingly, with the calibration procedure, there are two types of instruments in a laboratory: those whose calibration is absolute (measurement of time), and those whose calibration is relative to a specific standard (e.g. pH/mV-meters).

No analysis can be performed without utilization of balances and/or glassware. Therefore, the accuracy of analysis is directly dependent on weighing and volume measuring. Accurate weighing of a sample requires the use of an appropriate balance: macro-, micro-, or ultramicrobalance. When ultramicrobalances are used for weighing, corrections of temperature and pressure parameters are required. Also, special calibration procedures should be adopted for ultramicro- and microbalances. The analytical balance has evolved from a typical single-pan balance to an electronic balance, which has improved the quality of the weighing process (Jenemann, 1997). One of the parameters that affect the quality of the weighing is the temperature of the object: hot or cold objects must be brought to ambient temperature before being weighed. There are two types of weighing done in analytical chemistry: rough and accurate. A rough weighing to two or three significant figures is normally used when the amount of substances to be weighed needs only to be known within a few percentage points. Accurate weighings are reserved for obtaining the weight of a sample to be analyzed; these are performed only on an analytical balance, usually to the nearest 0.1–0.01 mg, and they need the best reliability. Calibration and maintenance of the balances are essential. If the laboratory has not got specialized personnel to perform the calibration, contract and outside balance calibration are required. The procedure used for the calibration should be specified in the final report, so that the audit can evaluate its correctness.

The four important concerns for the glassware are: class, washing, storage, and responsibility (Singer and Upton, 1993). Quantitative glassware should be Class A. Procedures for washing and drying glassware should be specified. The storage area must be appropriate to minimize wear and tear. The source of material should be appropriate to its intended use, the conditions of use must be clear, the storage conditions must be spelled out, and the responsibility for ordering, storage, and inventory must be clear. Glassware cannot be used for the determination of all substances, e.g., fluoride, boron.

Standard substances, reagents, columns, and electrodes have got a very important role in the calibration of laboratory apparatus for both qualitative and quantitative analysis. For example, an improvement of the selectivity can be achieved by increasing the S/N ratio when diamond paste electrodes are replacing glassy carbon and carbon paste electrodes (Stefan and Bairu, 2003).

Computer-driven instruments (e.g., diode-arrays, chromatographs, PGSTATs computer driven) are very easy to operate. All operational parameters can be precisely fixed accordingly with the type of analysis and with the selectivity and sensitivity of the required method. Use of an operational program can automatically do the recording and saving of the curves obtained. Data analysis and processing of data can be automatically done, as part of the operational program performed for a certain analysis. Updating of the programs is essential when such instruments are in use.

Preventative maintenance goes hand in hand with the calibration of the instruments. While the laboratory manager is responsible for the calibration of the instruments, somebody outside may get the responsibility for the maintenance of the entire facility. The key preventative maintenance activities to check are: the general procedure, the schedule, the specific procedure, and the records of preventive maintenance performed (Singer and Upton, 1993). Unscheduled maintenance is also part of preventative maintenance, as every time an instrument is repaired, cleaned, or adjusted, a record must be kept. The comparison between the scheduled and unscheduled maintenance activities gives an idea of the validity of the proposed schedule of maintenance as well as a real indication of the functioning of the instrument.

AUTOMATION OF ANALYTICAL PROCESS

The analytical process includes all the steps between sample collection and analytical information (Aboul-Enein et al., 2000). The main steps of an analytical process are: input, black box, and output (Baiulescu, 1987). The input represents all the steps between sample collection and sample measurement. The black box is the place where the measurement takes place, and the analytical data are obtained. The output represents data processing. The most laborious and time-consuming steps are those of the input and output.

A laboratory of quality control needs to be highly automated. By automation, a high number of analyses will be performed in a shorter time (e.g. these are already implemented in many clinical and water analysis laboratories), the contamination of the operators is minimum or eliminated (e.g. laboratories of radioanalysis), and the objectivity of the measurements increased due to elimination of the subjectivity of the operators. Comparing the recorded data obtained for the analysis of analyte(s) in a standard using both automated and manual standard methods, the audit should determine if by automation of the analytical process the quality and reliability of the analytical information increased or are at least the same as in the standard manual method. These parallel measurements should be performed at certain intervals of time. The data obtained (accuracy, precision, etc.) must be recorded in a special book.

The automation can be performed for the full analytical process (in this case, it is important to use computer software tailored to conduct automatically the steps of the analytical process) or only for certain steps of the analytical process. The flow system (continuous flow analysis, flow injection analysis, and sequential injection analysis) can be implemented in order to automate the analytical process. To be automated, an analytical process should have the following characteristics: rapidity, accuracy, reproducibility, flexibility, and reliability. Accordingly, not all the analytical processes can be automated. A proof of these qualities of the proposed analytical process before automation should be recorded in a book and presented to the audit.

The following parameters should be followed by the audit: objectivity (utilisation of a computer for driving the instrumentation and data processing), rapidity (the number of analytical information obtained increased), flexibility (the ability of automated or automatic device to optimise the operational parameters), and reliability (the quality of the analytical information is not changing in time). There are two types of automation equipment: automatic devices and automated devices. Automatic devices perform specific operations at a given point in an analysis, frequently the measurement step while automated devices can control and regulate a process without human intervention, an attribute that causes them to be used in process control systems (Christian, 1994).

ACQUISITION AND PROCESSING OF DATA

Acquisition and processing of data are characterized by a very high subjectivity if they are not performed automatically. Good records of data acquisition and methods of data processing must be presented for the audit, when a manual method is performed. The elimination of the subjectivity of these steps of the analytical process can be done by utilization either of a computer-driven instrument or of an automatic or automated system. In both cases, the acquisition of data is done automatically; the data are saved in the computer, and by using a "data analysis" button they can be processed in order to obtain the analytical information. For these computer-driven instruments, the updating of the programs is essential.

Presentation of analytical information is very important because "No quantitative experimental value is of any value unless it is accompanied by an estimate of the uncertainty involved in its measurement" (Pyell, 2001). The number of significant figures indicates the precision of the value. Standard deviation is a measure of the uncertainty of measurement. In the presentation of the analytical results, the average of at least 10 valid measurements should be followed by the standard deviation and the number of measurements should be always mentioned.

The role of the computer is to process the data, store the data, and transmit the data to the operator. The instrument–computer interface is very important for correct data acquisition. Instruments record analog signals whereas the computer is digital in nature and accepts only digital data. Accordingly, the interface should contain an analog to digital (A/D) converter. The data can be presented in the computer in digital form unless one wants to have them in the analog form when a D/A converter should be required for the computer. The software of the computer should be adequate for data acquisition, as its performance should be correlated with the number, speed, and complexity of the signals that are obtained from the instrument.

The utilization of chemometrics in data processing is a must in quality laboratories. Chemometrics has become an independent and very complex interdisciplinary science, which requires mathematical knowledge. It entails statistics, parameter optimisation strategy, neural networks, parameters estimation, and so on. A very important application of chemometrics is its utilization for multicomponent analysis. Computer programs are available for most of the calculations used in data processing. The training of the laboratory personnel to properly use these programs is essential and should be done at certain intervals of time. Signal processing is defined as "a discipline of chemometrics that is concerned with the manipulation of analytical data to make the information obtained in the data more accessible" (Brown and Bear, 1993).

A widely applied discipline of chemometrics is pattern recognition, which involves the classification and identification of samples. Its purpose is to develop a semiquantitative model that can be applied to the identification of unknown sample patterns. To assure the best reliability, pattern recognition requires the application of a minimum of two analytical methods.

There are analyses that required data banks. However, it is necessary for such a bank to accumulate data continuously and to be updated in order to operate efficiently. The incorporation of a data bank in a computer made the analytical information more reliable. By using chemometrics and data bank for data processing, the reliability of the analytical information will have the maximum value. Records of updating of data banks (materials used for updating and the interval of time when the updating is performed) should be kept. This updating can be done periodically or accordingly with the needs of the analyses that must be performed.

UNCERTAINTY OF MEASUREMENTS

Uncertainty (U) is a performance characteristic of the analytical process that must be estimated for both qualitative and quantitative analyses in order to determine the reliability of the analytical information (R_{AI}) (Aboul-Enein et al., 2000):

$$R_{AI} = f(U_S, U_M, U_I, U_{DP})$$

where U_S is the uncertainty of the sampling process, U_M is the uncertainty of the method, U_I is the uncertainty of the instrument, and U_{DP} is the uncertainty of data processing. Accordingly, the uncertainties associated with each step in the analytical process have a contribution to the reliability of the analytical information. To increase the value of the reliability of the analytical information, it is necessary to decrease the values of the uncertainties.

It has been demonstrated several times that the uncertainty arising from the sampling is often larger than that arising from method, instrument, and data processing. Indeed, the uncertainty of the sampling process increases with the complexity of the matrix. Selecting the most sensitive and selective method of analysis, accordingly with the complexity of the matrix as well as with the sampling procedure can minimize the uncertainty related to the method. By choosing the correct instrument for the proposed method of analysis, the uncertainty related to the instrument should be minimized. The value of uncertainty for data processing will be very small if the acquisition of data is done automatically using a computer program and if for data processing, the chemometrics is used.

According to Guide to the Expression of Uncertainty in Measurement, estimation of uncertainties is based on the identification and quantification of the effect of influence parameters, and requires an understanding of the measurement process, the factors influencing the result, and the uncertainties associated with those factors.

In quantitative analysis, the uncertainty can be expressed as standard deviation or relative standard deviation of the measurement. Contrary to quantitative results, uncertainty in qualitative analysis cannot be expressed as an interval around the predicted value (Pulido et al., 2003). The ways of estimating the uncertainty in qualitative or screen systems are: contingency tables, Bayes' theorem, statistical intervals, and performance curves. The selection of one of these procedures should be done accordingly with the analytical problem and the screen technique used. For example, when the screening technique gives positive or negative answers, the only procedure that cannot be used is the one based on statistical intervals; contingency tables can be used in the majority of applications; when an instrumental screening that gives continuous response is used, the recommended procedure is that based on statistical intervals.

ISO/IEC 17025 requested for the laboratories, the estimation of overall uncertainties associated to the analytical process. The laboratories must report the values of the uncertainties where specified by the method, where required by a client, and/or where the interpretation of the results could be compromised by the lack of knowledge of the uncertainty. The clients of the laboratories are usually using the values of uncertainties for comparative purposes. The clients may be interested in (Mueller, 2002): the reliability of the results and the possibility of their complementation by a statement about their uncertainty; knowledge of the certainty of the conformity statement made about the tested product; and if the test reports are correct, useful, and comprehensive for the laboratory's clients. Accordingly, the values of the uncertainties related to qualitative analysis or to each of the steps of the analytical process in the case of the quantitative analysis must be available.

RELIABILITY OF MEASUREMENTS

Reliability is the maintenance of quality through time (Aboul-Enein et al., 2000). For analytical chemistry, reliability is the correspondence of results (analytical information) obtained using different apparatus.

In the previous subchapter (Uncertainty of measurement), the reliability was defined as a function of the uncertainties of sampling process, method, instrument, and data processing. The reliability can be also defined as a function of the reliabilities of the sampling process (R_S), method (R_M), instrument (R_I) and data processing (R_{DP}) (Aboul-Enein et al., 2000):

$$R_{AI} = f(R_S, R_M, R_I, R_{DP})$$

But from a practical point of view it is easier to determine the reliability as a function of the uncertainties than as a function of the reliabilities of the same parameters. Using a smaller scale to express the reliability of the analytical information as a function of the uncertainties, one can consider the uncertainties of homogeneity of the sample, recovery, analysis blank, measurement standard, calibration, matrix effect and interferences, measuring instrument, and data processing in order to estimate the overall reliability. To obtain the best reliability for analytical information, it is essential to start with a reliable sample and to use a reliable sampling process, a reliable method connected with the type of analysis, reliable instrumentation, and the best software for data acquisition and processing.

Reliability is essential in quality control. It can be increased by automation. Three qualities are essential for the system analyst in order to obtain accurate and reliable analytical information: capability, correctness, and creativity (Aboul-Enein et al., 2000).

Taking into account the contribution of sample, sampling process, instrument, method, and data processing to the reliability of the analytical information, the reports of the audit on each of these parameters will reflect the overall reliability of the analytical information obtained in the laboratory.

STANDARDS AND STANDARDIZATION

Standards and standardization are essential for laboratories. Standard methods are requested for the quality control of a product in order to be introduced into the market. The standard methods are defined in terms of accuracy and precision.

Standardization refers to the compounds (standard compounds) and to analytical methods (standard methods). Primary standards are recognized to have the highest metrological qualities for their specified fields of application. The list of primary standards is established at the international level by the Bureau International des Poids et Mesures (BIPM). Every nation can select a few of the primary standards from this list or add to its primary standards in order to obtain a new list of primary standards that suit the nation's needs. The quality of the standards is essential.

The standards whose value is assigned by comparison with a primary standard of the same quantity are called secondary standards. The quality of the secondary standard can be determined using a primary standard and a standard method. The primary standard and the standard method used for the assessment of the quality of a secondary standard should be specified for each of the secondary standards used in the laboratory.

The qualities of a standard are: it must be illustrative from a quality and quantity point of view; it must possess good stability and homogeneity; and it must have a composition similar with the sample that will be assayed.

A standard method is the one that assures maximum reliability when it is used for the characterization of a material. The analytical methods used as standard methods must have the following characteristics: rapidity, reproducibility, flexibility, and reliability. Sometimes it is impossible to find a standard method that can be used for the determination of the quality of a certain sample. In this case, two or more methods with similar analytical performances in terms of sensitivity, detection limits, and selectivity may be used to certify the quality of the sample.

The standard methods and the standards are simple to be used only for simple samples. The increase of the complexity of the matrix makes the utilization of standard and standard methods difficult. In this case, one can find reliable methods of analysis that can be standardized and used for the assessment of the quality of samples with a given complexity. The parameters of the standardized method and the way adopted for its standardization should be recorded and presented to the audit.

A microdata bank containing standards, standard methods, and methods of standardization of the substances and analytical methods is requested for each laboratory and should be presented for the audit.

SUPPLIES ORGANIZATION

An efficient laboratory requires a defined organization of its supplies. A laboratory should have a documented procurement process to ensure that all supplies purchased meet requirements for their intended use. Microbiological media, for example, should not be accepted without completed certificates of analysis from the vendor. Also, sterile supplies must be inspected upon receipt and must have documentation of the sterilization cycle used. Vendor audits should be performed, where appropriate, to certify the manufacture of supplies.

Written procedures should describe how reagents, pure chemicals, and any combinations therefrom are inspected or tested to confirm their identities. A permanent record should be kept listing the name of each material received, the manufacturer's expiration date (if applicable), the laboratory's assigned expiration date, and any testing or inspection performed to confirm its identity. In assigning expiration dates, the laboratory should determine the criteria and have a written procedure describing the criteria, e.g. first use date, receipt date, or manufacturer's recommendations.

In addition to assigning expiration dates to materials, it is a strongly recommended practice that the laboratory follow a first-in, first-out (FIFO) inventory system to ensure that any supply having a limited shelf stability is used before its expiration date. A lack of control in this area can cause delays in testing and possibly the mistaken use of expired materials which can cause questionable test results.

An observation of the organization of supplies in the laboratory can be a significant indicator of the concern that laboratory personnel have for efficiency, inventory control, and laboratory cleanliness. When conducting an audit or when being given a preliminary tour of a facility, it is very useful to observe and ask about the organization and control of supplies. Organization of supplies is another area that reflects the concerns and attitudes of the laboratory management.

PROFICIENCY TESTING

Proficiency testing was developed as a tool to evaluate clinical laboratory accuracy (Belk and Sunderman, 1947). Concerns have been raised relating to the use of proficiency testing programs as criteria for accreditation (Daniel, 1990). Many laboratories, e.g. clinical laboratories and federally regulated environmental laboratories, are required to pass proficiency testing in order to become accredited and maintain accreditation.

Where proficiency testing is a requirement for laboratory accreditation, it is wise to inspect the records of the laboratory's proficiency testing program. The frequency of testing should at least meet the minimal requirements of the accreditation program. The laboratory should have a procedure which details how "check" samples are handled, stored, tested, and reported, and where results are filed.

Proficiency test records must show that "check" samples were handled according to routine sample control procedures, testing, and document review. This is to assure that the objective of the proficiency testing program was met (Daniel, 1990).

Where proficiency testing is not a requirement, it is strongly recommended as an evaluation tool to improve the accuracy of testing.

Some of the criteria that can be used to measure the usefulness of a "check" sample are: use of a national standard analyte, stability of an analyte in test sample during shipping, number of internal repeat tests performed, and number of external tests performed.

The recommended testing program should be corroborated with an external, unbiased laboratory. But a successful program can also be developed among multiple laboratories of an individual business, which provides discretion and confidentiality in this decade of corporate competitiveness. In this situation, one central laboratory acts as the "check" sample library and also utilizes an outside contract laboratory to confirm sample identification.

IV

OTHER AREAS TO CONSIDER

CUSTOMER SATISFACTION

One of the laboratory's most important responsibilities is to satisfy the needs of its customers. Competitiveness, technical competence, and growth are influenced by the concern given to satisfying customers.

Part of an audit should be to determine if a customer satisfaction program exists. If a program exists, determine if the employees are aware of the needs of the customers and whether the program is successful.

A laboratory should periodically solicit comments and requirements from all its internal and external customers. The responses should be evaluated and action should be taken where appropriate. The latter is a simplistic form of quality function deployment (Bossert, 1990; Day, 1993). In simple terms, the laboratory should ask its clients how service can be improved. Internally, employees should periodically be asked for feedback that can be used to help improve customer relations, conditions of the testing facility, employee development and training, and more.

It is sufficient to say there definitely is a place for total quality processes (customer satisfaction, quality improvement, quality function deployment, lean laboratory, statistical quality control, etc.) in the laboratory.

LABORATORY HEALTH AND SAFETY

Health and safety of the laboratory personnel are usually areas of concern to the laboratory management and not to the client who contracts testing from the laboratory. But, a smart auditor should investigate the design of the facility and any modifications to protect employees from chemical and biological contaminants. The laboratory should have a safety program for handling chemicals, biologicals, and any hazardous materials. It should be a well-documented program with information accessible to all employees at all times (not only working hours, but also after normal working hours and on weekends as well). Material Safety Data sheets should be readily accessible and familiar to all laboratory personnel who handle hazardous materials.

Why is the area of heath and safety important to an auditor? A well-run laboratory shows observable concern for its employees' health and safety. When health and safety concerns are given low priority, be sure that other areas will show signs of weak management.

V

COMMUNICATIONS AND FOLLOW-UP

CLOSING CONFERENCE, FINAL REPORT, AND FOLLOW-UP

At the conclusion of an audit, all findings should be summarized and discussed. It is very important that this is a summary. There should be no surprises at this closing conference or meeting. Each individual who accompanied the auditor(s) during the audit should have been made aware of concerns when they were made apparent. The auditor and the client's management who are intimately responsible for implementing any changes in the audited process should attend the meeting. In other words, the discussion should involve top management representation who have the authority to initiate and support any recommended changes. Personnel who are held accountable by management for the supervision of the audited areas should also attend this meeting, because they must understand the reasons for and actually implement any recommended changes.

At the closing meeting the auditor reiterates each observation made, assigns levels of concern to each observation, and listens to the client respond to each observation. This is the key to the success of the closing conference. The auditor must exhibit a balance in leadership by showing good listening skills and providing clear and concise details of the audit.

The closing conference is the first opportunity for the client to formally respond to the audit findings by prioritizing any problems observed and stating their plans for resolving problems. If contradictions develop in the understanding or interpretation of observations, they should be discussed at this time. It will take an auditor's strong skills to properly and professionally deal with any misunderstandings at that time. It is the auditor's responsibility, for his/her profession and for the organization he/she represents, to assure that observations are weighted in a reasonable and appropriate manner based on the reasons or requirements originally set forth in planning to perform the audit. The auditor must then assure that the client provides a schedule and plans to resolve each problem discussed. The most important point agreed upon at the closing conference should be that resolving the discussed problems will ensure continued accuracy and confidence in the laboratory's business.

The auditor should give a list of observations to the client's top management at the closing conference. A final report is written by the auditor or auditing team after the closing conference. The report should be written to the laboratory's top official and should contain the list of observations that were stated at the closing conference. The observations should be classified by importance. In addition to the summary of observations and their classifications, the final report should contain a statement of the time required for a response. Usually a period of 30 days is allowed for a client to respond. And additional 60–90 days is expected for the implementation of changes. The time for implementation should be based on both requirements of the auditor's firm or an agreement between the laboratory and the auditor.

Follow-up to an audit can be performed in two ways. A physical reinspection should be performed within 30 days, if possible, if any relevant change is made to facility design, housekeeping and maintenance practices, or changes in instrumentation. Most other areas where changes are made can be docu-

Table 1   Criteria for Determining Audit Frequency

Factor	Higher concern	Lower concern
1. *Quantity of tests performed*		
High or moderate number	X	
Low number		X
2. *Nature of testing*		
Health related	X	
Legal related	X	
Routine		X
Research	X	
3. *Laboratory performance history*		
Excellent		X
Average	X	
4. *Test methodology*		
Custom or new	X	
Standard		X
5. *Laboratory audit history*		
Accreditation		X
No accreditation	X	
6. *Regulatory audits (e.g., FDA, EPA)*		
Met requirements		X
Did not meet requirements	X	

mented, and a written statement of such changes should be required from the client's top official within the specified period of time. In many cases, procedural or documentation changes can be approved without an immediate physical reinspection, and in the next scheduled audit of the laboratory, particular attention should be first directed toward those changes.

FREQUENCY OF AUDIT

In general, a standard frequency for quality auditing of a laboratory does not exist, except where it is specified in the accreditation criteria by which a laboratory is evaluated. Planning an audit schedule is very individualistic, and it is customized to the requirements of the quality assurance program (if internal auditors are utilized) or of the firm contracted to perform the audit (if external auditors are contracted).

Table 1 in this chapter is an example of a list of criteria, which can be used to determine how often to audit.

It is prudent to perform a quality audit of each laboratory at least annually. Laboratory auditing can be prioritized by using the higher concern factors for increasing the frequency of audits and the lower concern factors for decreasing the frequency of audits. The sample table is only an example of prioritizing criteria. It is important that each firm develop its own selection criteria. They should use those criteria consistently to determine the frequency of quality audits of client laboratories.

EPILOG

A laboratory quality audit should always be performed with the primary intention of improving the laboratory. Since many areas of quality improvement exist, the audit becomes an important management tool for evaluating technical competence and competitiveness. Good communications can drive organizational improvement. When a continuous quality improvement process is in place, a laboratory can become more productive, more competitive, and a better place to be employed.

Audits performed by government agencies, accreditation groups, and international standards certification groups are usually very thorough. These audits should be incorporated into a total quality laboratory audit (TQLA) plan, which combines internal and external audits to evaluate the laboratory as often as possible. Add these audits to an effective customer satisfaction process and the laboratory will be well prepared for the journey toward continuous quality improvement.

Appendix A

SAMPLE LABORATORY QUALITY AUDIT DOCUMENT

A. Business Data

1. Name and address of firm _____
   _____
   _____

2. Lab Director, name _____
   Phone and fax number _____

3. Contact person (if other than Lab Director) _____

4. Name of tests performed for (hospital or company name) _____

5. Materials submitted for testing, annual requirements _____

6. Accreditations, organizational memberships _____

7. Date of last FDA inspection _____
   (State whether GMP or GLP inspection) _____
   Are results available? _____

8. Results of this audit, approval status _____

   Date of Audit _____ Auditor _____

The following audit questions are categorically separated to simplify information gathering. The grading scheme is based on three levels of competency, and the author should mark the appropriate number for the corresponding level of competency; 3 = excellent, 2 = average, 1 = unsatisfactory.

B. Personnel

All contract laboratory testing should be performed or supervised by a scientist(s) whose education, experience, and knowledge lend credence to the performance of testing and the analysis of results. The scientific understanding of the test methodology is an important qualification, so that discussion of analysis results is effective and efficient. Since microbiological testing can be affected by poor hygiene, it is necessary that an adequate health program be administered in the laboratory.

Questions:

Data

Who is responsible for supervision of the laboratory? _____

_____

What is their educational and training background? _____

_____

How many analysts/microbiologists are available in the laboratory?

_____

Information

Are the educational and training backgrounds of all
analysts sufficient to perform testing on submissions?   3   2   1

Is there a training program for GMPs, or GLPs?   3   2   1

Is there a training program to maintain job competency?   3   2   1

Is there an instrumentation/test procedure training
program with annual recertification?   3   2   1

Is the program offered internally?   3   2   1

Is there documentation of qualified training and
certification of each analyst who performs
tests on submissions?   3   2   1

Are analysts certified (e.g., Microbiology)?   3   2   1

Comments: _____
_____
_____

C. Facilities

Physical facilities should be adequate with respect to space, equipment, environmental control, and maintenance. Good housekeeping should be evident. Special requirements, such as clean rooms, should be adequately maintained for the respective use.

Questions:

Data

Who is responsible for administering the general housekeeping
program? _____
_____

Information

Is adequate space available for the type of testing performed?   3   2   1

Does the laboratory appear organized?   3   2   1

Is there a routine housekeeping schedule?   3   2   1

Comments: _____
_____
_____

D. Instrumentation and Calibration

The laboratory should follow a routine preventive maintenance and calibration program. Instrumentation used in the preparation and analysis of samples should be maintained and calibrated at appropriate time intervals and with sufficient frequency to assure high quality performance. Program procedures should be written, and there should be documentation of each maintenance and calibration performed on an instrument.

Each instrument that is calibrated must be labeled with the date when the most recent calibration was performed. Calibrations should be performed by a person experienced with the instrument using appropriate certified standards.

Questions:

Data

List the main instrument(s) used to perform testing on samples
submitted. _____

_____

Who are the manufacturers of the listed instruments? _____

_____

List the names of any outside firms that perform calibration of any
of the listed instruments. _____

_____

What standards are used for calibration? _____

_____

_____

Information

Does a preventive maintenance (PM) program exist for the instrument(s)?	3	2	1
Is the person who performs PM qualified?	3	2	1
Are there written procedures for the operation and calibration of the instrument(s)?	3	2	1
Are there written procedures for the preventive maintenance program?	3	2	1
Is there a written schedule for calibration?	3	2	1
Are all calibrations documented?	3	2	1
Is each instrument visibly labeled with its most recent calibration date?	3	2	1
Are standards used to calibrate an instrument?	3	2	1
If reference standards are used, is there a written procedure which specifies the vendor of the standard and its required handling?	3	2	1
If nonreference standards are used, is there a written procedure detailing the required preparation of the standard?	3	2	1

Are thermometers, if used, calibrated against standard
NITS thermometers? 3 2 1

Are all instruments calibrated by qualified personnel? 3 2 1

Are all instruments operated by experienced personnel? 3 2 1

Are instruments calibrated frequently enough to ensure
reproducibility and reliability? 3 2 1

Comments: _____

_____

_____

E. Quality Assurance

1. Program

A formal quality program should exist with appropriate documents that
state the responsibilities and objectives of the programs. If a formal program
does not exist, the requirements of this audit in the specific areas that
follow must be met, in addition to the acquisition of a statement by the
lab director to initiate a formal quality program within 6 months of the
audit.

Quality control must be performed in the laboratory. Appropriate
protocols should be written and available for review. The laboratory must
provide a signed statement which says that they comply with GMP and,
where applicable, with GLP regulations.

Questions:

Data

Who administers the quality assurance program? _____

Who is responsible for laboratory document control? _____

_____

Information

Does the laboratory follow a quality assurance program? 3 2 1

Are there written protocols for the quality assurance program? 3 2 1

Is there a written procedure for document control? 3 2 1

Comments: _____

_____

_____

2. Test Procedures

Procedures used for testing of materials should be specified by (company name) Quality Assurance. Any modifications to the procedures must be reviewed and approved by (company name) Quality Assurance. All procedures must be in written format, reviewed and signed by personnel with appropriate levels of authority, on periodic basis. Procedures should state or reference the required materials and standards necessary to perform the test(s). Each modified procedure must be validated, reviewed, and approved by (company name) Quality Assurance.

Where appropriate, a routine qualification of the testing capability of a laboratory can be performed. (Company name) can submit to the contract laboratory a combination of standard, spiked samples along with placebo samples. Routine qualification should be performed at least on an annual basis.

Questions:

Data

What compendial methods are followed? _____

From where are reference standards purchased for procedures? _____

_____

Information

Do written procedures exist for testing all materials submitted?	3	2	1
Are written procedures reviewed, and are they up-to-date?	3	2	1
Do procedures represent compendial methods?	3	2	1
Does a protocol exist which requires that client be notified and asked to review any changes made to test procedures?	3	2	1
Have procedures been validated for precision and reproducibility?	3	2	1
Does a protocol exist that states what changes would cause a revalidation to occur?	3	2	1

Are procedures qualified periodically after the
initial validation?   3 2 1

Are reference standards specified in the procedure(s)?   3 2 1

When testing a material submitted, if a procedure
is repeated for any reason, is (company name) formally notified? 3 2 1

Does the laboratory take part in a proficiency testing
program on a routine basis?   3 2 1

3. **Chemicals, Biologicals, References Standards**

a. *Chemicals and Reagents, Purchased*

Chemicals and reagents should be used only within labeled or posted expiration dates. They should be purchased from reputable manufacturers, and inspected periodically to confirm their identification. They should be stored appropriately.

A satisfactory program should employ a system that alerts users to obtain new supply before the expiration date of the chemical or reagent is reached.

Questions:

Data

From where are chemicals and reagents purchased?

_____
_____
_____
_____

What manufacturers' brands are purchased?

_____
_____
_____
_____

Information

Are vendors approved through a vendor certification program? 3 2 1

Is a first-in-first (FIFO) system followed?   3 2 1

Is yes, for what materials?

_____

_____

Are chemicals inspected for proper labeling and expiration
dating when they are received?   3   2   1

Are expiration dates, where applicable, logged-in?   3   2   1

Are chemicals tested to confirm their identification
when they are received?   3   2   1

Is there a written procedure for each identification test?   3   2   1

Are chemicals and reagents stored appropriately?   3   2   1

If refrigeration is required, is the temperature monitored?   3   2   1

Is there a DEA approved program for narcotic storage?   3   2   1

b.   *Chemical or Biological Reference Standards, Purchased*

All chemical or biological standards should be purchased from a qualified
source. Each standard should be accompanied by a certificate or analysis
stating the source of testing, date of testing, and the standard to which
the product was compared. All standards must be stored under conditions
stated by the supplier. Expiration dating must be present on the standard,
as well as on any further products prepared by these standards.

Questions:

Data

From whom are Reference Standards purchased? _____

_____

_____

Information

Is a certificate of analysis obtained with each Standard?   3   2   1

Is each Standard tested internally to confirm its quality?   3   2   1

Are Standards stored appropriately?   3   2   1

Does each Standard have a legible expiration date?   3   2   1

c. *Chemical or Biological Reference Standards, Internally Prepared*

All chemical or biological standards that are prepared internally should be tested to confirm their quality with known, purchased standards or with standards supported by highly reliable data and testing. All standards must be stored under the best conditions to support their stability. All standards must be dated for their expiration. Water used in test procedures and for cleaning equipment/supplies must be of acceptable purity for its intended use.

Questions:

Data

What Standards used in testing materials are prepared internally? _____

What is the purity (or grade) of water used for analyses and preparation of standards? _____

What is the purity (or grade) of water used for cleaning glassware and other equipment? _____

What tests are performed on the water to determine its purity (or grade)?

Information

Are standards prepared frequently?	3 2 1
Are internally prepared standards tested routinely to confirm their quality?	3 2 1
Are they tested frequently?	3 2 1
Are the standards properly stored?	3 2 1
Are the standards dated for expiration?	3 2 1
Have the expiration dates been validated?	3 2 1

Are water purity tests performed frequently?   3   2   1

Are tests on water adequate to determine its purity?   3   2   1

If a treatment system is used to prepare water for
laboratory use, does a written procedure exist for
monitoring that system?   3   2   1

4.   Sample Receipt, Handling, Storage, Documentation

A written standard procedure should exist for documentation and labeling
of samples and for storage conditions. The procedure should assure that
all samples can be tracked from receipt to testing. Samples should be stored
under conditions which are specified by (company name), where appropriate. Storage facilities should be controlled and monitored for temperature
and humidity where necessary. The disposition of all samples after testing
should be agreed upon by client and the contract laboratory.

Questions:

Data

Who is responsible for receiving samples for testing? _____
_____

What documentation and coding is used to track a sample from receipt to
test? _____
_____
_____
_____

Where are samples stored before and after testing? _____
_____

How long are samples kept after test results are reported? _____
_____

What happens to a sample after testing and final report are complete?
_____

Information

Is there a written procedure for documentation, labeling,
and storage of samples?   3   2   1

Is the storage facility monitored and controlled
for temperature? For humidity?   3   2   1

Comments: _____

_____

_____

5. Date Recording and Reporting

All raw data should preferably be recorded in bound notebooks. Data
should have proper identification to track original data to final report.
Data and calculations should be reviewed by a second analyst or by a
supervisor before the final report is written. A formal written explanation
should be provided whenever a retest is performed. A single person should
be named by the contract laboratory as the contact if any questions arise
regarding testing and data. Confidentiality must be assured for all client-
related data.

Questions:

Data

Who should be contacted for any questions regarding results? _____

_____

Who transfers raw data to the final report forms? _____

_____

Information

Is raw data recorded in bound books?   3   2   1

Is the instrument used identified with the raw data?   3   2   1

Is there a way to track a final report to the original raw data?   3   2   1

Is the format of a final report complete and thorough?   3   2   1

Is there a review of the data, calculations, and final report?   3   2   1

Do all reviewers sign the data or final report?   3   2   1

Are all test data reported to client?   3   2   1

If a retest is performed, is a formal written explanation
of the reason for the retest provided automatically?   3   2   1

F. Audit Considerations for Special Testing

This section will be used by an auditor experienced in the scientific discipline appropriate to the methodology performed by the laboratory. Specific review questions should be developed by a person directly responsible for and experienced in the type of testing being audited. (The following sections contain a list of sample questions for specific areas of testing. The list is not all inclusive.)

1. Microbiological Bioburden Testing

Have sterilizer cycles been validated for media preparation?
Have test methodologies been validated to assure that no inhibitory or cidal activity is present?
Can microbiological media prepared internally be tracked to the respective lot number and sterilizer cycle used?
How are specimens collected, transported, and preserved?
Where is microbial testing performed?
Is there an environmental monitoring program for the area where testing is performed?
What controls are used to validate media, methods, and test environment?
When was the last proficiency test performed? Results?

2. Sterility Testing

Where is testing performed? (hood, clean room, both, and neither)
Is testing performed manually, semiautomatically, or is the system totally automated?
What is the frequency of environmental testing in the test area?
What are the environmental action levels?
What actions are taken responding to a result which is above the action level?
Are there written procedures for environmental testing and action level response?
Are personnel tested routinely for hand sanitation?
What gowning is used by testing personnel?
What procedure is followed when a positive test occurs in test sample?
What procedure is followed when a positive test occurs in negative control or if positive controls do not grow adequately?
If a test is repeated at any time, is client notified?

3. Bacterial Endotoxins Testing

Where is the testing performed?
Is testing performed manually, semiautomatically, or is the system totally automated?

What method is used to depyrogenate testing supplies?
Are any supplies purchased with depyrogenation certification?
If water is required to reconstitute sample, what is the source and quality of the water?
Is water for injection tested routinely? How often?
If a negative control results in a positive endotoxin test, what procedure is followed?
If inhibition or enhancement occurs in testing, what procedure is followed?

4. Pyrogen Testing, and Other Safety Tests

Is the animal facility managed to comply to GLPs?
Are animals cared for by experienced personnel?
Does a veterinarian routinely check the health of the animals?
How often are new animals purchased?
What time intervals are allowed between testing on the same animals?
How often are the thermometers calibrated?
Is the animal facility inspected frequently for sanitation?
What standard of reference is used for absorption from site of injection?

References

1. Garfield FM. Quality Assurance Principles for Analytical Laboratories. AOAC, 1984.
2. Taylor JK. Quality Assurance of Chemical Measurements. Lewis Publishers, Inc. 1980.

Appendix B

NATIONAL VOLUNTARY LABORATORY ACCREDITATION PROGRAM, NIST HANDBOOK 150

Editors
V. R. White, D. F. Alderman,
and C. D. Faison

Procedures and General Requirements

1. GENERAL INFORMATION

1.1. Purpose and Scope

1.1.1. The NIST Handbook 150 sets forth the procedures and general requirements under which the NVLAP operates as an unbiased third party to accredit both testing and calibration laboratories. Supplementary technical and administrative requirements are provided in supporting handbooks (NIST Handbook 150 series) and documents, as needed, depending on the criteria established for specific Laboratory Accreditation Programs (LAPs).

1.1.2. This handbook is to be used by laboratories in developing their quality, administrative, and technical systems that govern their operations. Laboratory clients, regulatory authorities, and accreditation bodies may also use it in confirming or recognizing the competence of laboratories.

1.1.3. Notes are given to provide clarification of the text, examples, and guidance. They do not contain requirements and do not form an integral part of the accreditation criteria.

In Sec. 5 of the handbook, the term *NVLAP Note* indicates NVLAP reference to or clarification of a particular requirement.

1.1.4. Compliance with regulatory and safety requirements on the operation of laboratories is not covered by this handbook.

1.1.5.   If testing and calibration laboratories comply with the requirements of this handbook, they will operate a quality system for their testing and calibration activities that also meets the requirements of ISO 9001:1994 when they engage in the design/development of new methods, and/or develop test programs combining standard and nonstandard test and calibration methods, and ISO 9002:1994 when they only use standard methods.

1.2.   Organization of Handbook

Section 1 describes considerations, which relate in general to all aspects of NVLAP. Section 2 describes how LAPs are requested, developed, announced, and terminated. Section 3 describes the process for accrediting laboratories. Sections 4 and 5 provide the criteria for NVLAP accreditation found in clauses 4 and 5 of ISO/IEC 17025:1999. Annexes A and B present requirements for referencing NVLAP accreditation and achieving traceability, respectively.

1.3.   Description of NVLAP

1.3.1.   The NVLAP is a U.S. government entity administered by the National Institute of Standards and Technology (NIST), an agency of the Department of Commerce. The NVLAP accredits testing and calibration laboratories found competent to perform specific tests or calibrations.

1.3.2.   The NVLAP operates a quality system that is compliant with ISO/IEC Guide 58:1993.

1.3.3.   The NVLAP is a voluntary system which:

   a.   provides a mechanism for the recognition of testing and calibration laboratories based on internationally accepted standards and procedures;
   b.   provides laboratory management with documentation for use in the development and implementation of their quality systems;
   c.   identifies competent laboratories for use by regulatory agencies, purchasing authorities, and product certification systems;
   d.   provides laboratories with a process to aid them in reaching a higher level of performance, resulting in the generation of improved engineering and product information; and
   e.   promotes the acceptance of test and calibration results between economies and accreditors to support trade facilitation activities.

1.3.4.   The NVLAP is comprised of a series of LAPs established on the basis of requests and demonstrated need. The Chief of NVLAP does not unilaterally propose or decide the scope of a LAP. The specific tests and calibrations, types of tests and calibrations, or standards to be included in a LAP are determined by an open process during the establishment of the LAP (Sec. 2).

1.3.5. The NVLAP programs are established:

   a. for public and private testing and calibration laboratories, including commercial laboratories, manufacturers' in-house laboratories, university laboratories, and federal, state, and local government laboratories;

   b. to meet legal requirements, regulations or codes, and contract specifications, or to recognize laboratories found competent to meet the needs of their clients; and

   c. as the basis for guidance to facilitate agreements on mutual recognition of accreditation of laboratories between NVLAP and other accreditation systems.

1.3.6. The NVLAP accreditation is:

   a. based on evaluation of a laboratory's management and technical qualifications and competence for conducting specific test methods, measurements, and services in specified fields of testing or calibration;

   b. granted only after thorough evaluation of an applicant has demonstrated that all NVLAP criteria have been met;

   c. acknowledged by the issuance of two documents to attest to that compliance: 1) a Certificate of Accreditation, and 2) a scope of accreditation, which details the specific test methods, measurements, and services for which a laboratory has been accredited; and

   d. administered in a nondiscriminatory manner, and not conditional on the size of a laboratory or on its membership in any association or group.

Accreditation does not relieve a laboratory from complying with applicable federal, state, and local laws and regulations.

1.4. References

The following documents are referenced in the text or notes of this handbook:

- ANSI/NCSL Z540-2-1997, *U.S. Guide to the Expression of Uncertainty in Measurement.*
- BIPM/IEC/IFCC/ISO/IUPAC/IUPAP/OIML: 1993, *International Vocabulary of Basic and General Terms in Metrology* (VIM).
- BIPM/IEC/IFCC/ISO/IUPAC/IUPAP/OIML:1993, *Guide to the Expression of Uncertainty in Measurement* (GUM).
- ISO 8402:1994, *Quality management and quality assurance—Vocabulary.*
- ISO 9001:1994, *Quality systems—Model for Quality Assurance in Design, Development, Production, Installation, and Servicing.*
- ISO 9002:1994, *Quality systems—Model for Quality Assurance in Production, Installation, and Servicing.*

- ISO Guide 43-1:1997, *Proficiency testing by interlaboratory comparisons*—Part 1: *Development and operation of proficiency testing schemes.*
- ISO Guide 43-2:1997, *Proficiency testing by interlaboratory comparisons*—Part 2.: *Selection and use of proficiency testing schemes by laboratory accreditation bodies.*
- ISO/IEC 17025:1999, *General requirements for the competence of testing and calibration laboratories.*
- ISO/IEC Guide 2:1996, *Standardization and related activities— General vocabulary.*
- ISO/IEC Guide 58:1993, *Calibration and testing laboratory accreditation systems—General requirements for operation and recognition.*
- NIST Technical Note 1297: *Guidelines for Evaluating and Expressing the Uncertainty of NIST Measurement Results,* 1994 Edition.

1.5. Definitions

For the purposes of this handbook, the following definitions apply:

1.5.1. Accreditation: Formal recognition that a laboratory is competent to carry out specific tests or calibrations or types of tests or calibrations.

1.5.2. Accreditation Criteria: Set of requirements used by an accrediting body, which a laboratory must meet in order to be accredited.

1.5.3. Accuracy of Measurement: Closeness of the agreement between the result of a measurement and a true value of the measurand (See also uncertainty of measurement).

NOTE 1. "Accuracy" is a qualitative concept.

NOTE 2. The term *precision* should not be used for "accuracy."

[VIM: 1993, 3.5].

1.5.4. Approved Signatory: Individual who is recognized by NVLAP as competent to sign accredited laboratory test or calibration reports.

NOTE. The Approved Signatory is responsible for the technical content of the report and is the person to be contacted by NVLAP, laboratory clients, or others in case of questions or problems with the report. Approved signatories shall be persons with responsibility, authority, and technical capability within the organization for the results produced.

1.5.5. Assessment, On-site: On-site examination of a testing or calibration laboratory to evaluate its compliance with the conditions and criteria for accreditation.

1.5.6. Authorized Representative: Individual who is authorized by the laboratory or the parent organization to sign the NVLAP application forms and commits the laboratory to fulfill the NVLAP requirements.

1.5.7. Best Measurement Capability: Smallest uncertainty of measurement of a laboratory can achieve within its scope of accreditation, when performing more-or-less routine calibrations of nearly ideal measurement standards intended to define, realize, conserve or reproduce a unit of that quantity or one or more of its values, or when performing more-or-less routine calibrations of nearly ideal measuring instruments designed for the measurement of that quantity.

1.5.8. Calibration: Set of operations that establish, under specified conditions, the relationship between values of quantities indicated by a measuring instrument or measuring system, or values represented by a material, measure or a reference material (RM), and the corresponding values realized by standards.

NOTE 1.   The result of a calibration permits either the assignment of values of measurands to the indications or the determination of corrections with respect to indications.

NOTE 2.   A calibration may also determine other metrological properties such as the effect of influence quantities.

NOTE 3.   The result of a calibration may be recorded in a document, sometimes called a *calibration certificate* or a *calibration report*.

[VIM: 1993, 6.11].

NOTE 4.   The result of a calibration is sometimes expressed as a calibration factor, or as a series of calibration factors in the form of a calibration curve.

1.5.9.   Calibration Method: (*See* method of measurement.)

1.5.10.   Calibration Procedure: (*See* measurement procedure.)

1.5.11.   Certificate of Accreditation: Document issued by NVLAP to a laboratory that has met the criteria and conditions for accreditation. A current Certificate of Accreditation may be used as proof of accredited status. A Certificate of Accreditation is always accompanied by a Scope of Accreditation.

1.5.12.   Certified Reference Material (CRM): RM, accompanied by a certificate, one or more of whose property values are certified by a procedure which establishes traceability to an accurate realization of the unit in which the property values are expressed, and for which each certified value is accompanied by an uncertainty at a stated level of confidence. (See also RM.)

NOTE 1.   The definition of a "RM certificate" is given in VIM: 1993, 4.2.

NOTE 2.   The CRMs are generally prepared in batches for which the property values are determined within stated uncertainty limits by measurements on samples representative of the whole batch.

NOTE 3.   The certified properties of CRMs are sometimes conveniently and reliably realized when the material is incorporated into a specially fabricated device, e.g., a substance of known triple-point into a triple-point cell, a glass of known optical density into a transmission filter, spheres of uniform particle size mounted on a microscope slide. Such devices may also be considered as CRMs.

NOTE 4.   All CRMs lie within the definition of "measurement standards" or "etalons" given in the VIM.

NOTE 5.   Some RMs and CRMs have properties, which, because they cannot be correlated with an established chemical structure or for other reasons, cannot be determined by exactly defined physical and chemical measurement methods. Such materials include certain biological materials such as vaccines to which an international unit has been assigned by the World Health Organization.

[VIM: 1993, 6.14].

1.5.13.   Client: Any person or organization that engages the services of a testing or calibration laboratory.

1.5.14.   Competence: Ability of a laboratory to meet the NVLAP conditions and to conform to the criteria in NVLAP publications for specific test and calibration methods.

1.5.15.   Deficiency: Nonfulfillment of NVLAP conditions and/or criteria for accreditation; sometimes referred to as a *nonconformance*.

1.5.16.   Error (of measurement): Result of a measurement minus a true value of the measurand.

NOTE 1.   Since a true value cannot be determined, in practice a conventional true value is used.

NOTE 2.   When it is necessary to distinguish "error" from "relative error," the former is sometimes called *absolute error of measurement*. This should not be confused with *absolute value of error*, which is the modulus of the error.

[VIM: 1993, 3.10].

1.5.17. Influence Quantity: Quantity that is not the measurand but that affects the result of the measurement.

EXAMPLES
1. Temperature of a micrometer used to measure length;
2. Frequency in the measurement of the amplitude of an alternating electric potential difference;
3. Bilirubin concentration in the measurement of hemoglobin concentration in a sample of human blood plasma.

[VIM: 1993., 2.7].

1.5.18. Interlaboratory comparisons: Organization, performance, and evaluation of tests or calibrations on the same or similar items or materials by two or more laboratories in accordance with predetermined conditions.

NOTE. In some circumstances, one of the laboratories involved in the intercomparison may be the laboratory, which provided the assigned value for the test item.

[ISO/IEC Guide 43-1:1997, 3.7 expanded].

1.5.19. Internal Audit: Systematic and documented process for obtaining evidence and evaluating it objectively to verify that a laboratory's operations comply with the requirements of its quality system.

1.5.20. Laboratory: Organization that performs tests and/or calibrations. When a laboratory is part of an organization that carries out activities additional to testing and calibration, the term *laboratory* refers only to those parts of that organization that are involved in the testing and calibration process. A laboratory's activities may be carried out at a permanent, temporary, or remote location.

NOTE. The NVLAP further defines *laboratory* as being a physical entity— that is, a testing or calibration facility that is separate and apart physically from any other laboratory whether or not sharing common ownership, management, or quality systems with any other laboratory(s).

The NVLAP previously differentiated between *main facilities* and *subfacilities*. This distinction is no longer recognized (Exception: As long as there is no break in accreditation, any laboratory previously accredited as a *subfacility* may request to be "grandfathered" in its accreditation renewal under the former classification as a *subfacility,* including the unique conditions associated with that classification.).

1.5.21.   Laboratory Accreditation Body: Body that conducts and administers a laboratory accreditation system and grants accreditation.

1.5.22.   Laboratory Accreditation System: System that has its own rules of procedure and management for carrying out laboratory accreditation.

1.5.23.   LAP: Laboratory accreditation program established and administered under NVLAP, consisting of test methods or calibrations relating to specific products or fields of testing or calibration.

1.5.24.   Limits of Permissible Error (of a Measuring Instrument): Extreme values of an error permitted by specifications, regulations, etc., for a given measuring instrument.

[VIM: 1993, 5.21].

NOTE   This term is frequently referred to as *tolerance* in the United States.

1.5.25.   Management Review: Formal evaluation by top management of the status and adequacy of the quality system in relation to quality policy and objectives.

NOTE 1.   Management review may include review of the quality policy.

NOTE 2.   Quality audit results are one of the possible inputs to management review.

NOTE 3.   The term *top management* refers to the management of the organization whose quality system is being reviewed.

[ISO 8402:1994, 3.9].

1.5.26.   Measurand: Particular quantity subject to measurement.

EXAMPLE
Vapor pressure of a given sample of water at 20°C.

NOTE 1.   The specification of a measurand may require statements about quantities such as time, temperature, and pressure.

[VIM: 1993, 2.6].

NOTE 2.   As appropriate, this may be the *measured quantity* or the *quantity to be measured.*

1.5.27. Measurement: Set of operations having the object of determining a value of a quantity.

[VIM: 1993, 2.1].

1.5.28. Measurement Assurance: Process to ensure adequate measurement results that may include, but is not limited to: 1) use of good experimental design principles so that the entire measurement process, its components, and relevant influence factors can be well characterized, monitored, and controlled; 2) complete experimental characterization of the measurement process uncertainty including statistical variations, contributions from all known or suspected influence factors, imported uncertainties, and the propagation of uncertainties throughout the measurement process; and 3) continuously monitoring the performance and state of statistical control of the measurement process with proven statistical process control techniques including the measurement of well-characterized check standards along with the normal workload and the use of appropriate control charts.

1.5.29. Measurement Procedure: Set of operations, described specifically, used in the performance of particular measurements according to a given method.

NOTE   A measurement procedure is usually recorded in a document that is sometimes itself called a *measurement procedure* (or a *measurement method)* and is usually in sufficient detail to enable an operator to carry out a measurement without additional information.

[VIM: 1993, 2.5].

1.5.30. Measuring and Test Equipment (M & TE): All of the measuring instruments, measurement standards, RMs, auxiliary apparatus, and instructions that are necessary to perform a measurement. This term includes measuring equipment used in the course of testing and inspection, as well as that used in calibration.

NOTE   In the context of this handbook, the term *measuring and test equipment* is taken to encompass *measuring instruments* and *measurement standards*. Moreover, an *RM* is considered to be a type of *measurement standard.*

1.5.31. Measuring Instrument: Device intended to be used to make measurements, alone or in conjunction with supplementary device(s).

[VIM: 1993, 4.1].

1.5.32. Method of measurement: Logical sequence of operations, described generically, used in the performance of measurements.

NOTE   Methods of measurement may be qualified in various ways such as:

- substitution method
- differential method
- null method

[VIM: 1993, 2.4].

1.5.33.   NVLAP Lab Code: Unique numeric identifier assigned by NVLAP to each applicant laboratory; e.g., 101000–0. It is used for identification, record-keeping, and database management, and appears on formal accreditation documents.

1.5.34.   Precision: Repeatability of measurement data; the similarity of successive independent measurements of a single magnitude generated by repeated applications of a process under specified conditions.

1.5.35.   Proficiency testing (laboratory):   Determination of laboratory testing performance by means of interlaboratory comparisons.

[ISO/IEC Guide 2:1996, 13.5].

NOTE   For the purposes of this handbook, the term *laboratory proficiency testing* is taken in its widest sense and includes, for example:

- a.   Qualitative schemes—for example, where laboratories are required to identify a component of a test item.
- b.   Data transformation exercises—for example, where laboratories are furnished with sets of data and are required to manipulate the data to provide further information.
- c.   Single item testing—where one item is sent to a number of laboratories sequentially and returned to the organizer at intervals.
- d.   One-off exercises—where laboratories are provided with a test item on a single occasion.
- e.   Continuous schemes—where laboratories are provided with test items at regular intervals on a continuing basis.
- f.   Sampling—for example, where individuals or organizations are required to take samples for subsequent analysis.

[ISO/IEC Guide 43-1:1997, 3.6].

1.5.36.   Quality Manual: Document stating the quality policy and describing the quality system of an organization.

[ISO 8402:1994, 3.12].

1.5.37.   Quality System: Organizational structure, procedures, processes, and resources needed to implement quality management.

NOTE 1. The quality system should be as comprehensive as needed to meet the quality objectives.

NOTE 2. The quality system of an organization is designed primarily to satisfy the internal managerial needs of the organization. It is broader than the requirements of a particular customer who evaluates only the relevant part of the quality system.

NOTE 3. For contractual or mandatory quality assessment purposes, demonstration of the implementation of identified quality system elements may be required.

[ISO 8402:1994, 3.6].

1.5.38. Reference Material (RM): Material or substance one or more of whose property values are sufficiently homogeneous and well established to be used for the calibration of an apparatus, for the assessment of a measurement method, or for assigning values to materials.

NOTE An RM may be in the form of a pure or mixed gas, liquid, or solid. Examples are water for the calibration of viscometers, sapphire as a heat-capacity calibrant in calorimetry, and solutions used for calibration in chemical analysis.

[VIM: 1993, 6.13].

1.5.39. Reference Standard: Standard, generally having the highest metro-logical quality available at a given location or in a given organization, from which measurements made there are derived.

[VIM: 1993, 6.6].

1.5.40. Requirement: Provision that conveys criteria to be fulfilled.

[ISO/IEC Guide 2:1996, 7.5].

1.5.41. Resolution (of a displaying device): Smallest difference between indications of a displaying device that can be meaningfully distinguished.

NOTE 1. For a digital displaying device, this is the change in the indication when the least significant digit changes by one step.

NOTE 2. This concept applies also to a recording device.

[VIM: 1993, 5.12].

1.5.42. Revocation: Removal of the accredited status of a laboratory if the laboratory is found to have violated the terms of its accreditation.

1.5.43. Scope of accreditation: Document issued by NVLAP to a laboratory that lists the test methods or services, or calibration services, for which the laboratory is accredited. A scope of accreditation is always accompanied by a Certificate of Accreditation.

1.5.44. Stability: Ability of a measuring instrument to maintain its constant metrological characteristics with time.

NOTE 1. Where stability with respect to a quantity other than time is considered, this should be stated explicitly.

NOTE 2. Stability may be quantified in several ways, for example:

- in terms of the time over which a metrological characteristic changes by a stated amount, or
- in terms of the change in a characteristic over a stated time

[VIM: 1993, 5.14].

1.5.45. Standard, International (Measurement): Standard recognized by an international agreement to serve internationally as the basis for assigning values to other standards of the quantity concerned.

[VIM: 1993, 6.2].

1.5.46. Standard, Measurement: Material measure, measuring instrument, RM, or measuring system intended to define, realize, conserve or reproduce a unit or one or more values of a quantity to serve as a reference.

EXAMPLES
- a.   1. kg mass standard;
- b. $1.00\,\Omega$ standard resistor;
- c.   standard ammeter;
- d.   cesium frequency standard;
- e.   standard hydrogen electrode;
- f.   reference solution of cortisol in human serum having a certified concentration.

NOTE 1. A set of similar material measures or measuring instruments that, through their combined use, constitutes a standard is called a collective standard.

NOTE 2. A set of standards of chosen values that, individually or in combination, provides a series of values of quantities of the same kind is called a group standard.

[VIM: 1993, 6.1].

1.5.47. Standard, National (Measurement): Standard recognized by a national decision to serve, in a country, as the basis for assigning values to other standards of the quantity concerned.

[VIM: 1993, 6.3].

1.5.48. Standard, Primary: Standard that is designated or widely acknowledged as having the highest metrological qualities and whose value is accepted without reference to other standards of the same quantity.

NOTE   The concept of primary standard is equally valid for base quantities and derived quantities.

[VIM: 1993, 6.4].

1.5.49.   Standard, Reference: (*See* Reference Standard.)

1.5.50.   Standard, Secondary: Standard whose value is assigned by comparison with a primary standard of the same quantity.

[VIM: 1993, 6.5].

1.5.51.   Standard, transport (or transfer):   Standard used as an intermediary to compare standards.

NOTE The term *transfer device* should be used when the intermediary is not a standard.

[VIM: 1993, 6.8].

1.5.52.   Standard, Working: Standard that is used routinely to calibrate or check material measures, measuring instruments or RMs.

NOTE 1.   A working standard is usually calibrated against a reference standard.

NOTE 2.   A working standard used routinely to ensure that measurements are being carried out correctly is called a *check standard*.

[VIM: 1993, 6.7].

1.5.53.   Subfacility: Laboratory operating under the technical direction and quality system of an accredited main facility. (See note under *laboratory*.)

1.5.54.   Suspension: Temporary removal of the accredited status of a laboratory when the laboratory is found to be out of compliance with the terms of its accreditation.

1.5.55. Test: Technical operation that consists of the determination of one or more characteristics of a given product, process, or service according to a specified procedure.

[ISO/IEC Guide 2:1996, 13.1].

1.5.56. Test Method: Specified technical procedure for performing a test.

[ISO/IEC Guide 2:1996, 13.2].

1.5.57. Traceability: Property of the result of a measurement or the value of a standard whereby it can be related to stated references, usually national or international standards, through an unbroken chain of comparisons all having stated uncertainties.

NOTE 1. The concept is often expressed by the adjective *traceable*.

NOTE 2. The unbroken chain of comparisons is called a *traceability chain*.

[VIM: 1993, 6.10].

1.5.58. Uncertainty of measurement: Parameter, associated with the result of a measurement, that characterizes the dispersion of the values that could reasonably be attributed to the measurand.

NOTE 1. The parameter may be, for example, a standard deviation (or a given multiple of it), or the half-width of an interval having a stated level of confidence.

NOTE 2. Uncertainty of measurement comprises, in general, many components. Some of these components may be evaluated from the statistical distribution of the results of series of measurements and can be characterized by experimental standard deviations. The other components, which can also be characterized by standard deviations, are evaluated from assumed probability distributions based on experience or other information.

NOTE 3 It is understood that the result of the measurement is the best estimate of the value of the measurand, and that all components of uncertainty, including those arising from systematic effects, such as components associated with corrections and reference standards, contribute to the dispersion.

This definition is that of the *Guide to the expression of uncertainty in measurement* in which its rationale is detailed (see, in particular, and Annex D [10]).

[VIM:1993, 3.9].

1.5.59. Uncertainty, Type A (Evaluation of): Method of evaluation of uncertainty by the statistical analysis of series of observations.

[GUM: 1993, 2.3.2].

1.5.60. Uncertainty, Type B (evaluation of): Method of evaluation of uncertainty by means other than the statistical analysis of series of observations.

[GUM: 1993, 2.3.3].

1.5.61. Verification: Confirmation by examination and provision of objective evidence that specified requirements have been fulfilled.

NOTE 1. In design and development, verification concerns the process of examining the result of a given activity to determine conformity with the stated requirements for that activity.

NOTE 2. The term *verified* is used to designate the corresponding status.

[ISO 8402:1994, 2.17].

NOTE 3. In the United States, verification is frequently referred to as *calibration*.

1.6. NVLAP Publications

The NVLAP publishes a variety of documents in paper and electronic formats for use by testing and calibration laboratories and others needing information about the NVLAP program. Accredited and applicant laboratories are routinely sent new and revised NVLAP publications. Many of these publications are available on the NVLAP web site, < www.nist.gov/nvlap > .

1.6.1. NIST Handbook 150, *NVLAP Procedures and General Requirements*

This handbook sets forth the procedures under which NVLAP operates and the general requirements for accreditation as prescribed in Title 15 of the U.S. Code of Federal Regulations, Part 285—the primary document describing the legal basis for NVLAP. Sections 4 and 5 of the handbook include all requirements of ISO/IEC 17025, *General requirements for the competence of testing and calibration laboratories.* Annexes A and B contain additional requirements specific to NVLAP.

1.6.2. NIST Handbook 150 Series (Program-Specific Handbooks)

This series of handbooks contains guidance, interpretive information, and technical requirements for the LAPs. A separate handbook is published for each LAP or unique field of testing or calibration, and tailors the general criteria in NIST Handbook 150 to the specific tests, calibrations, or types of tests or calibrations covered by a LAP. A program-specific handbook(s) and NIST Handbook 150 constitute the collective body of requirements that must be met by a laboratory seeking NVLAP accreditation.

1.6.3. NIST Special Publication 810, *NVLAP Directory*

This publication is updated annually and lists the following information for each accredited laboratory: NVLAP Lab Code, laboratory name and address, Authorized Representative, phone and fax numbers, e-mail and URL addresses (if available), accreditation expiration date, and scope of accreditation. The directory is distributed worldwide to participating laboratories, manufacturers, suppliers, retailers, professional and trade associations, standards groups, and government agencies.

The NVLAP web site, <www.nist.gov/nvlap>, also includes a listing of NVLAP-accredited laboratories; it is updated on a periodic basis.

1.6.4. Other Publications

The NVLAP may publish other documents from time to time, as necessary to ensure communication of program information to laboratories and interested parties, such as policy guides, laboratory bulletins, technical briefs, and newsletters.

1.7. Confidentiality

To the extent permitted by applicable laws, NVLAP will protect the confidentiality of all information obtained relating to the application, on-site assessment, proficiency testing, evaluation, and accreditation of laboratories.

1.8. Referencing NVLAP Accreditation (See Also Annex A, p. 112)

1.8.1. The term *NVLAP* (represented by the NVLAP logo) is a federally registered certification mark of the NIST and the federal government, who retain exclusive rights to control the use thereof. Permission to use the term and/or logo is granted to NVLAP-accredited laboratories for the limited purposes of announcing their accredited status, and for use on reports that describe only testing and calibration within the scope of accreditation. The NIST reserves the right to control the quality of the use of the term *NVLAP* and of the logo itself.

1.8.2. The NVLAP's policy is to control the use of the term and logo and to ensure that accredited laboratories express their accredited status in a manner that is clear and accurate, and not misleading. This policy applies to test and calibration reports, letterheads, contracts, business cards, brochures, advertising, web sites, and any other use not specified herein.

1.8.3. The NVLAP-accredited laboratories are authorized to use the term and logo to reference their accredited status, subject to the conditions presented in Annex A. Failure to comply with the conditions may result in suspension or revocation of a laboratory's accreditation.

1.8.4. Use of the term and logo by other persons and organizations shall be authorized in writing by NVLAP on a case-by-case basis.

1.8.5. Photographic and electronic copies of the logo are available from NVLAP upon request.

1.8.6. Use of the term and logo by a laboratory whose status is suspended, revoked, or voluntarily terminated is specified in Secs. 3.9–3.11.

1.9. Mutual Recognition

Consistent with applicable, laws, and regulations, the Director of NIST may negotiate and conclude agreements for NVLAP with other laboratory accreditation entities. These agreements, realized through MRAs, serve to support trade facilitation activities and promote harmonization of laboratory accreditation criteria. At a minimum, any agreement must provide that accredited laboratories of other accreditation bodies meet conditions for accreditation comparable to and consistent with those set out in this handbook.

1.10. Information Collection Requirements

The information collection requirements contained in these procedures have been approved by the Office of Management and Budget under the Paperwork Reduction Act and have been assigned OMB control number 0693–0003.

2. LAP ESTABLISHMENT, DEVELOPMENT, AND IMPLEMENTATION

2.1. Bases for Establishment

The NVLAP establishes LAPs in response to legislative or administrative actions or to requests from private sector entities and government agencies.

2.1.1. LAPs Established Through Legislative or Administrative Actions: Upon receipt of a mandate for a LAP based on legislative or administrative action, the Chief of NVLAP shall publish a *Federal Register* notice:

    a. stating the purpose of the LAP, including the national or international need;
    b. describing the general scope of the LAP;
    c. identifying government agencies having oversight; and
    d. providing information to any interested party wishing to receive routine information on the development of the LAP.

2.1.2. LAPs established by request

2.1.2.1. A request to establish a LAP must be made in writing to the Chief of NVLAP. Each request must include:

a. the scope of the LAP in terms of products, testing services, or calibration services proposed for inclusion;

b. specific identification of the applicable standards and test methods, including appropriate designations, and the organizations or standards-writing bodies having responsibility for them;

c. a statement of the perceived need for the LAP including:

- technical and economic reasons why the LAP would benefit the public interest;
- evidence of a national need to accredit testing or calibration laboratories for the specific scope beyond that served by an existing laboratory accreditation program in the public or private sector;
- an estimate of the number of laboratories that are likely to seek accreditation; and
- an estimate of the number and nature of the users of such laboratories; and

d. a statement of the extent to which the requestor is willing to support necessary developmental aspects of the LAP with funding and personnel.

2.1.2.2. If the requestor is a private sector entity, then the request must include a description of the ways in which the following conditions have been met:

a. public notice of meetings and other activities related to the LAP request is provided in a timely fashion and is distributed in a manner designed to reach the attention of interested persons;

b. meetings are open and participation in activities is available to interested persons;

c. decisions reached by the private sector entity in the development of a request for an LAP represent substantial agreement of the interested persons;

d. prompt consideration is given to the expressed views and concerns of interested persons;

e. adequate and impartial mechanisms for handling substantive and procedural complaints and appeals are in place; and

f. appropriate records of all meetings are maintained and the official procedures used by the private sector entity to make a formal request for an LAP are made available upon request to any interested person.

2.1.2.3. If the requestor is a federal, state, or local government agency, then the request must include a description of the procedures followed or a citation of the specific authority used to identify a need for the LAP. For state and local government agencies, the request must also include a statement explaining why the LAP should be of national scope.

2.1.2.4.   The NVLAP may request clarification of the information submitted in the request.

2.1.2.5.   The Chief of NVLAP shall analyze the request and any supporting information received, and after consultation with interested parties through public workshops and other means, shall determine if there is need for the requested LAP. In making this determination, the Chief of NVLAP shall consider the following:

    a.   the needs and scope of the LAP requested;
    b.   the needs and scope of the user population;
    c.   the nature and content of other relevant public and private sector LAPs;
    d.   compatibility with the criteria referenced in Sections 4 and 5;
    e.   the importance of the requested LAP to commerce, consumer well-being, or the public health and safety; and
    f.   the economic and technical feasibility of accrediting laboratories for the tests or calibrations, types of tests or calibrations, or standards requested.

2.1.2.6.   The Chief of NVLAP shall make the decision to either:

    a.   develop the LAP, if a need has been demonstrated and resources are available for the LAP's development;
    b.   defer development of the LAP until resources become available, if a need has been demonstrated and there are no resources for development; or
    c.   not develop the LAP, if a need has not been demonstrated.

2.1.2.7.   The Chief of NVLAP shall inform the requestor and other interested parties of the LAP decision.

2.2.   Development of Technical Requirements

2.2.1.   Technical requirements for accreditation are specific for each LAP. The requirements tailor the criteria referenced in Sections 4 and 5 to the tests or calibrations, types of tests or calibrations, or standards covered by the LAP.

2.2.2.   The NVLAP shall develop the technical requirements based on relevant and impartial expert advice. This advice may be obtained through one or more public workshops or other suitable means.

2.2.3.   The NVLAP shall make every reasonable effort to ensure that the affected testing or calibration community within the scope of the LAP is informed of any planned workshop. A summary of each workshop shall be prepared and made available upon request.

2.3. Coordination with Federal Agencies

As a means of ensuring effective and meaningful cooperation, input, and participation by those federal agencies that may have an interest in and may be affected by established LAPs, NVLAP shall communicate and consult with appropriate officials within those agencies.

2.4. Announcing the Establishment of a LAP

When NVLAP has completed the development of the technical requirements, it shall publish a notice in *the Federal Register* announcing the establishment of the LAP. The notice will identify the scope of the LAP and advise laboratories how to apply for accreditation.

2.5. Adding to or Modifying a LAP

2.5.1. An LAP may be added to, modified, or realigned based on either a written request or a need identified by NIST. Any person wishing to add or delete specific tests or calibrations, types of tests or calibrations, or standards may submit a request to NVLAP.

2.5.2. The NVLAP may choose to make the additions or modifications available for accreditation under a LAP when:

    a. the additional tests or calibrations, types of tests or calibrations, or standards requested are directly relevant to the LAP;

    b. it is feasible and practical to accredit testing or calibration laboratories for the additional tests or calibrations, types of tests or calibrations, or standards; and

    c. it is likely that laboratories will seek accreditation for the additional tests or calibrations, types of tests or calibrations, or standards.

2.6. Termination of a LAP

2.6.1. The Chief of NVLAP may terminate a LAP when he/she determines that a need no longer exists to accredit laboratories for the services covered under the scope of the LAP. In the event that the Chief of NVLAP proposes to terminate a LAP, a notice will be published in the *Federal Register* setting forth the basis for that determination.

2.6.2. When a LAP is terminated, NVLAP will no longer grant or renew accreditations following the effective date of termination. Accreditations previously granted shall remain effective until their expiration date unless terminated voluntarily by the laboratory or revoked by NVLAP. Technical expertise will be maintained by NVLAP while any accreditation remains effective.

3. ACCREDITATION PROCESS

3.1. Application for Accreditation

3.1.1. Required Information

3.1.1.1. A laboratory may apply for accreditation in any of the established LAPs. The applicant laboratory shall provide a completed application to NVLAP, pay all required fees, agree to conditions for accreditation, and provide a quality manual to NVLAP (or a designated NVLAP assessor) prior to the assessment process.

3.1.1.2. Required information for accreditation includes, but is not limited to:

   a. legal name and full address of the laboratory;
   b. ownership of the laboratory;
   c. Authorized Representative's name and contact information;
   d. names, titles, and contact information for laboratory staff nominated to serve as Approved Signatories of test or calibration reports that reference NVLAP accreditation;
   e. organization chart defining relationships that are relevant to performing testing and calibrations covered in the accreditation request;
   f. general description of the laboratory, including its facilities and scope of operation; and
   g. requested scope of accreditation.

3.1.1.3. The General Application shall be signed by the laboratory's Authorized Representative who commits the laboratory to comply with the conditions of accreditation (see Sec. 3.1.2) and with the requirements contained in Sections 4 and 5 of this handbook. Before signing the application, the Authorized Representative should review all documents provided with the application package and become familiar with NVLAP requirements. Only the Authorized Representative can authorize a change in the scope or nature of a laboratory's application.

3.1.1.4. The laboratory shall provide a copy of its quality manual and related documentation, where appropriate, prior to the on-site assessment. The NVLAP will review the quality system documentation and discuss any noted deficiencies with the Authorized Representative before the assessment is performed.

3.1.4 Conditions for Accreditation

To become accredited and maintain accreditation, a laboratory shall agree in writing to:

   a. comply at all times with the NVLAP criteria for accreditation as set forth in this handbook and relevant technical documents;

b. fulfill the accreditation procedure, especially to receive the assessment team, to pay the fees charged to the applicant laboratory whatever the result of the assessment may be, and to accept the charges of subsequent maintenance of the accreditation of the laboratory;

c. participate in proficiency testing as required;

d. follow NVLAP conditions for referencing accreditation status (see Annex A);

e. resolve all deficiencies;

f. report to NVLAP within 30 days any major changes that affect the laboratory's:

  − legal, commercial, organizational, or ownership status;
  − organization and management; e.g., key managerial staff;
  − policies or procedures, where appropriate;
  − location;
  − personnel, equipment, facilities, working environment, or other resources, where significant;
  − Authorized Representative or Approved Signatories; or
  − other such matters that may affect the laboratory's capability, or scope of accredited activities, or compliance with the requirements of this handbook and relevant technical documents;

g. return to NVLAP the Certificate of Accreditation and the scope of accreditation for revision or other action should it be requested to do so by NVLAP.

3.1.3.   Fees for Accreditation

3.1.3.1. General: The NVLAP operates on a cost-reimbursable basis from fees paid by participating laboratories that apply for accreditation in specific NVLAP fields of testing or calibration. For fee calculation purposes, a field is considered to be any area of accreditation that is a separate line item on the NVLAP Fee Schedule.

3.1.3.2. Fee Structure: The fee structure is reviewed annually and revised, as necessary. The current structure incorporates four major fee categories:

a. The *Initial Application Fee* covers costs associated with processing an applicant for the first time. It is paid only one time per laboratory and is due with the initial application for accreditation.

b. The *Administrative/Technical Support Fee* covers costs associated with NVLAP and other NIST staff conducting the program in all areas for which accreditation is offered and for providing these services to participating laboratories. A discount is available if multiple fields of accreditation are selected by a laboratory.

  This fee is due annually regardless of the accreditation status of a laboratory. Laboratories that have been enrolled in a program for more than 1 year and are not yet accredited will be invoiced annually for the Administrative/Technical Support Fee, based on the date the laboratory's initial application was accepted by NVLAP.

c. The *On-Site Assessment Fee* covers costs incurred for on-site assessment visits and is due only for a year in which an assessment is scheduled.

NOTE   The optional use of a preassessment visit will be considered if it is decided that such a visit would result in a better definition of the scope of accreditation which has been requested by the laboratory. In such cases, the preassessment costs will be charged to the laboratory in addition to the actual On-Site Assessment Fee.
A laboratory will be charged for an additional assessment visit if required as the result of deficiencies in meeting NVLAP criteria. The fee for this additional assessment visit is the same as the On-Site Assessment Fee on the NVLAP Fee Schedule.

A laboratory will not be charged separately for a monitoring visit, which may be initiated by NVLAP at any time during the accreditation period for cause or on a random selection basis (see Sec. 3.8).

d. The *Proficiency Testing Fee* covers costs relating to the provision of proficiency test samples and artifacts, the collection and analysis of laboratory results, and reports to NVLAP. Laboratories participating in required proficiency testing not covered by initial or renewal accreditation fees will be invoiced.

3.1.3.3.   Fee Refund Policy:

3.1.3.3.1 This refund policy applies to laboratories that withdraw from the NVLAP program.

3.1.3.3.2 The *Initial Application Fee* is nonrefundable.

3.1.3.3.3   The amount of the *Administrative/Technical Support Fee* to be refunded depends upon the length of time that has elapsed between the laboratory's renewal date and the date NVLAP was notified of the decision to withdraw (see Table 1). If a laboratory is seeking initial accreditation (i.e., has never been accredited for a specific program), the time of withdrawal will be counted as the number of months after the date the initial application was received.

3.1.3.3.4.   The *On-Site Assessment Fee* is refundable only if no on-site related costs have been incurred. Otherwise, costs incurred will be deducted from the *On-Site Assessment Fee.*

3.1.3.3.5.   The portion of the *Proficiency Testing Fee* for any proficiency testing planned but not sent to the laboratory, or for any proficiency testing that was not initiated, will be refunded. No refund will be given for artifacts sent but returned unmeasured by the laboratory.

Table 1   Amount of Administrative/Technical Support Fee
Refund

Time of withdrawal (No. of months)	Refund amount
Less than 3 months	3/4
3 months to less than 6 months	1/2
6 months to less than 9 months	1/4
9 months or greater	No refund

3.1.4.   Receipt of Application

3.1.4.1. Upon receipt of a laboratory's application for accreditation,
NVLAP shall:

    a.   assign a NVLAP Lab Code to the applicant laboratory;
    b.   acknowledge receipt of the application in writing;
    c.   request further information, if necessary;
    d.   confirm payment of fees before proceeding with the assessment
        process; and
    e.   specify the next step(s) in the accreditation process.

3.1.4.2. The information received shall be used for the preparation of the
on-site assessment and shall be treated with appropriate confidentiality
(see Sec. 1.7).

3.1.5.   Laboratories Located Outside of the United States

3.1.5.1.   In cases where laboratory documents are not in English, or laboratory personnel do not speak English, it is the responsibility of the laboratory
to provide an interpreter(s), subject to NVLAP approval, to assist the NVLAP
assessor(s) during the on-site assessment. The interpreter will assist the
assessor(s) with conversing directly with laboratory management and technical staff and with reviewing laboratory documentation. Documents such as
quality manuals, procedures, standards, and test reports sent to NVLAP prior
to on-site assessments or reviewed during assessments may be required to be
provided in English to verify compliance with NVLAP requirements.

3.1.5.2.   Some of the fees listed on the NVLAP Fee Schedule may be insufficient to cover the costs incurred by an applicant laboratory located outside of
the United States. In such cases, the laboratory will be responsible for all
additional costs incurred. Additional fees will be charged, if necessary, for
travel by NVLAP assessor(s) outside the United States, for shipment of proficiency testing materials to the laboratories, and for any additional administrative expenses. To ensure that the initial or renewal application is
processed without delay, payment (in U.S. currency) of the appropriate listed
fees should accompany the application. When all the additional costs asso-

ciated with the application have been identified, an invoice for any additional fee amount owed will be sent to the laboratory.

3.1.5.3. Pursuant to U.S. Department of Commerce Export Regulations and/or U.S. Department of State International Traffic in Arms Regulations, certain technologies, equipment, data, and software may not be exported from the United States to certain foreign destinations without first obtaining an export license or official approval. If a laboratory uses such technologies, NVLAP requires that the laboratory possesses, and shows upon request the appropriate license or official U.S. Government approval. For export and license information for the Department of Commerce's regulations, contact the Bureau of Export Administration, Washington, DC, telephone 202–482–4811, fax 202–482–3617, or see the Bureau's web site < http://www.bxa.doc.gov >. For export and license information regarding the State Department's International Traffic in Arms Regulations, please contact the Department of State, telephone 202–663–2980, or see the Department's Defense Trade Controls web site < http://www.pmdtc.org >.

3.2. Assessment

3.2.1. Frequency and Scheduling

3.2.1.1. Before initial accreditation, during the first renewal year, and every 2 years thereafter, an on-site assessment of each laboratory is conducted to determine compliance with the NVLAP criteria.

3.2.1.2. After payment of the required fees, the laboratory will be contacted to schedule a mutually acceptable date for the on-site assessment.

3.2.1.3. An assessment normally takes 1–5 days depending on the scope of the laboratory's application. Every effort is made to conduct an assessment with as little disruption as possible to the normal operations of the laboratory.

3.2.2 Assessors

3.2.2.1. The NVLAP shall select qualified assessors to evaluate all information collected from an applicant laboratory and to conduct the assessment on its behalf at the laboratory and any other sites where activities to be covered by the accreditation are performed.

3.2.2.2. Assessors are selected on the basis of their professional and academic achievements, experience in the field of testing or calibration, management experience, training, technical knowledge, and communications skills. For example, they may be engineers or scientists currently active in the field, consultants, or college professors.

3.2.2.3. An assessor is assigned to conduct an on-site assessment of a particular laboratory on the basis of how well his or her experience matches the type of testing or calibration to be assessed, as well as the absence of conflict of interest. The NVLAP provides the laboratory with a short biographical sketch of the assessor(s). A lead assessor will be assigned if needed. A laboratory may request an alternate assessor if a conflict of interest or prior business relationship exists.

3.2.3   Conduct of Assessment

3.2.3.1.   Assessors use checklists provided by NVLAP so that each laboratory receives an assessment comparable to that received by others.

3.2.3.2.   During the Assessment, the assessor meets with management and laboratory personnel, examines the quality system, reviews staff information, examines equipment and facilities, observes demonstrations of testing or calibrations, and examines tests or calibration reports.

3.2.3.3.   The assessor reviews laboratory records, including resumes, job descriptions of key personnel, training, and competency, evaluations for all staff members who routinely perform, or affect the quality of the testing or calibration for which accreditation is sought. The assessor need not be given information, which violates individual privacy, such as salary, medical information, or performance reviews outside the scope of the accreditation program. The staff information may be kept in the laboratory's official personnel folders or in separate folders that contain only the information that the NVLAP assessor needs to review.

3.2.3.4.   At the conclusion of the assessment, the assessor conducts an exit briefing to discuss observations and any deficiencies with the Authorized Representative and other responsible laboratory staff.

3.2.4.   Assessment Report

3.2.4.1. At the exit briefing, the assessor submits a written report on the compliance of the laboratory with the accreditation requirements, together with the completed checklists, where appropriate. The report shall include as a minimum:

   a.   date(s) of assessment;
   b.   the names of the assessor(s) responsible for the report;
   c.   the names and addresses of all the laboratory sites assessed;
   d.   the assessed scope of accreditation or reference thereto; and
   e.   comments and/or deficiencies cited by the assessor(s) on the compliance of the laboratory with the accreditation requirements.

3.2.4.2.   The report must be signed by the laboratory's Authorized Representative to acknowledge the discussion and receipt of the report.

3.2.4.3. The assessor forwards the original report to NVLAP and leaves a copy with the laboratory.

3.2.5 Deficiency Notification and Resolution

3.2.5.1. Laboratories are informed of deficiencies during the on-site assessment, and deficiencies are documented in the assessment report (Sec. 3.2.4.1.(e)).

3.2.5.2. A laboratory shall respond in writing to NVLAP within 30 days of the date of the assessment report. The response shall be signed by the Authorized Representative and includes documentation that the specified deficiencies have either been corrected and/or a plan of corrective actions. A corrective action plan must include a list of actions, target completion dates, and names of persons responsible for discharging those actions.

3.2.5.3. If substantial deficiencies have been cited, NVLAP may require an additional on-site assessment, at additional cost to the laboratory, prior to granting accreditation. All deficiencies and resolutions will be subject to thorough review and evaluation prior to an accreditation decision (see Sec. 3.4).

3.2.5.4. An on-site assessment review is conducted after a laboratory has undergone an on-site assessment, whether or not deficiencies are found, to determine if the laboratory has met all of the on-site assessment requirements.

3.3. Proficiency Testing

3.3.1 General

3.3.1.1. Proficiency testing is an integral part of the NVLAP accreditation process. The performance of tests or calibrations and reporting of results from proficiency testing assists NVLAP in determining the overall effectiveness of the laboratory. Information obtained from proficiency testing helps to identify problems in a laboratory; if problems are found, NVLAP works with the laboratory staff to solve them.

3.3.1.2. The NVLAP proficiency testing is consistent with the provisions contained in ISO/IEC Guide 43: 1997 (Parts 1 and 2), where applicable. Proficiency testing may be organized by NVLAP itself or by a NVLAP-approved provider of services.

3.3.2 Types of Proficiency Testing

Proficiency testing requirements are associated with most fields of accreditation. Proficiency testing techniques vary depending on the nature of the test item, the method in use, and the number of laboratories participating.

3.3.2.1. Proficiency Testing using *Interlaboratory* comparisons may utilize randomly selected specimens from a batch of uniform material, selected specimens with known properties and results, artifacts with similar properties that have not been characterized, and one-of-a-kind artifacts.

3.3.2.2. Proficiency testing may use such *intralaboratory* techniques as comparisons of computer software implementations to reference implementations, use of standard RMs, and use of fundamental physical laws.

3.3.2.3. Proficiency testing for calibration laboratories may involve comparison of the results of measurements made by the laboratory on selected instruments or artifacts with calibration results obtained independently by NIST/NVLAP.

3.3.3. Analysis and Reporting

Proficiency testing data are analyzed by NVLAP and the participants' own results are reported to them. Summary results are available upon request to other interested parties; e.g., professional societies and standards writing bodies. The identity and performance of individual laboratories are kept confidential.

3.3.4. Proficiency Testing Deficiencies

3.3.4.1. Unsatisfactory participation in any NVLAP proficiency testing program is a technical deficiency which must be resolved in order to obtain initial accreditation or maintain accreditation.

3.3.4.2. Proficiency testing deficiencies are defined as, but not limited to, one or more of the following:

    a. failure to meet specified proficiency testing performance requirements prescribed by NVLAP;

    b. failure to participate in a regularly scheduled "round" of proficiency testing for which the laboratory has received instructions and/or materials;

    c. failure to submit laboratory control data as required; and

    d. failure to produce acceptable test or calibration results when using NIST Standard Reference Materials or special artifacts whose properties are well characterized and known to NIST/NVLAP.

3.3.4.3. The NVLAP will notify the laboratory of proficiency testing deficiencies and actions to be taken to resolve the deficiencies. Denial or suspension of accreditation will result from failure to resolve deficiencies.

3.4. Accreditation Decision

3.4.1. The Chief of NVLAP is responsible for all NVLAP accreditation actions, including granting, renewing, suspending, and revoking any NVLAP accreditation.

3.4.2. The accreditation decision is based on NVLAP review of information gathered during the accreditation process and a determination of whether or not all requirements for accreditation have been fulfilled.

3.4.3.   The evaluation process considers the laboratory's record as a whole, including:

   a.   information provided on the application;
   b.   results of quality system documentation review;
   c.   on-site assessment reports;
   d.   actions taken by the laboratory to correct deficiencies; and
   e.   results of proficiency testing, if required.

3.4.4.   Based on this evaluation, NVLAP determines whether or not a laboratory should be accredited. If the evaluation reveals deficiencies, NVLAP shall inform the laboratory in writing of the deficiencies, and the laboratory must respond as specified in 3.2.5. All deficiencies must be resolved to NVLAP's satisfaction before accreditation can be granted.

3.5. Granting Accreditation

3.5.1.   Initial accreditation is granted when a laboratory has met all NVLAP requirements. One of four accreditation renewal dates (January 1, April 1, July 1, or October 1) is assigned to the laboratory and is usually retained as long as the laboratory remains in the program. The renewal period is 1 year; accreditation expires and is renewable on the assigned date.

3.5.2.   Renewal dates may be reassigned to provide benefits to the laboratory and/or NVLAP. If a renewal date is changed, the laboratory will be notified in writing of the change and any related adjustment in fees.

3.5.3.   When accreditation is granted, NVLAP shall provide to the laboratory a Certificate of Accreditation and a scope of accreditation, which permit identification of:

   a.   the name and address of the laboratory that has been accredited;
   b.   the scope of the accreditation, including:

      –   the tests or calibrations, or types of tests or calibrations, for which accreditation has been granted;
      –   for calibrations, the type of measurement performed, the measurement range, and best measurement uncertainty;
      –   for tests, the materials or products tested, the methods used, and the tests performed;
      –   for specific tests and calibrations for which accreditation has been granted, the methods used defined by written standards or reference documents that have been accepted by the accreditation body;

   c.   the laboratory's Authorized Representative;
   d.   the expiration date of the accreditation; and
   e.   the NVLAP Lab Code.

3.6. Renewal of Accreditation

3.6.1.   Each accredited laboratory shall be sent a renewal application package before the expiration date of its accreditation to allow sufficient time to complete the renewal process.

3.6.2.   Fees for renewal are charged according to services required as listed on the NVLAP Fee Schedule.

3.6.3.   Both the application and fees must be received by NVLAP prior to expiration of the laboratory's current accreditation to avoid a lapse in accreditation.

3.6.4.   On-site assessments of currently accredited laboratories are performed in accordance with the procedures in Section 3.2. If deficiencies are found during the assessment of an accredited laboratory, the laboratory must submit a satisfactory response concerning resolution of deficiencies within 30 days of notification or face possible suspension or revocation of accreditation.

3.6.5.   Undue delay in the resolution of deficiencies may necessitate another on-site assessment at additional cost to the laboratory.

3.7. Changes to Scope of Accreditation

A laboratory may request in writing changes to its scope of accreditation. If the laboratory requests additions to its scope, it must meet all NVLAP criteria for the additional tests or calibrations, types of tests or calibrations, or standards. The need for an additional on-site assessment and/or proficiency testing will be determined on a case-by-case basis.

3.8. Monitoring Visits

3.8.1.   In addition to regularly scheduled assessments, monitoring visits may be conducted by NVLAP at any time during the accreditation period. They may occur for cause or on a random selection basis. While most monitoring visits will be scheduled in advance with the laboratory, NVLAP may conduct unannounced monitoring visits.

3.8.2.   The scope of a monitoring visit may range from checking a few designated items to a complete review. The assessors may review deficiency resolutions, verify reported changes in the laboratory's personnel, facilities, or operations, or administer proficiency testing, when appropriate.

3.9. Suspension of Accreditation

3.9.1.   If NVLAP finds that an accredited laboratory has violated the terms of its accreditation or the provisions of these procedures, NVLAP may suspend the laboratory's accreditation, or advise of NVLAP's intent to revoke accreditation (see Sec. 3.10). The determination by NVLAP whether to suspend the laboratory or to propose revocation of a laboratory's accreditation will depend on the nature of the violation(s) of the terms of its accreditation.

3.9.2.   If a laboratory's accreditation is suspended, NVLAP shall notify the laboratory of that action stating the reasons for and conditions of the suspension and specifying the action(s) the laboratory must take to have its accreditation reinstated. A reassessment of the laboratory may also be required for reinstatement. Conditions of suspension will include prohibiting the laboratory from using the NVLAP logo on its test or calibration reports, correspondence, or advertising during the suspension period in the area(s) affected by the suspension.

3.9.3.   The NVLAP will not require a suspended laboratory to return its Certificate and scope of accreditation, but the laboratory must refrain from using the NVLAP logo in the area(s) affected until such time as the problem(s) leading to the suspension has been resolved. When accreditation is reinstated, NVLAP will authorize the laboratory to resume testing or calibration activities in the previously suspended area(s) as an accredited laboratory.

3.10   Denial and Revocation of Accreditation

3.10.1   If NVLAP proposes to deny or revoke accreditation of a laboratory, NVLAP shall inform the laboratory of the reasons for the proposed denial or revocation and the procedure for appealing such a decision.

3.10.2   The laboratory will have 30 days from the date of receipt of the proposed denial or revocation letter to appeal the decision to the Director of NIST. If the laboratory appeals the decision to the Director of NIST, the proposed denial or revocation will be stayed pending the outcome of the appeal. The proposed denial or revocation will become final through the issuance of a written decision to the laboratory in the event that the laboratory does not appeal the proposed denial or revocation within the 30-day period.

3.10.3   If accreditation is revoked, the laboratory may be given the option of voluntarily terminating the accreditation (see Sec. 3.11).

3.10.4   A laboratory whose accreditation has been revoked must cease use of the NVLAP logo on any of its reports, correspondence, or advertising related to the area(s) affected by the revocation. If the revocation is total, NVLAP will instruct the laboratory to return its Certificate and scope of accreditation and to remove the NVLAP logo from all test or calibration reports, correspondence, or advertising. If the revocation affects only some, but not all of the items listed on a laboratory's scope of accreditation, NVLAP will issue a revised scope that excludes the revoked area(s) in order that the laboratory might continue operations in accredited areas.

3.10.5 A laboratory whose accreditation has been denied or revoked may reapply (see Sec. 3.1) and be accredited if the laboratory:

  a. completes the assessment and evaluation process; and
  b. meets the NVLAP conditions and criteria for accreditation.

3.11. Voluntary Termination of Accreditation

3.11.1. A laboratory may at any time terminate its participation and responsibilities as an accredited laboratory by advising NVLAP in writing of its desire to do so.

3.11.2. Upon receipt of a request for termination, NVLAP shall terminate the laboratory's accreditation, notify the laboratory that its accreditation has been terminated, and instruct the laboratory to return its Certificate and scope of accreditation and to remove the NVLAP logo from all test and calibration reports, correspondence, and advertising.

3.11.3. A laboratory whose accreditation has been voluntarily terminated may reapply (see Sec. 3.1) and be accredited if the laboratory:

  a. completes the assessment and evaluation process; and
  b. meets the NVLAP conditions and criteria for accreditation.

4. MANAGEMENT REQUIREMENTS FOR ACCREDITATION

4.1. Organization

4.1.1. The laboratory or the organization of which it is part shall be an entity that can be held legally responsible.

4.1.2. It is the responsibility of the laboratory to carry out its testing and calibration activities in such a way as to meet the requirements of this handbook and to satisfy the needs of the client, the regulatory authorities or organizations providing recognition.

4.1.3. The laboratory management system shall cover work carried out in the laboratory's permanent facilities, at sites away from its permanent facilities, or in associated temporary or mobile facilities.

4.1.4. If the laboratory is part of an organization performing activities other than testing and/or calibration, the responsibilities of key personnel in the organization that have an involvement or influence on the testing arid/or calibration activities of the laboratory shall be defined in order to identify potential conflicts of interest.

NOTE 1. Where a laboratory is part of a larger organization, the organizational arrangements should be such that departments having conflicting interests, such as production, commercial marketing, or financing, do not adversely influence the laboratory's compliance with the requirements of this handbook.

NOTE 2. If the laboratory wishes to be recognized as a third-party laboratory, it should be able to demonstrate that it is impartial and that it and its personnel are free from any undue commercial, financial, and other pressures, which might influence their technical judgement. The third-party testing or calibration laboratory should not engage in any activities that may endanger the trust in its independence of judgement and integrity in relation to its testing or calibration activities.

4.1.5. The laboratory shall

a. have managerial and technical personnel with the authority and resources needed to carry out their duties and to identify the occurrence of departures from the quality system or from the procedures for performing tests and/or calibrations, and to initiate actions to prevent or minimize such departures (see also Sec. 5.2);

b. have arrangements to ensure that its management and personnel are free from any undue internal and external commercial, financial and other pressures, and influences that may adversely affect the quality of their work;

c. have policies and procedures to ensure the protection of its clients' confidential information and proprietary rights, including procedures for protecting the electronic storage and transmission of results;

d. have policies and procedures to avoid involvement in any activities that would diminish confidence in its competence, impartiality, judgement, or operational integrity;

e. define the organization and management structure of the laboratory, its place in any parent organization, and the relationships between quality management, technical operations, and support services;

f. specify the responsibility, authority, and interrelationships of all personnel who manage, perform, or verify work affecting the quality of the tests and/or calibrations;

g. provide adequate supervision of testing and calibration staff, including trainees, by persons familiar with methods and procedures, purpose of each test and/or calibration, and with the assessment of the test or calibration results;

h. have technical management which has overall responsibility for the technical operations and the provision of the resources needed to ensure the required quality of laboratory operations;

i. appoint a member of staff as quality manager (however named) who, irrespective of other duties and responsibilities, shall have defined responsibility and authority for ensuring that the quality system is implemented and followed at all times; the quality manager shall have direct access to the highest level of management at which decisions are made on laboratory policy or resources;

j. appoint deputies for key managerial personnel (see NOTE below).

NOTE   Individuals may have more than one function and it may be impractical to appoint deputies for every function.

4.2.   Quality System

4.2.1.   The laboratory shall establish, implement, and maintain a quality system appropriate to the scope of its activities. The laboratory shall document its policies, systems, programs, procedures, and instructions to the extent necessary to assure the quality of the test and/or calibration results. The system's documentation shall be communicated to, understood by, available to, and implemented by the appropriate personnel.

4.2.2.   The laboratory's quality system policies and objectives shall be defined in a quality manual (however named). The overall objectives shall be documented in a quality policy statement. The quality policy statement shall be issued under the authority of the chief executive. It shall include at least the following:

a. the laboratory management's commitment to good professional practice and to the quality of its testing and calibration in servicing its clients;

b. the management's statement of the laboratory's standard of service;

c. the objectives of the quality system;

d. a requirement that all personnel concerned with testing and calibration activities within the laboratory familiarize themselves with the quality documentation and implement the policies and procedures in their work; and

e. the laboratory management's commitment to compliance with this handbook.

NOTE   The quality policy statement should be concise and may include the requirement that tests and/or calibrations shall always be carried out in accordance with stated methods and clients' requirements. When the test and/or calibration laboratory is part of a larger organization, some quality policy elements may be in other documents.

4.2.3.   The quality manual shall include or make reference to the supporting procedures including technical procedures. It shall outline the structure of the documentation used in the quality system.

4.2.4.   The roles and responsibilities of technical management and the quality manager, including their responsibility for ensuring compliance with this handbook, shall be defined in the quality manual.

4.3. Document Control

4.3.1 General

The laboratory shall establish and maintain procedures to control all documents that form part of its quality system (internally generated or from external sources), such as regulations, standards, other normative documents, test and/or calibration methods, as well as drawings, software, specifications, instructions, and manuals.

NOTE 1. In this context "document" could be policy statements, procedures, specifications, calibration tables, charts, text books, posters, notices, memoranda, software, drawings, plans, etc. These may be on various media, whether hard copy or electronic, and they may be digital, analog, photographic, or written.

NOTE 2. The control of data related to testing and calibration is covered in Section 5.4.7. The control of records is covered in Section 4.12.

4.3.2 Document Approval and Issue

4.3.2.1. All documents issued to personnel in the laboratory as part of the quality system shall be reviewed and approved for use by authorized personnel prior to issue. A master list or an equivalent document control procedure identifying the current revision status and distribution of documents in the quality system shall be established and be readily available to preclude the use of invalid and/or obsolete documents.

4.3.2.2. The procedure(s) adopted shall ensure that:

a. authorized editions of appropriate documents are available at all locations where operations essential to the effective functioning of the laboratory are performed;
b. documents are periodically reviewed and, where necessary, revised to ensure continuing suitability and compliance with applicable requirements;
c. invalid or obsolete documents are promptly removed from all points of issue or use, or otherwise assured against unintended use;
d. obsolete documents retained for either legal or knowledge preservation purposes are suitably marked.

4.3.2.3. Quality system documents generated by the laboratory shall be uniquely identified. Such identification shall include the date of issue and/or revision identification, page numbering, the total number of pages or a mark to signify the end of the document, and the issuing authority(ies).

4.3.3 Document Changes

4.3.3.1. Changes to documents shall be reviewed and approved by the same function that performed the original review unless specifically designated otherwise. The designated personnel shall have access to pertinent background information upon which to base their review and approval.

4.3.3.2.   Where practicable, the altered or new text shall be identified in the document or the appropriate attachments.

4.3.3.3.   If the laboratory's documentation control system allows for the amendment of documents by hand pending the reissue of the documents, the procedures and authorities for such amendments shall be defined. Amendments shall be clearly marked, initialed, and dated. A revised document shall be formally reissued as soon as practicable.

4.3.3.4.   Procedures shall be established to describe how changes in documents maintained in computerized systems are made and controlled.

4.4.   Review of Requests, Tenders, and Contracts

4.4.1.   The laboratory shall establish and maintain procedures for the review of requests, tenders, and contracts. The policies and procedures for these reviews leading to a contract for testing and/or calibration shall ensure that:

   a.   the requirements, including the methods to be used, are adequately defined, documented, and understood (see Sec. 5.4.2);
   b.   the laboratory has the capability and resources to meet the requirements;
   c.   the appropriate test and/or calibration method is selected and capable of meeting the clients' requirements (see Sec. 5.4.2).

Any differences between the request or tender and the contract shall be resolved before any work commences. Each contract shall be acceptable both to the laboratory and the client.

NOTE 1.   The request, tender, and contract review should be conducted in a practical and efficient manner, and the effect of financial, legal, and time schedule aspects should be taken into account. For internal clients, reviews of requests, tenders, and contracts can be performed in a simplified way.

NOTE 2.   The review of capability should establish that the laboratory possesses the necessary physical, personnel, and information resources, and that the laboratory's personnel have the skills and expertise necessary for the performance of the tests and/or calibrations in question. The review may also encompass results of earlier participation in interlaboratory comparisons or proficiency testing and/or the running of trial test or calibration programs using samples or items of known value in order to determine uncertainties of measurement, limits of detection, confidence limits, etc.

NOTE 3.   A contract may be any written or oral agreement to provide a client with testing and/or calibration services.

4.4.2. Records of reviews, including any significant changes, shall be maintained. Records shall also be maintained of pertinent discussions with a client relating to the client's requirements or the results of the work during the period of execution of the contract.

NOTE   For review of routine and other simple tasks, the date and the identification (e.g., the initials) of the person in the laboratory responsible for carrying out the contracted work are considered adequate. For repetitive routine tasks, the review need be made only at the initial enquiry stage or on granting of the contract for ongoing routine work performed under a general agreement with the client, provided that the client's requirements remain unchanged. For new, complex, or advanced testing and/or calibration tasks, a more comprehensive record should be maintained.

4.4.3.   The review shall also cover any work that is subcontracted by the laboratory.

4.4.4.   The client shall be informed of any deviation from the contract.

4.4.5.   If a contract needs to be amended after work has commenced, the same contract review process shall be repeated and any amendments shall be communicated to all affected personnel.

4.5.   Subcontracting of Tests and Calibrations

4.5.1.   When a laboratory subcontracts work whether because of unforeseen reasons (e.g., workload, need for further expertise, or temporary incapacity) or on a continuing basis (e.g., through permanent subcontracting, agency, or franchising arrangements), this work shall be placed with a competent subcontractor. A competent subcontractor is one that, for example, complies with this handbook for the work in question.

4.5.2.   The laboratory shall advise the client of the arrangement in writing and, when appropriate, gain the approval of the client, preferably in writing.

4.5.3.   The laboratory is responsible to the client for the subcontractor's work, except in the case where the client or a regulatory authority specifies which subcontractor is to be used.

4.5.4.   The laboratory shall maintain a register of all subcontractors that it uses for tests and/or calibrations and a record of the evidence of compliance with this handbook for the work in question.

4.6.   Purchasing Services and Supplies

4.6.1.   The laboratory shall have a policy and procedure(s) for the selection and purchasing of services and supplies it uses that affect the quality of the tests and/or calibrations. Procedures shall exist for the purchase, reception, and storage of reagents and laboratory consumable materials relevant for the tests and calibrations.

4.6.2. The laboratory shall ensure that purchased supplies and reagents and consumable materials that affect the quality of tests and/or calibrations are not used until they have been inspected or otherwise verified as complying with standard specifications or requirements defined in the methods for the tests and/or calibrations concerned. These services and supplies used shall comply with specified requirements. Records of actions taken to check compliance shall be maintained.

4.6.3. Purchasing documents for items affecting the quality of laboratory output shall contain data describing the services and supplies ordered. These purchasing documents shall be reviewed and approved for technical content prior to release.

NOTE   The description may include type, class, grade, precise identification, specifications, drawings, inspection instructions, other technical data including approval of test results, the quality required, and the quality system standard under which they were made.

4.6.4. The laboratory shall evaluate suppliers of critical consumables, supplies, and services which affect the quality of testing and calibration, and shall maintain records of these evaluations and list those approved.

4.7. Service to the Client

The laboratory shall afford clients or their representatives' cooperation to clarify the client's request and to monitor the laboratory's performance in relation to the work performed, provided that the laboratory ensures confidentiality to other clients.

NOTE 1.   Such cooperation may include:

a. providing the client or the client's representative reasonable access to relevant areas of the laboratory for the witnessing of tests and/or calibrations performed for the client;
b. preparation, packaging, and dispatch of test and/or calibration items needed by the client for verification purposes.

NOTE 2.   Clients value the maintenance of good communication, advice, and guidance in technical matters, and opinions and interpretations based on results. Communication with the client, especially in large assignments, should be maintained throughout the work. The laboratory should inform the client of any delays or major deviations in the performance of the tests and/or calibrations.

NOTE 3   Laboratories are encouraged to obtain other feedback, both positive and negative, from their clients (e.g., client surveys). The feedback should be used to improve the quality system, testing and calibration activities, and client service.

4.8.   Complaints

The laboratory shall have a policy and procedure for the resolution of complaints received from clients or other parties. Records shall be maintained of all complaints and of the investigations and corrective actions taken by the laboratory (see also Sec. 4.10).

4.9.   Control of Nonconforming Testing and/or Calibration Work

4.9.1.   The laboratory shall have a policy and procedures that shall be implemented when any aspect of its testing and/or calibration work, or the results of this work, do not conform to its own procedures or the agreed requirements of the client. The policy and procedures shall ensure that:

a.   the responsibilities and authorities for the management of nonconforming work are designated and actions (including halting of work and withholding of test reports and calibration certificates, as necessary) are defined and taken when nonconforming work is identified;
b.   an evaluation of the significance of the nonconforming work is made;
c.   corrective actions are taken immediately, together with any decision about the acceptability of the nonconforming work;
d.   where necessary, the client is notified and work is recalled;
e.   the responsibility for authorizing the resumption of work is defined.

NOTE   Identification of nonconforming work or problems with the quality system or with testing and/or calibration activities can occur at various places within the quality system and technical operations. Examples are customer complaints, quality control, instrument calibration, checking of consumable materials, staff observations or supervision, test report and calibration certificate checking, management reviews, and internal or external audits.

4.9.2.   Where the evaluation indicates that the nonconforming work could recur or that there is doubt about the compliance of the laboratory's operations with its own policies and procedures, the corrective action procedures given in Sec. 4.10 shall be promptly followed.

4.10   Corrective Action

4.10.1   General

The laboratory shall establish a policy and procedure and shall designate appropriate   authorities   for   implementing   corrective   action   when nonconforming work or departures from the policies and procedures in the quality system or technical operations have been identified.

NOTE A problem with the quality system or with the technical operations of the laboratory may be identified through a variety of activities, such as control of nonconforming work, internal or external audits, management reviews, feedback from clients or staff observations.

4.10.2 Cause Analysis

The procedure for corrective action shall start with an investigation to determine the root cause(s) of the problem.

NOTE   Cause analysis is the key and sometimes the most difficult part in the corrective action procedure. Often the root cause is not obvious and thus a careful analysis of all potential causes of the problem is required. Potential causes could include client requirements, the samples, sample specifications, methods and procedures, staff skills and training, consumables, or equipment and its calibration.

4.10.3 Selection and Implementation of Corrective Actions

Where corrective action is needed, the laboratory shall identify potential corrective actions. It shall select and implement the action(s) most likely to eliminate the problem and to prevent recurrence.

Corrective actions shall be to a degree appropriate to the magnitude and the risk of the problem.

The laboratory shall document and implement any required changes resulting from corrective action investigations.

4.10.4 Monitoring of Corrective Actions

The laboratory shall monitor the results to ensure that the corrective actions taken have been effective.

4.10.5 Additional Audits

Where the identification of nonconformances or departures casts doubts on the laboratory's compliance with its own policies and procedures, or on its compliance with this handbook, the laboratory shall ensure that the appropriate areas of activity are audited in accordance with Sec. 4.13 as soon as possible.

NOTE Such additional audits often follow the implementation of the corrective actions to confirm their effectiveness. An additional audit should be necessary only when a serious issue or risk to the business is identified.

4.11. Preventive Action

4.11.1. Needed improvements and potential sources of nonconformances, either technical or concerning the quality system, shall be identified. If preventive action is required, action plans shall be developed, implemented, and monitored to reduce the likelihood of the occurrence of such nonconformances and to take advantage of the opportunities for improvement.

4.11.2. Procedures for preventive actions shall include the initiation of such actions and application of controls to ensure that they are effective.

NOTE 1. Preventive action is a proactive process to identify opportunities for improvement rather than a reaction to the identification of problems or complaints.

NOTE 2. Apart from the review of the operational procedures, the preventive action might involve analysis of data, including trend and risk analyses and proficiency-testing results.

4.12. Control of Records

4.12.1. General

4.12.1.1. The laboratory shall establish and maintain procedures for identification, collection, indexing, access, filing, storage, maintenance, and disposal of quality and technical records. Quality records shall include reports from internal audits and management reviews as well as records of corrective and preventive actions.

4.12.1.2. All records shall be legible and shall be stored and retained in such a way that they are readily retrievable in facilities that provide a suitable environment to prevent damage or deterioration and to prevent loss. Retention times of records shall be established.

NOTE   Records may be in any media, such as hard copy or electronic media.

4.12.1.3. All records shall be held secure and in confidence.

4.12.1.4. The laboratory shall have procedures to protect and back up records stored electronically and to prevent unauthorized access to or amendment of these records.

4.12.2. Technical Records

4.12.2.1. The laboratory shall retain records of original observations, derived data, and sufficient information to establish an audit trail, calibration records, staff records, and a copy of each test report or calibration certificate issued for a defined period. The records for each test or calibration shall contain sufficient information to facilitate, if possible, identification of factors affecting the uncertainty and to enable the test or calibration to be repeated under conditions as close as possible to the original. The records shall include the identity of personnel responsible for the sampling, performance of each test, and/or calibration and checking of results.

NOTE 1.   In certain fields, it may be impossible or impracticable to retain records of all original observations.

NOTE 2.   Technical records are accumulations of data (see Sec. 5.4.7) and information which result from carrying out tests and/or calibrations and which indicate whether specified quality or process parameters are achieved. They may include forms, contracts, work sheets, workbooks, check sheets, work notes, control graphs, external and internal test reports and calibration certificates, clients' notes, papers, and feedback.

4.12.2.2. Observations, data, and calculations shall be recorded at the time they are made and shall be identifiable to the specific task.

4.12.2.3. When mistakes occur in records, each mistake shall be crossed out, not erased, made illegible or deleted, and the correct value entered alongside. All such alterations to records shall be signed or initialed by the person making the correction. In the case of records stored electronically, equivalent measures shall be taken to avoid loss or change of original data.

4.13.   Internal Audits

4.13.1.   The laboratory shall periodically, and in accordance with a predetermined schedule and procedure, conduct internal audits of its activities to verify that its operations continue to comply with the requirements of the quality system and this handbook. The internal audit program shall address all elements of the quality system, including the testing and/or calibration activities. It is the responsibility of the quality manager to plan and organize audits as required by the schedule and requested by management. Such audits shall be carried out by trained and qualified personnel who are, wherever resources permit, independent of the activity to be audited.

NOTE   The cycle for internal auditing should normally be completed in 1 year.

4.13.2.   When audit findings cast doubt on the effectiveness of the operations or on the correctness or validity of the laboratory's test or calibration results, the laboratory shall take timely corrective action, and shall notify clients in writing if investigations show that the laboratory results may have been affected.

4.13.3. The area of activity audited, the audit findings, and corrective actions that arise from them shall be recorded.

4.13.4. Follow-up audit activities shall verify and record the implementation and effectiveness of the corrective action taken.

4.14. Management Reviews

4.14.1. In accordance with a predetermined schedule and procedure, the laboratory's executive management shall periodically conduct a review of the laboratory's quality system and testing and/or calibration activities to ensure their continuing suitability and effectiveness, and to introduce necessary changes or improvements. The review shall take account of:

- the suitability, of policies, and procedures;
- reports from managerial and supervisory personnel;
- the outcome of recent internal audits;
- corrective and preventive actions;
- assessments by external bodies;
- the results of interlaboratory comparisons or proficiency tests;
- changes in the volume and type of the work;
- client feedback;
- complaints;
- other relevant factors, such as quality control activities, resources, and staff training.

NOTE 1. A typical period for conducting a management review is once in every 12 months.

NOTE 2. Results should feed into the laboratory planning system and should include the goals, objectives, and action plans for the coming year.

NOTE 3 A management review includes consideration of related subjects at regular management meetings.

4.14.2. Findings from management reviews and the actions that arise from them shall be recorded. The management shall ensure that those actions are carried out within an appropriate and agreed timescale.

5. TECHNICAL REQUIREMENTS FOR ACCREDITATION

5.1. General

5.1.1. Many factors determine the correctness and reliability of the tests and/or calibrations performed by a laboratory. These factors include contributions from:

- human factors (Sec. 5.2);
- accommodation and environmental conditions (Sec. 5.3);
- test and calibration methods and method validation (Sec. 5.4);
- equipment (Sec. 5.5);
- measurement traceability (Sec. 5.6 and Annex B);
- sampling (Sec. 5.7);
- the handling of test and calibration items (Sec. 5.8).

5.1.2.   The extent to which the factors contribute to the total uncertainty of measurement differs considerably between (types of) tests and between (types of) calibrations. The laboratory shall take account of these factors in developing test and calibration methods and procedures, in the training and qualification of personnel, and in the selection and calibration of the equipment it uses.

5.2. Personnel

5.2.1.   The laboratory management shall ensure the competence of all who operate specific equipment, perform tests and/or calibrations, evaluate results, and sign test reports, and calibration certificates. When using staffs who are undergoing training, appropriate supervision shall be provided. Personnel performing specific tasks shall be qualified on the basis of appropriate education, training, experience, and/or demonstrated skills, as required.

NOTE 1.   In some technical areas (e.g., nondestructive testing), it may be required that the personnel performing certain tasks hold personnel certification. The laboratory is responsible for fulfilling specified personnel certification requirements. The requirements for personnel certification might be regulatory, included in the standards for the specific technical field, or required by the client.

NOTE 2.   The personnel responsible for the opinions and interpretation included in test reports should, in addition to the appropriate qualifications, training, experience, and satisfactory knowledge of the testing carried out, also have:

- relevant knowledge of the technology used for the manufacturing of the items, materials, products, etc., tested, or the way they are used or intended to be used, and of the defects or degradations which may occur during or in service;
- knowledge of the general requirements expressed in the legislation and standards; and
- an understanding of the significance of deviations found with regard to the normal use of the items, materials, products, etc., concerned.

5.2.2.   The management of the laboratory shall formulate the goals with respect to the education, training, and skills of the laboratory personnel.

The laboratory shall have a policy and procedures for identifying training needs and providing training of personnel. The training program shall be relevant to the present and anticipated tasks of the laboratory.

5.2.3.   The laboratory shall use personnel who are employed by, or under contract to, the laboratory. Where contracted and additional technical and key support personnel are used, the laboratory shall ensure that such personnel are supervised and competent and that they work in accordance with the laboratory's quality system.

5.2.4.   The laboratory shall maintain current job descriptions for managerial, technical, and key support personnel involved in tests and/or calibrations.

NOTE   Job descriptions can be defined in many ways. As a minimum, the following should be defined:

- the responsibilities with respect to performing tests and/or calibrations;
- the responsibilities with respect to the planning of tests and/or calibrations and evaluation of results;
- the responsibilities for reporting, opinions, and interpretations;
- the responsibilities with respect to method modification and development and validation of new methods;
- expertise and experience required;
- qualifications and training programs;
- managerial duties.

5.2.5.   The management shall authorize specific personnel to perform particular types of sampling, test and/or calibration, to issue test reports and calibration certificates, to give opinions and interpretations and to operate particular types of equipment. The laboratory shall maintain records of the relevant authorization(s), competence, educational and professional qualifications, training, skills and experience of all technical personnel, including contracted personnel. This information shall be readily available and shall include the date on which authorization and/or competence is confirmed.

NVLAP Note: This requirement also applies to Approved Signatories (see Sec. 1.5.4).

5.3.   Accommodation and Environmental Conditions

5.3.1.   Laboratory facilities for testing and/or calibration, including but not limited to energy sources, lighting, and environmental conditions, shall be such as to facilitate correct performance of the tests and/or calibrations.

The laboratory shall ensure that the environmental conditions do not invalidate the results or adversely affect the required quality of any measurement. Particular care shall be taken when sampling and tests and/or calibrations are undertaken at sites other than a permanent laboratory facility. The technical requirements for accommodation and environmental conditions that can affect the results of tests and calibrations shall be documented.

5.3.2. The laboratory shall monitor, control, and record environmental conditions as required by the relevant specifications, methods, and procedures or where they influence the quality of the results. Due attention shall be paid, for example, to biological sterility, dust, electromagnetic disturbances, radiation, humidity, electrical supply, temperature, and sound and vibration levels, as appropriate to the technical activities concerned. Tests and calibrations shall be stopped when the environmental conditions jeopardize the results of the tests and/or calibrations.

5.3.3. There shall be effective separation between neighboring areas in which there are incompatible activities. Measures shall be taken to prevent cross-contamination.

5.3.4. Access to and use of areas affecting the quality of the tests and/or calibrations shall be controlled. The laboratory shall determine the extent of control based on its particular circumstances.

5.3.5. Measures shall be taken to ensure good housekeeping in the laboratory. Special procedures shall be prepared where necessary.

5.4. Test and Calibration Methods and Method Validation

5.4.1. General

The laboratory shall use appropriate methods and procedures for all tests and/or calibrations within its scope. These include sampling, handling, transport, storage, and preparation of items to be tested and/or calibrated, and, where appropriate, an estimation of the measurement uncertainty as well as statistical techniques for analysis of test and/or calibration data.

The laboratory shall have instructions on the use and operation of all relevant equipment, and on the handling and preparation of items for testing and/or calibration, or both, where the absence of such instructions could jeopardize the results of tests and/or calibrations. All instructions, standards, manuals, and reference data relevant to the work of the laboratory shall be kept up to date and shall be made readily available to personnel (see Sec. 4.3). Deviation from test and calibration methods shall occur only if the deviation has been documented, technically justified, authorized, and accepted by the client.

NOTE   International, regional or national standards, or other recognized specifications that contain sufficient and concise information on how to perform the tests and/or calibrations do not need to be supplemented or rewritten as internal procedures if these standards are written in a way that they can be used as published by the operating staff in a laboratory. It may be necessary to provide additional documentation for optional steps in the method or additional details.

5.4.2. Selection of Methods

The laboratory shall use test and/or calibration methods, including methods for sampling, which meet the needs of the client and which are appropriate for the tests and/or calibrations it undertakes. Methods published in international, regional, or national standards shall preferably be used. The laboratory shall ensure that it uses the latest valid edition of a standard unless it is not appropriate or possible to do so. When necessary, the standard shall be supplemented with additional details to ensure consistent application.

When the client does not specify the method to be used, the laboratory shall select appropriate methods that have been published either in international, regional or national standards, or by reputable technical organizations, or in relevant scientific texts or journals, or as specified by the manufacturer of the equipment. Laboratory-developed methods or methods adopted by the laboratory may also be used if they are appropriate for the intended use and if they are validated. The client shall be informed as to the method chosen. The laboratory shall confirm that it can properly operate standard methods before introducing the tests or calibrations. If the standard method changes, the confirmation shall be repeated.

The laboratory shall inform the client when the method proposed by the client is considered to be inappropriate or out of date.

5.4.3. Laboratory-Developed Methods

The introduction of test and calibration methods developed by the laboratory for its own use shall be a planned activity and shall be assigned to qualified personnel equipped with adequate resources.
Plans shall be updated as development proceeds and effective communication amongst all personnel involved shall be ensured.

5.4.4. Nonstandard Methods

When it is necessary to use methods not covered by standard methods, these shall be subject to agreement with the client and shall include a clear specification of the client's requirements and the purpose of the test and/or calibration. The method developed shall have been validated appropriately before use.
NOTE For new test and/or calibration methods, procedures should be developed prior to the tests and/or calibrations being performed and should contain at least the following information:

   a. appropriate identification;
   b. scope;
   c. description of the type of item to be tested or calibrated;
   d. parameters or quantities and ranges to be determined;
   e. apparatus and equipment, including technical performance requirements;
   f. reference standards and RMs required;

g. environmental conditions required and any stabilization period
    needed;
h. description of the procedure, including

  – affixing of identification marks, handling, transporting, storing,
    and preparation of items;
  – checks to be made before the work is started;
  – checks that the equipment is working properly and, where
    required, calibration and adjustment of the equipment before
    each use;
  – the method of recording the observations and results;
  – any safety measures to be observed;

i. criteria and/or requirements for approval/rejection;
j. data to be recorded and method of analysis and presentation;
k. the uncertainty or the procedure for estimating uncertainty.

5.4.5. Validation of Methods

5.4.5.1. Validation is the confirmation by examination and the provision of
objective evidence that the particular requirements for a specific intended
use are fulfilled.

5.4.5.2. The laboratory shall validate nonstandard methods, laboratory-
designed/developed methods, standard methods used outside their intended
scope, and amplifications and modifications of standard methods to confirm
that the methods are fit for the intended use. The validation shall be as
extensive as is necessary to meet the needs of the given application or field
of application. The laboratory shall record the results obtained, the procedure used for the validation, and a statement as to whether the method is
fit for the intended use.

NOTE 1.   Validation may include procedures for sampling, handling, and
transportation.

NOTE 2.   The techniques used for the determination of the performance of a
method should be one of, or a combination of, the following:

  – calibration using reference standards or RMs;
  – comparison of results achieved with other methods;
  – interlaboratory comparisons;
  – systematic assessment of the factors influencing the result;
  – assessment of the uncertainty of the results based on scientific under-
    standing of the theoretical principles of the method and practical
    experience.

NOTE 3.   When some changes are made in the validated non-standard
methods, the influence of such changes should be documented and, if appropriate, a new validation should be carried out.

5.4.5.3. The range and accuracy of the values obtainable from validated methods (e.g., the uncertainty of the results, detection limit, selectivity of the method, linearity, limit of repeatability and/or reproducibility, robustness against external influences, and/or cross-sensitivity against interference from the matrix of the sample/test object), as assessed for the intended use, shall be relevant to the clients' needs.

NOTE 1.   Validation includes specification of the requirements, determination of the characteristics of the methods, a check that the requirements can be fulfilled by using the method, and a statement on the validity.

NOTE 2.   As method-development proceeds, regular review should be carried out to verify that the needs of the client are still being fulfilled. Any change in requirements requiring modifications to the development plan should be approved and authorized.

NOTE 3.   Validation is always a balance between costs, risks, and technical possibilities. There are many cases in which the range and uncertainty of the values (e.g., accuracy, detection limit, selectivity, linearity, repeatability, reproducibility, robustness and cross-sensitivity) can only be given in a simplified way due to lack of information.

5.4.6.   Estimation of Uncertainty of Measurement

5.4.6.1. A calibration laboratory, or a testing laboratory performing its own calibrations, shall have and shall apply a procedure to estimate the uncertainty of measurement for all calibrations and types of calibrations.

5.4.6.2. Testing laboratories shall have and shall apply procedures for estimating uncertainty of measurement. In certain cases, the nature of the test method may preclude rigorous, metrologically and statistically valid, calculation of uncertainty of measurement. In these cases, the laboratory shall at least attempt to identify all the components of uncertainty and make a reasonable estimation, and shall ensure that the form of reporting of the result does not give a wrong impression of the uncertainty. Reasonable estimation shall be based on knowledge of the performance of the method and on the measurement scope and shall make use of, for example, previous experience and validation data.

NOTE 1.   The degree of rigor needed in an estimation of uncertainty of measurement depends on factors such as:

- the requirements of the test method;
- the requirements of the client;
- the existence of narrow limits on which decisions on conformance to a specification are based.

NOTE 2.   In those cases where a well-recognized test method specifies limits to the values of the major sources of uncertainty of measurement and specifies the form of presentation of calculated results, the laboratory is considered to have satisfied this clause by following the test method and reporting instructions (see Sec. 5.10).

5.4.6.3.   When estimating the uncertainty of measurement, all uncertainty components, which are of importance in the given situation shall be taken into account using appropriate methods of analysis.

NOTE 1.   Sources contributing to the uncertainty include, but are not necessarily limited to, the reference standards and RMs used, methods and equipment used, environmental conditions, properties and condition of the item being tested or calibrated, and the operator.

NOTE 2.   The predicted long-term behavior of the tested and/or calibrated item is not normally taken into account when estimating the measurement uncertainty.

NOTE 3   For further information, see ISO 5725 and the Guide to the Expression of Uncertainty in Measurement (see Sec. 1.4).

NVLAP Note: ANSI/NCSL Z540–2-1997 and NIST Technical Note 1297, 1994 edition, are considered to be equivalent to the Guide to the Expression of Uncertainty in Measurement (GUM).

5.4.7.   Control of Data

5.4.7.1. Calculations and data transfers shall be subject to appropriate checks in a systematic manner.

5.4.7.2. When computers or automated equipment are used for the acquisition, processing, recording, reporting, storage or retrieval of test, or calibration data, the laboratory shall ensure that:

    a. computer software developed by the user is documented in sufficient detail and is suitably validated as being adequate for use;

    b. procedures are established and implemented for protecting the data; such procedures shall include, but not be limited to, integrity and confidentiality of data entry or collection, data storage, data transmission, and data processing;

    c. computers and automated equipment are maintained to ensure proper functioning and are provided with the environmental and operating conditions necessary to maintain the integrity of test and calibration data.

NOTE   Commercial off-the-shelf software (e.g., word processing, database and statistical programs) in general use within their designed application range may be considered to be sufficiently validated. However, laboratory software configuration/modifications should be validated as in Sec. 5.4.7.2(a).

5.5. Equipment

5.5.1. The laboratory shall be furnished with all items of sampling, measurement, and test equipment required for the correct performance of the tests and/or calibrations (including sampling, preparation of test and/or calibration items, processing and analysis of test, and/or calibration data). In those cases where the laboratory needs to use equipment outside its permanent control, it shall ensure that the requirements of this handbook are met.

5.5.2. Equipment and its software used for testing, calibration, and sampling shall be capable of achieving the accuracy required and shall comply with specifications relevant to the tests and/or calibrations concerned. Calibration programs shall be established for key quantities or values of the instruments where these properties have a significant effect on the results. Before being placed into service, equipment (including that used for sampling) shall be calibrated or checked to establish that it meets the laboratory's specification requirements and complies with the relevant standard specifications. It shall be checked and/or calibrated before use (see Sec. 5.6).

5.5.3. Equipment shall be operated by authorized personnel. Up-to-date instructions on the use and maintenance of equipment (including any relevant manuals provided by the manufacturer of the equipment) shall be readily available for use by the appropriate laboratory personnel.

5.5.4. Each item of equipment and its software used for testing and calibration and significant to the result shall, when practicable, be uniquely identified.

5.5.5. Records shall be maintained of each item of equipment and its software significant to the tests and/or calibrations performed. The records shall include at least the following:

a. the identity of the item of equipment and its software;
b. the manufacturer's name, type identification, and serial number or other unique identification;
c. checks that equipment complies with the specification (see Sec. 5.5.2);
d. the current location, where appropriate;
e. the manufacturer's instructions, if available, or reference to their location;
f. dates, results, and copies of reports and certificates of all calibrations, adjustments, acceptance criteria, and the due date of next calibration;
g. the maintenance plan, where appropriate, and maintenance carried out to date;
h. any damage, malfunction, modification, or repair to the equipment.

5.5.6. The laboratory shall have procedures for safe handling, transport, storage, use, and planned maintenance of measuring equipment to ensure proper functioning and in order to prevent contamination or deterioration.

NOTE   Additional procedures may be necessary when measuring equipment is used outside the permanent laboratory for tests, calibrations, or sampling.

5.5.7. Equipment that has been subjected to overloading or mishandling gives suspect results, or has been shown to be defective or outside specified limits, shall be taken out of service. It shall be isolated to prevent its uses or clearly labeled or marked as being out of service until it has been repaired and shown by calibration or test to perform correctly. The laboratory shall examine the effect of the defect or departure from specified limits on previous tests and/or calibrations and shall institute the "Control of nonconforming work" procedure (see Sec. 4.9).

5.5.8. Whenever practicable, all equipment under the control of the laboratory and requiring calibration shall be labeled, coded, or otherwise identified to indicate the status of calibration, including the date when last calibrated and the date or expiration criteria when recalibration is due.

5.5.9. When, for whatever reason, equipment goes outside the direct control of the laboratory, the laboratory shall ensure that the function and calibration status of the equipment are checked and shown to be satisfactory before the equipment is returned to service.

5.5.10. When intermediate checks are needed to maintain confidence in the calibration status of the equipment, these checks shall be carried out according to a defined procedure.

5.5.11. Where calibrations give rise to a set of correction factors, the laboratory shall have procedures to ensure that copies (e.g., in computer software) are correctly updated.

5.5.12. Test and calibration equipment, including both hardware and software, shall be safeguarded from adjustments, which would invalidate the test and/or calibration results.

5.6.   Measurement Traceability

5.6.1   General

All equipment used for tests and/or calibrations, including equipment for subsidiary measurements (e.g., for environmental conditions) having a significant effect on the accuracy or validity of the result of the test, calibration, or sampling shall be calibrated before being put into service. The laboratory shall have an established program and procedure for the calibration of its equipment.

NOTE   Such a program should include a system for selecting, using, cali-brating, checking, controlling and maintaining measurement standards, RMs used as measurement standards, and measuring and test equipment used to perform tests and calibrations.

NVLAP Note: See Annex B for requirements for the implementation of trace-ability policy in NVLAP-accredited laboratories.

5.6.2   Specific Requirements
5.6.2.1.   Calibration

5.6.2.1.1 For calibration laboratories, the program for calibration of equip-ment shall be designed and operated so as to ensure that calibrations and measurements made by the laboratory are traceable to the International System of Units (SI) *(Système international d'unités)*.

A calibration laboratory establishes traceability of its own measurement standards and measuring instruments to the SI by means of an unbroken chain of calibrations or comparisons linking them to relevant primary stan-dards of the SI units of measurement. The link to SI units may be achieved by reference to national measurement standards. National measurement stan-dards may be primary standards, which are primary realizations of the SI units or agreed representations of SI units based on fundamental physical constants, or they may be secondary standards which are standards cali-brated by another national metrology institute. When using external calibra-tion services, traceability of measurement shall be assured by the use of calibration services from laboratories that can demonstrate competence, measurement capability, and traceability. The calibration certificates issued by these laboratories shall contain the measurement results, including the measurement uncertainty and/or a statement of compliance with an identi-fied metrological specification (see also Sec. 5.10.4.2).

NOTE 1.   Calibration laboratories fulfilling the requirements of this hand-book are considered to be competent. A calibration certificate bearing an accreditation body logo from a calibration laboratory accredited to this hand-book, for the calibration concerned, is sufficient evidence of traceability of the calibration data reported.

NOTE 2.   Traceability to SI units of measurement may be achieved by refer-ence to an appropriate primary standard (see VIM: 1993, 6.4) or by reference to a natural constant, the value of which in terms of the relevant SI unit is known and recommended by the General Conference of Weights and Measures (CGPM) and the International Committee for Weights and Measures (CIPM).

NOTE 3   Calibration laboratories that maintain their own primary standard or representation of SI units based on fundamental physical constants can claim traceability to the SI system only after these standards have been compared, directly or indirectly, with other similar standards of a national metrology institute.

NOTE 4   The term "identified metrological specification" means that it must be clear from the calibration certificates, which, specifying the measurements, have been compared with, by including the specification or by giving an unambiguous reference to the specification.

NOTE 5   When the terms "international standard" or "national standard" are used in connection with traceability, it is assumed that these standards fulfill the properties of primary standards for the realization of SI units.

NOTE 6   Traceability to national measurement standards does not necessarily require the use of the national metrology institute of the country in which the laboratory is located.

NOTE 7   If a calibration laboratory wishes or needs to obtain traceability from a national metrology institute other than in its own country, this laboratory should select a national metrology institute that actively participates in the activities of BIPM either directly or through regional groups.

NOTE 8   The unbroken chain of calibrations or comparisons may be achieved in several steps carried out by different laboratories that can demonstrate traceability.

5.6.2.1.2.   There are certain calibrations that currently cannot be strictly made in SI units. In these cases, calibration shall provide confidence in measurements by establishing traceability to appropriate measurement standards such as:

- the use of CRMs provided by a competent supplier to give a reliable physical or chemical characterization of a material;
- the use of specified methods and/or consensus standards that are clearly described and agreed by all parties concerned.

Participation in a suitable program of interlaboratory comparisons is required where possible.

5.6.2.2. Testing:
5.6.2.2.1.   For testing laboratories, the requirements given in Sec. 5.6.2.1. apply for measuring and test equipment with measuring functions used, unless it has been established that the associated contribution from the calibration contributes little to the total uncertainty of the test result. When this situation arises, the laboratory shall ensure that the equipment used can provide the uncertainty of measurement needed.

NOTE   The extent to which the requirements in Sec. 5.6.2.1 should be followed depends on the relative contribution of the calibration uncertainty to the total uncertainty. If calibration is the dominant factor, the requirements should be strictly followed.

5.6.2.2.2 Where traceability of measurements to SI units is not possible and/or not relevant, the same requirements for traceability to, for example, CRMs, agreed methods, and/or consensus standards are required as for calibration laboratories (see Sec. 5.6.2.1.2).

5.6.3. Reference Standards and RMs

5.6.3.1. Reference Standards: The laboratory shall have a program and procedure for the calibration of its reference standards. Reference standards shall be calibrated by a body that can provide traceability as described in Sec. 5.6.2.1. Such reference standards of measurement held by the laboratory shall be used for calibration only and for no other purpose, unless it can be shown that their performance as reference standards would not be invalidated. Reference standards shall be calibrated before and after any adjustment.

5.6.3.2. Reference Materials: The RMs shall, where possible, be traceable to SI units of measurement or to CRMs. Internal RMs shall be checked as far as is technically and economically practicable.

5.6.3.3. Intermediate Checks:

Checks needed to maintain confidence in the calibration status of reference, primary, transfer or working standards, and RMs shall be carried out according to defined procedures and schedules.

5.6.3.4. Transport and Storage:

The laboratory shall have procedures for safe handling, transport, storage, and use of reference standards and RMs in order to prevent contamination or deterioration and in order to protect their integrity.

NOTE Additional procedures may be necessary when reference standards and RMs are used outside the permanent laboratory for tests, calibrations, or sampling.

5.7. Sampling

5.7.1. The laboratory shall have a sampling plan and procedures for sampling when it carries out sampling of substances, materials, or products for subsequent testing or calibration. The sampling plan as well as the sampling procedure shall be available at the location where sampling is undertaken. Sampling plans shall, whenever reasonable, be based on appropriate statistical methods. The sampling process shall address the factors to be controlled to ensure the validity of the test and calibration results.

NOTE 1. Sampling is a defined procedure whereby a part of a substance, material, or product is taken to provide for testing or calibration of a representative sample of the whole. Sampling may also be required by the appropriate specification for which the substance, material, or product is to be tested or calibrated. In certain cases (e.g., forensic analysis), the sample may not be representative but is determined by availability.

NOTE 2. Sampling procedures should describe the selection, sampling plan, withdrawal, and preparation of a sample or samples from a substance, material, or product to yield the required information.

5.7.2. Where the client requires deviations, additions, or exclusions from the documented sampling procedure, these shall be recorded in detail with the appropriate sampling data and shall be included in all documents containing test and/or calibration results, and shall be communicated to the appropriate personnel.

5.7.3. The laboratory shall have procedures for recording relevant data and operations relating to sampling that forms part of the testing or calibration that is undertaken. These records shall include the sampling procedure used, the identification of the sampler, environmental conditions (if relevant) and diagrams, or other equivalent means to identify the sampling location as necessary and, if appropriate, the statistics the sampling procedures are based upon.

5.8. Handling of Test and Calibration Items

5.8.1. The laboratory shall have procedures for the transportation, receipt, handling, protection, storage, retention and/or disposal of test and/or calibration items, including all provisions necessary to protect the integrity of the test or calibration item, and to protect the interests of the laboratory and the client.

5.8.2. The laboratory shall have a system for identifying test and/or calibration items. The identification shall be retained throughout the life of the item in the laboratory. The system shall be designed and operated so as to ensure that items cannot be confused physically or when referred to in records or other documents. The system shall, if appropriate, accommodate a subdivision of groups of items and the transfer of items within and from the laboratory.

5.8.3. Upon receipt of the test or calibration item, abnormalities or departures from normal or specified conditions, as described in the test or calibration method, shall be recorded. When there is doubt as to the suitability of an item for test or calibration, or when an item does not conform to the description provided, or the test or calibration required is not specified in sufficient detail, the laboratory shall consult the client for further instructions before proceeding and shall record the discussion.

5.8.4. The laboratory shall have procedures and appropriate facilities for avoiding deterioration, loss or damage to the test or calibration item during storage, handling, and preparation. Handling instructions provided with the item shall be followed. When items have to be stored or conditioned under specified environmental conditions, these conditions shall be maintained, monitored, and recorded. Where a test or calibration item or a portion of an item is to be held secure, the laboratory shall have arrangements for storage and security that protect the condition and integrity of the secured items or portions concerned.

NOTE 1. Where test items are to be returned into service after testing, special care is required to ensure that they are not damaged or injured during the handling, testing, or storing/waiting processes.

NOTE 2. A sampling procedure and information on storage and transport of samples, including information on sampling factors influencing the test or calibration result, should be provided to those responsible for taking and transporting the samples.

NOTE 3. Reasons for keeping a test or calibration item secure can be for reasons of record, safety or value, or to enable complementary tests and/or calibrations to be performed later.

5.9. Assuring the Quality of Test and Calibration Results

The laboratory shall have quality control procedures for monitoring the validity of tests and calibrations undertaken. The resulting data shall be recorded in such a way that trends are detectable and, where practicable, statistical techniques shall be applied to the reviewing of the results. This monitoring shall be planned and reviewed and may include, but not be limited to, the following:

   a. regular use of CRMs and/or internal quality control using secondary RMs;
   b. participation in interlaboratory comparison or proficiency-testing programs;
   c. replicate tests or calibrations using the same or different methods;
   d. retesting or recalibration of retained items;
   e. correlation of results for different characteristics of an item.

NOTE The selected methods should be appropriate for the type and volume of the work undertaken.

5.10. Reporting the Results

5.10.1. General

The results of each test, calibration, or series of tests or calibrations carried out by the laboratory shall be reported accurately, clearly, unambiguously and objectively, and in accordance with any specific instructions in the test or calibration methods.

The results shall be reported usually in a test report or a calibration certificate (see NOTE 1. below), and shall include all the information requested by the client and necessary for the interpretation of the test or calibration results and all information required by the method used. This information is normally that required by Secs. 5.10.2 and 5. 10. 3, or 5.10.4.

In the case of tests or calibrations performed for internal clients, or in the case of a written agreement with the client, the results may be reported in a simplified way. Any information, listed in Secs. 5.10.2–5.10.4, which is not reported to the client shall be readily available in the laboratory which carried out the tests and/or calibrations.

NOTE 1. Test reports and calibration certificates are sometimes called test certificates and calibration reports, respectively.

NOTE 2. The test reports or calibration certificates may be issued as hard copy or by electronic data transfer provided that the requirements of this handbook are met.

5.10.2. Test Reports and Calibration Certificates

Each test report or calibration certificate shall include at least the following information, unless the laboratory has valid reasons for not doing so:

    a. a title (e.g., "Test Report" or "Calibration Certificate");

    b. the name and address of the laboratory, and the location where the tests and/or calibrations were carried out, if different from the address of the laboratory;

    c. unique identification of the test report or calibration certificate (such as the serial number), and on each page an identification in order to ensure that the page is recognized as a part of the test report or calibration certificate, and a clear identification of the end of the test report or calibration certificate;

    d. the name and address of the client;

    e. identification of the method used;

    f. a description of, the condition of, and unambiguous identification of the item(s) tested or calibrated;

    g. the date of receipt of the test or calibration item(s) where this is critical to the validity and application of the results, and the date(s) of performance of the test or calibration;

    h. reference to the sampling plan and procedures used by the laboratory or other bodies where these are relevant to the validity or application of the results;

    i. the test or calibration results with, where appropriate, the units of measurement;

    j. the name(s), function(s) and signature(s), or equivalent identification of person(s) authorizing the test report or calibration certificate;

k. where relevant, a statement to the effect that the results relate only to the items tested or calibrated.

NVLAP Note: NVLAP defines the person(s) who authorizes the test report or calibration certificate as the Approved Signatory (see Sec. 1.5.4).

NOTE 1. Hard copies of test reports and calibration certificates should also include the page number and total number of pages.

NOTE 2. It is recommended that laboratories include a statement specifying that the test report or calibration certificate shall not be reproduced except in full, without written approval of the laboratory.

5.10.3. Test Reports

5.10.3.1. In addition to the requirements listed in Sec. 5.10.2, test reports shall, where necessary for the interpretation of the test results, include the following:

a. deviations from, additions to, or exclusions from the test method, and information on specific test conditions, such as environmental conditions;
b. where relevant, a statement of compliance/noncompliance with requirements and/or specifications;
c. where applicable, a statement on the estimated uncertainty of measurement; information on uncertainty is needed in test reports when it is relevant to the validity or application of the test results, when a client's instruction so requires, or when the uncertainty affects compliance to a specification limit;
d. where appropriate and needed, opinions and interpretations (see Sec. 5.10.5);
e. additional information which may be required by specific methods, clients, or groups of clients.

5.10.3.2. In addition to the requirements listed in Secs. 5.10.2 and 5.10.3.1, test reports containing the results of sampling shall include the following, where necessary for the interpretation of test results:

a. the date of sampling;
b. unambiguous identification of the substance, material or product sampled (including the name of the manufacturer, the model or type of designation and serial numbers as appropriate);
c. the location of sampling, including any diagrams, sketches or photographs;
d. a reference to the sampling plan and procedures used;
e. details of any environmental conditions during sampling that may affect the interpretation of the test results;

    f. any standard or other specification for the sampling method or procedure, and deviations, additions to or exclusions from the specification concerned.

5.10.4. Calibration Certificates

5.10.4.1 In addition to the requirements listed in Sec. 5.10.2, calibration certificates shall include the following, where necessary for the interpretation of calibration results:

    a. the conditions (e.g., environmental) under which the calibrations were made that have an influence on the measurement results;
    b. the uncertainty of measurement and/or a statement of compliance with an identified metrological specification or clauses, thereof;
    c. evidence that the measurements are traceable (see NOTE 2. in Sec. 5.6.2.1.1).

5.10.4.2. The calibration certificate shall relate only to quantities and the results of functional tests. If a statement of compliance with a specification is made, this shall identify which clauses of the specification are met or not met.

When a statement of compliance with a specification is made omitting the measurement results and associated uncertainties, the laboratory shall record those results and maintain them for possible future reference.

When statements of compliance are made, the uncertainty of measurement shall be taken into account.

5.10.4.3. When an instrument for calibration has been adjusted or repaired, the calibration results before and after adjustment or repair, if available, shall be reported.

5.10.4.4. A calibration certificate (or calibration label) shall not contain any recommendation on the calibration interval except where this has been agreed with the client. This requirement may be superseded by legal regulations.

5.10.5. Opinions and Interpretations

When opinions and interpretations are included, the laboratory shall document the basis upon which the opinions and interpretations have been made. Opinions and interpretations shall be clearly marked as such in a test report.

NOTE 1. Opinions and interpretations should not be confused with inspections and product certifications as intended in ISO/IEC 17020 and ISO-/IEC Guide 65.

NOTE 2. Opinions and interpretations included in a test report may comprise, but not be limited to, the following:

– an opinion on the statement of compliance/noncompliance of the
  results with requirements;
– fulfillment of contractual requirements;
– recommendations on how to use the results;
– guidance to be used for improvements.

NOTE 3.   In many cases, it might be appropriate to communicate the opinions and interpretations by direct dialogue with the client. Such dialogue should be written down.

5.10.6.   Testing and Calibration Results Obtained From Subcontractors

When the test report contains results of tests performed by subcontractors, these results shall be clearly identified. The subcontractor shall report the results in writing or electronically.

When a calibration has been subcontracted, the laboratory performing the work shall issue the calibration certificate to the contracting laboratory.

5.10.7.   Electronic Transmission of Results

In the case of transmission of test or calibration results by telephone, telex, facsimile, or other electronic or electromagnetic means, the requirements of this handbook shall be met (see also Sec. 5.4.7).

5.10.8.   Format of Reports and Certificates

The format shall be designed to accommodate each type of test or calibration carried out and to minimize the possibility of misunderstanding or misuse.

NOTE 1.   Attention should be given to the layout of the test report or calibration certificate, especially with regard to the presentation of the test or calibration data and ease of assimilation by the reader.

NOTE 2.   The headings should be standardized as far as possible.

5.10.9   Amendments to Test Reports and Calibration Certificates

Material amendments to a test report or calibration certificate after issue shall be made only in the form of a further document, or data transfer, which includes the statement:

"Supplement to Test Report [or Calibration Certificate], serial
number ... [or as otherwise identified],"

or an equivalent form of wording.

Such amendments shall meet all the requirements of this handbook.

When it is necessary to issue a complete new test report or calibration certificate, this shall be uniquely identified and shall contain a reference to the original that it replaces.

ANNEX A (NORMATIVE)

Referencing NVLAP Accreditation

The following conditions, which pertain to the use of the term *NVLAP* and the NVLAP logo, shall be met by a laboratory in order to become and remain accredited. Failure to comply with these conditions may result in suspension or revocation of a laboratory's accreditation.

    a. The laboratory shall have a policy and procedure for controlling the use of the term *NVLAP* and the NVLAP logo.

    b. The term and logo shall not be used in a manner that brings NVLAP into disrepute or misrepresents a laboratory's scope of accreditation or accredited status.

    c. When the term *NVLAP* is used to reference a laboratory's accredited status, it shall be accompanied by the NVLAP Lab Code.

    d. Reference to the NVLAP Lab Code by an applicant laboratory that has not yet achieved accreditation shall include a statement accurately reflecting the laboratory's status.

    e. When the NVLAP logo is used to reference a laboratory's accredited status, it shall be accompanied by the NVLAP Lab Code in an approved caption. The caption shall appear below and in close proximity to the logo. The following captions have been approved by NVLAP:

    – "For the scope of accreditation under NVLAP Lab Code 000000-0" (Fig. 1)

    – "NVLAP Lab Code 000000-0" (Fig. 2).

    f. The form of the NVLAP logo must conform to the following guidelines:

FOR THE SCOPE OF ACCREDITATION UNDER NVLAP LAB CODE 000000-0

Figure 1   NVLAP logo and caption 1.

NVLAP LAB CODE 000000-0

Figure 2   NVLAP logo and caption 2.

- The logo shall stand by itself and shall not be combined with any other logo, symbol, or graphic.
- The aspect ratio (height to width) shall be 1–2.25 (Fig. 3).
- The logo and caption shall be of a size that allows the caption to be easily read. The size of the caption shall not exceed the size of the logo itself.
- The logo shall appear in black, blue, or other color approved by NVLAP, and may be filled or unfilled. In the case of a filled logo, the same color shall be used for the outline and the fill.

g. The name of at least one Approved Signatory shall appear on a test or calibration report that displays the NVLAP logo or references NVLAP accreditation. A computer-generated report may have the Approved Signatory's name printed along with the test or calibration results, as long as there is evidence that there is a system in place to ensure that the report cannot be generated without the review and consent of the Approved Signatory. There may be legal or contractual requirements for original signatures to appear on the report.

Width = 2.25 (does not include registration mark)

Figure 3   Aspect ratio of the NVLAP logo (width to height).

h. When the term and logo are used on test or calibration reports, such use shall be limited to reports in which some or all of the data are from tests or calibrations performed by the laboratory under its scope of accreditation.

   A test or calibration report that contains both data covered by the accreditation and data not covered by the accreditation shall clearly identify the data that are not covered by the accreditation. The report must prominently display the following statement at the beginning of the report: "This report contains data that are not covered by the NVLAP accreditation."

i. When the term and logo are used on test or calibration reports that also include work done by subcontracted laboratories, such use shall be limited to reports in which some or all of the data are from tests or calibrations performed by the laboratory under its scope of accreditation.

   A test or calibration report that contains both data covered by the accreditation and data provided by a subcontractor shall clearly identify the data that were provided by the subcontracted laboratory. The report must prominently display the following statement at the beginning of the report: "This report contains data that were produced under subcontract by Laboratory X." If the subcontracted laboratory is accredited by NVLAP, then its Lab Code should also be stated. If the subcontracted laboratory is accredited by a body other than NVLAP, then the name of the accreditation body and the laboratory's number or other unique identifier should also be stated. If the subcontracted laboratory is not accredited, then this must be stated.

j. Each test and calibration report bearing the term or logo shall include a statement that the report must not be used by the client to claim product certification, approval, or endorsement by NVLAP, NIST, or any agency of the federal government.

k. When used in a contract or proposal, the term and logo shall be accompanied by a description of the laboratory's scope of accreditation and current accreditation status.

ANNEX B (NORMATIVE)

Implementation of Traceability Policy in Accredited Laboratories

It is a fundamental requirement that the results of all accredited calibrations and the results of all calibrations required to support accredited tests shall be traceable to national and international standards of measurement. The NIST Handbook 150 (and ISO/IEC 17025) details the specific requirements for traceability to be met by testing and calibration laboratories. This annex provides guidance as to how these requirements may be met and how traceability of measurement can be assured by an accredited laboratory.

B.1. General

Laboratories shall be able to demonstrate proper use of traceable standards and test and measurement equipment by competent laboratory personnel in a suitable environment in performing the tests for which accreditation is desired or held. This demonstration will include the determination of the appropriate measurement uncertainty.

Calibration certificates received by NVLAP-accredited testing and calibration laboratories with new or recalibrated equipment shall meet the requirements of ISO/IEC 17025. The certificates must include the uncertainty of measurement and/or a statement of compliance with an identified metrological specification or clauses thereof.

B.2. Demonstration of Traceability

B.2.1. The NVLAP-accredited laboratories may submit appropriate physical standards and test and measurement equipment directly to NIST or, when appropriate, to another national metrology institute. Accredited laboratories may obtain CRMs from NIST (called Standard Reference Materials under copyright) or from another national metrology institute. The use of a national metrology institute other than NIST shall be documented and will be assessed by NVLAP.

B.2.2. Testing laboratories that perform calibrations only for themselves do not need to be accredited as calibration laboratories. Calibration laboratories that perform specific calibrations only for themselves to support their accredited services do not need to be accredited for those calibrations. For the purpose of assuring traceability, an accredited laboratory may calibrate its own equipment if the appropriate requirements of NIST Handbook 150 have been met.

B.2.3. The NVLAP-accredited laboratories that do not demonstrate traceability as described in Secs. B.2.1 or B.2.2 shall use accredited calibration laboratory services wherever available. Accredited calibration laboratories are those accredited by NVLAP or by any accrediting body with which NVLAP has a mutual recognition arrangement. A listing of NVLAP-accredited calibration laboratories and of accreditation bodies with which NVLAP currently has agreements is available from NVLAP.

B.2.4. If an NVLAP-accredited laboratory submits physical standards or test and measurement equipment to a calibration service provider that is not accredited by NVLAP or by an accrediting body with which NVLAP has a mutual recognition arrangement, the laboratory shall:

    a. document that an appropriate accredited calibration service provider is not available;
    b. audit the claim of traceability of the provider of the calibration service and document the following areas related to the calibration

and claim of traceability of its standards and test and measurement equipment:

1. information regarding assessment of the quality system used by the calibration service provider;
2. the calibration procedure(s) used by the calibration service provider;
3. the physical standards or other test and measurement equipment used by the calibration service provider (including evidence of traceability to standards maintained by NIST or an appropriate national metrology institute and copies of relevant calibration certificates);
4. information regarding the calibration intervals of relevant standards or other test and measurement equipment;
5. the environmental conditions of the laboratory;
6. the method(s) by which uncertainties are determined (e.g., Guide to the Expression of Uncertainty in Measurement (GUM); and
7. the relative uncertainties achieved at all steps of the process;

c. pursue the traceability chain until traceability to appropriate stated references is completely validated, when a calibration service provider submits physical standards and/or test and measurement equipment used in the calibration to another laboratory(s) not accredited by NVLAP;
d. enter the audit documentation, including all findings of nonconformance and resolutions of those findings, into the laboratory's quality management record-keeping system.

NOTE   An on-site visit to the provider of the calibration service is encouraged, but is not required as long as the information listed above is obtained and otherwise verified. Self-declaration of compliance to ISO/IEC 17025   or other relevant standards by a calibration service provider is not acceptable evidence of verification of traceability. Citation of a NIST Test Number by the calibration service provider likewise is not acceptable evidence of verification, of traceability.

B.2.5.   If traceable calibration is not available or appropriate, laboratories may demonstrate comparison to a widely used standard that is clearly specified and mutually agreeable to all parties concerned, particularly in measurements where NIST does not maintain a U.S. national standard. For example, NIST does not maintain a standard for all hardness testing scales. There are several widely used commercial standards available for hardness. However, these standards may not all give equivalent measurement results; therefore, it is important to specify which standard is used and to obtain agreement among all parties involved that the choice made is acceptable.

Appendix C

GUIDE TO QUALITY IN ANALYTICAL CHEMISTRY (EURACHEM/CITAC)

1. AIMS AND OBJECTIVES

1.1.   The aim of this guide is to provide laboratories with guidance on best practice for the analytical operations they carry out. The guidance covers both qualitative and quantitative analysis carried out on a routine or nonroutine basis. A separate guide covers research and development work (CITAC/EURACHEM Guide reference A1).

1.2.   The guidance is intended to help those implementing quality assurance (QA) in laboratories. For those working towards accreditation, certification, or other compliance with particular quality requirements, it will help explain what these requirements mean. The guidance will also be useful to those involved in the quality assessment of analytical laboratories against those quality requirements. Cross-references to ISO/IEC 17025, ISO 9000 and OECD Good Laboratory Practice (GLP) requirements are provided.

1.3.   This document has been developed from the previous CITAC Guide 1 (which in turn was based on the EURACHEM/WELAC Guide), and updated to take account of new material and developments, particularly the new requirements of the standard, ISO/IEC 17025.

1.4.   This guide has been produced by a working group comprising David Holcombe, LGC, UK; Bernard King, NARL, Australia; Alan Squirrell, NATA, Australia, and Maire Walsh, State Laboratory, Ireland. In addition, over the years leading to the drafting of this and earlier versions of the guide, there has been extensive input from a large number of individuals and organizations, including CITAC, EURACHEM, EA, ILAC, AOACI, IUPAC, CCQM, and others (Refer Acronyms list).

1.5. This guide concentrates on the technical issues of QA, with emphasis on those areas where there is a particular interpretation required for chemical testing or related measurements. There are a number of additional aspects of QA where no guidance is given as these are fully addressed in other documents, such as ISO/IEC 17025. These include records; reports; quality systems; subcontracting; complaints; supplier's requirements; contract review; confidentiality; and data handling.

2. INTRODUCTION

2.1. The value of chemical measurements depends upon the level of confidence that can be placed in the results. Increasingly, the chemical testing community is adopting QA principles which, whilst not actually guaranteeing the quality of the data produced, increases the likelihood of it being soundly based and fit for its intended purpose.

2.2. Appropriate QA can enable a laboratory to show that it has adequate facilities and equipment for carrying out chemical analysis and that the work was carried out by competent staff in a controlled manner, following a documented validated method. QA should focus on the key issues which determine quality results, costs, and timeliness and avoid diversion of energies into less important issues.

2.3. Good QA practice, including its formal recognition by accreditation, certification, etc., help to ensure that results are valid and fit for purpose. However, it is important for both laboratories and their customers to realize that QA cannot guarantee that 100% of the individual results will be reliable. There are two reasons for this:

1. Mistakes/gross errors can occur, where, for example, the results for two samples are mixed-up. In a well-run laboratory, the frequency of mistakes will be small, but not zero.
2. Random and systematic errors also occur, leading to uncertainty in a measured result. The probability of a result lying within the stated uncertainty range depends on the level of confidence employed, but again, even in a well-ordered laboratory, deviant results will occasionally occur and very occasionally, the deviation will be large.

The business of QA is to manage the frequency of quality failures. The greater the effort taken, the smaller the number of quality failures that can be expected. It is necessary to balance the cost of QA against the benefit in reducing quality failures to an acceptable (nonzero) level.

2.4. The principles of QA have been formalized in a number of published protocols or standards. Those most widely recognized and used in chemical testing fall into three groups and are applied according to a laboratory's individual needs. The three groups are:

2.4.1. ISO/IEC 17025:1999: (Ref. B1) This standard addresses the technical competence of laboratories to carry out specific tests and calibrations and is used by laboratory accreditation bodies worldwide as the core requirements for the accreditation of laboratories.

2.4.2. ISO 9001:2000: (Ref. B2) and its national and international equivalents. This standard relates primarily to quality management, for facilities carrying out production, or providing services, including chemical analysis.

2.4.3. OECD Principles of Good Laboratory Practice (GLP): 1998 (Ref. B3) and its national and sectorial equivalents. These guidelines are concerned with the organizational processes and conditions under which laboratory studies related to certain regulatory work are carried out.

2.5. In addition, there are Total Quality Management (TQM) approaches to QA which place emphasis on continuous improvement (the new ISO 9001:2000 gives more emphasis here). Central to this guide is the contention that, at the technical level, good practice in analytical QA is independent of the formal QA system adopted.

2.6. A laboratory may decide to design its own QA procedures or it may follow one of the established protocols. In the latter case, it may claim informal compliance against the protocol or ideally may undergo independent assessment from an official expert body, with the aim of gaining independent endorsement of its quality system. Such independent assessment/endorsement is variously known as accreditation, registration, or certification depending on which standard the assessment is made against. In particular areas of analysis, accreditation is sometimes mandatory, however, in most cases, the laboratory is free to decide what sort of QA measures it wishes to adopt. The independent assessment route has recognized advantages, particularly where the laboratory's customers require objective evidence of the technical competence of the laboratory. For clarification of the term "accreditation" as used in this guide (see Secs. 3.2 and 4 below).

3. DEFINITIONS AND TERMINOLOGY

There are a number of important terms used in quality management and conformity assessment whose meaning may vary according to the context in which they are used. It is important to understand the distinction between the various terms. A few are presented here. The key reference is ISO Guide 2:1996—Ref. B4. Other terms can be found in ISO 9000:2000—Ref. B5 (Note: ISO 8402:1994—Quality—Vocabulary—has been withdrawn).

3.1. *QUALITY:* Degree to which a set of inherent characteristics fulfils requirements (ISO 9000:2000)

3.2. *ACCREDITATION:* 'Procedure by which an authoritative body gives formal recognition that a body or person is competent to carry out specific tasks' (ISO Guide 2:1996).

3.2.1. In the context of a laboratory making measurements, accreditation is a formal recognition that a laboratory is competent to carry out specific calibrations or tests or specific types of calibrations or tests. The mechanism under which accreditation is granted is described below in Sec. 4 and the core requirements document is ISO/IEC 17025:1999.

3.2.2. Accreditation is also used in the context of ISO 9000 based activities to describe the process whereby a national organization formally recognizes certification bodies as competent to assess and certify organizations as being compliant with the ISO 9000 series of standards ("quality management systems").

3.3. *CERTIFICATION:* 'Procedure by which a third party gives written assurance that a product, process, or service conforms to specified requirements' (ISO Guide 2:1996). Certification (sometimes known as registration) primarily differs from accreditation in that technical competence is not specifically addressed.

3.4. *QUALITY ASSURANCE* (QA): QA describes the overall measures that a laboratory uses to ensure the quality of its operations. Typically, this might include:

> A quality system
> Suitable laboratory environment
> Educated, trained, and skilled staff
> Training procedures and records
> Equipment suitably maintained and calibrated
> Quality control procedures
> Documented and validated methods
> Traceability and measurement uncertainty
> Checking and reporting procedures
> Preventative and corrective actions
> Proficiency testing
> Internal audit and review procedures
> Complaints procedures
> Requirements for reagents, calibrants, measurement standards, and reference materials

3.5. *QUALITY CONTROL* (QC): 'The operational techniques and activities that are used to fulfil requirements for quality.'

Quality control procedures relate to ensuring the quality of specific samples or batches of samples and include:

Analysis of reference materials/measurement standards
Analysis of blind samples
Use of quality control samples and control charts
Analysis of blanks
Analysis of spiked samples
Analysis in duplicate

Proficiency testing

More details on quality control and proficiency testing are given in Sec. 21.

3.6. *AUDIT AND REVIEW*:   In practice, quality audits take two forms. An audit carried out by an independent external body as part of the accreditation process is more usually known as an *assessment*. "Quality audits" carried out within the laboratory are sometimes subdivided into *audit*, often called 'internal audit' (which checks that the quality procedures are in place, and fully implemented), and *review* (which checks to ensure that the quality system is effective and achieves objectives). The review is carried out by senior management with responsibility for the quality policy and work of the laboratory.

In this guide the term *audit* refers to internal audit; *assessment* refers to external audit.

3.7. *STANDARD*: This word has a number of different meanings in the English language. In the past, it has been used routinely to refer firstly to written standards, i.e. widely adopted procedures, specifications, technical recommendations, etc., and secondly, to chemical or physical standards used for calibration purposes. In this guide, to minimize confusion, *standard* is used only in the sense of *written standards*. The term *measurement standard* is used to describe *chemical* or *physical standards*, used for calibration or validation purposes such as: chemicals of established purity and their corresponding solutions of known concentration; UV filters; weights, etc. Reference materials are one (important) category of measurement standards.

3.8. *REFERENCE MATERIAL (RM)*: 'Material or substance one or more of whose property values are sufficiently homogeneous and well established to be used for the calibration of an apparatus, the assessment of a measurement method, or for assigning values to materials.' (ISO Guide 30—Ref. C1)

3.9. *CERTIFIED REFERENCE MATERIAL (CRM)*:   'Reference material, accompanied by a certificate, one or more of whose property values are certified by a procedure, which establishes its traceability to an accurate realization of the units in which the property values are expressed, and for which each certified value is accompanied by an uncertainty at a stated level of confidence' (ISO Guide 30:1992—Ref. C1).

3.10.   *TRACEABILITY*: 'Property of the result of a measurement or the value of a standard whereby it can be related to stated references, usually national or international standards, through an unbroken chain of comparisons all having stated uncertainties.' (VIM 1993—Ref. B6).

3.11.   *MEASUREMENT UNCERTAINTY*: a parameter associated with the result of a measurement that characterizes the dispersion of the values that could reasonably be attributed to the measurand (VIM 1993—Ref. B6).

4.   ACCREDITATION

4.1.   The references to accreditation in this and successive sections refer to ISO/IEC 17025: 1999 (Ref. B1). Its requirements will be implemented by laboratories and accredited by accreditation bodies over a 3-year transition period ending December 2002. The standard is substantially longer than its predecessor and contains some new or enhanced requirements, as summarized below, but much of the new material was previously contained in supplementary guidance documents. Thus, the scale of the new requirements is not as great as might first appear. A table comparing the clauses of ISO/IEC 17025:1999 and its predecessor, ISO/IEC Guide 25: 1990 is found in Appendix C.

4.2.   Briefly, ISO/IEC 17025 includes new or enhanced requirements concerning the following:

- Contract review—precontract communications to ensure that the requirements are adequately specified and the services fully meet customer requirements.
- Purchasing services and supplies—a policy and procedures are required to ensure that they are fit for purpose.
- Sampling—a sampling plan and procedures are required where sampling is part of the work of the laboratory.
- Preventative action—proactively seeking to improve the processes thus minimizing the need for corrective action.
- Method validation, traceability, and measurement uncertainty— significantly enhanced emphasis on these requirements.
- Opinion and interpretation—this is now allowed in test reports.

4.3.   The requirements of the leading quality standards/protocols have many common or similar elements. For example, ISO/IEC 17025 incorporates the ISO 9001 (1994) quality system elements that are applicable to laboratories. A comparison of the major standards/protocols is given below:

Title	ISO/IEC 17025:1999	ISO 9001:2000	OECD GLP 1998 Organization for Economic Cooperation and Development
Scope	1	1	Section I–1
Normative references	2	2	
Terms and definitons	3	3→ISO 9000:2000	Section I–2
Management requirements	4	Various	Section II–1.1
Organization	4.1		
Study director			Section I–1.2
Quality manager	4.1.5	5.5.2	QM # GLP personnel Section II–2
Quality system	4.2	4	
Quality policy	4.2.2	5.3	
Quality manual	4.2.2	4.2.2	
Management commitment to quality	4.2.2	5.1	
Document control	4.3	4.2.3	
Document approval and issue	4.3.2	4.2.3	
Document changes	4.3.3	4.2.3	Section II–7.1
Review of requests, tenders, and contracts	4.4	7.2	
Subcontraction	4.5		
Purchasing services and supplies	4.6	7.4	
Verification of supplies	4.6.2	7.4.3	Section II–6.2.3 (test item only)
Customer focus		5.2, 8.2.1	
Service to the client	4.7	7.2.3	
Complaints	4.8	7.2.3	
Control of nonconforming work	4.9	8.3	
Improvement		8.5	
Cause analysis	4.10.2	8.5.2	
Corrective actions	4.10.3, 4.10.4	8.5.2	
Preventive action	4.11	8.5.3	
Control of records	4.12	4.2.4	Section II–10
Internal audits	4.13, 4.10.5	8.2.2	Section II–2.2

Continued

Title	ISO/IEC 17025:1999	ISO 9001:2000	OECD GLP 1998 Organization for Economic Cooperation and Development
Management reviews	4.14	5.6	
General technical requirements	5.1		
Personnel	5.2	6.2	Section II–1.3
Accommodation and environmental conditions	5.3	6.3, 6.4	Section II–3
Test and calibration methods	5.4	7.5.1	Section II–7
Method validation	5.4.5	7.5.2	
Measurement uncertainty	5.4.6		
Calculation and transcription checks	5.4.7.1		Section II–8.3
IT validation	5.4.7.2	6.3	Section II–1.1.2 (q)
Equipment	5.5	7.5.1	Section II–4
Equipment qualification	5.5.2	7.5.1, 7.5.2	Section II–5.1
Measurement traceability	5.6	7.6	
Calibration	5.6	7.6	Section II–4.2
Reference standards and reference materials	5.6.3	7.6	Section II–6
Sampling	5.7		
Handling of test or calibration items (transport/storage/identification/disposal)	5.8	7.5.5	
Sample identification	5.8.2	7.5.3	Section II–8.3.1
Assuring the quality of measurement results	5.9	7.5.1, 7.6, 8.2.3, 8.2.4	Section II–2
Reporting results	5.10		Section II–9
Opinions and interpretations	5.10.5		
Electronic transmission	5.10.7		
Amendments to reports	5.10.9	8.3	Section II–9.1.4

Note: Consideration is being given to the alignment of ISO/IEC 17025:1999 to bring the quality management system requirements in Sec. 4 (based on ISO 9001:1994) in line with ISO 9001:2000.

4.4.   Accreditation is granted to a laboratory for a specified set of activities (i.e. tests or calibrations) following assessment of that laboratory. Such assessments will typically include an examination of the analytical procedures in use, the quality system, and the quality documentation. The analytical procedures will be examined to ensure they are technically appropriate for the intended purpose and that they have been validated. The performance of tests may be witnessed to ensure documented procedures are being followed, and indeed can be followed. The laboratory's performance in external proficiency testing schemes may also be examined. Assessment may additionally include a "performance audit," where the laboratory is required to analyse samples supplied by the accrediting body and achieve acceptable levels of accuracy. This performance audit is effectively a form of proficiency testing (see Sec. 21).

4.5.   It is the responsibility of the laboratory to ensure that all procedures used are appropriate for their intended purpose. The assessment process examines this "fitness-for-purpose" aspect.

4.6.   Each accreditation body has established procedures against which it operates, assesses laboratories, and grants accreditation. For example, the laboratory accreditation bodies themselves work to requirements based on ISO/IEC Guide 58 (Ref. C8). Similarly, bodies offering certification schemes work to requirements of ISO/IEC Guide 62 (Ref. C9).

4.7.   Likewise, assessors are chosen against specified criteria. For example, the selection criteria for assessors appointed to assess for laboratory accreditation bodies are specified in ISO/IEC Guide 58. These include the requirement for technical expertise in the specific areas of operation being assessed.

4.8.   The benefit of accreditation is that it enables potential customers of the laboratory to have confidence in the quality of the work performed by the laboratory. Various international developments mean that the endorsement conferred by accreditation and other assessments have worldwide recognition. Many laboratory accreditation bodies (who have been evaluated and found to satisfy relevant requirements—see Sec. 4.6 above) have signed a multilateral agreement (The ILAC Arrangement) to recognize the equivalence of laboratory accreditation schemes. Similar international agreements have been developed for bodies associated with certification schemes.

4.9.   The guidance given below will be of use to laboratories seeking accreditation against ISO/IEC 17025, certification against ISO 9001, or compliance/registration with GLP principles.

5. SCOPE

5.1.   A laboratory may apply QA to all or part of its operations. Where a laboratory claims compliance against, or certification or accreditation to, a particular standard, it is important to be clear to what this compliance, certification, or accreditation applies. The formal statement of the activities which have been certified against ISO 9001, or accredited against ISO 17025 is known as the "scope." ISO 9000 and GLP require only a brief description of the activities covered, but with ISO/IEC 17025, a detailed description of the specific work covered by the accreditation is usually required.

5.2.   Quality management is aided by a clear statement of activities, which ideally should define the range of work covered, but without restricting the laboratory's operation. Different quality standards have different rules, but for ISO/IEC 17025, the scope may typically be defined in terms of:

    i.   the range of products, materials or sample types tested or analyzed;
    ii.   the measurements (or types of measurements) carried out;
    iii.   the specification or method/equipment/technique used;
    iv.   the concentration, range, and measurement uncertainty as appropriate.

5.3.   Definition of scope in specific terms is clearly most easily applied to laboratories carrying out routine testing to established procedures. Where nonroutine testing is carried out, a more flexible approach to scope is desirable. The scope must, however, be as specific as is feasible and the QA system maintained by the laboratory must ensure that the quality of the results is under control.

5.4.   A laboratory wishing to change its scope, either by adding additional tests or changing the methodology of existing tests, will require the approval of the accreditation body, which will have specified policy for such situations. Typically, it is possible to grant simple changes by examination of documentation. For more complex changes, particularly where new techniques are involved, additional assessment may be required.

6.   THE ANALYTICAL TASK

6.1.   Analysis is a complex multistage investigation which may be summarized by the following subtasks. Where appropriate the corresponding section in this guide is also listed. Not every step will be required each time a routine measurement is performed. Also, in reality, measurement is often an iterative process rather than the linear series of steps shown below:

- Specification of requirements—c.f. Sec. 7
- Information review*
- Creative thought*
- Sampling plan*—c.f. Sec. 8
- Sample—c.f. Sec. 22
- Sample preparation

- Preliminary analysis*
- Identification/confirmation of composition
- Quantitative analysis
- Data collection and review
- Data interpretation/problem solving
- Reporting/advice

Those marked* are of more significance in the context of nonroutine analysis.

The process is described in the form of a flow diagram in Fig. 1 in Sec. 19.

6.2.   Although different standards emphasize different aspects of QA and some of the above steps are not specifically covered, it is important that the QA of each stage is considered, and where relevant addressed.

7.   SPECIFICATION OF ANALYTICAL REQUIREMENT

7.1.   The laboratory has a duty to provide an analytical service for its customers that is appropriate to solving the customers problems.

7.2.   The key to good analysis is a clear and adequate specification of the requirement. This will need to be produced in co-operation with the customer who may need considerable help to translate their functional requirements into a technical analytical task. The analytical requirement may also develop during the course of a commission but should not drift. Any changes are likely to be customer driven but should have the agreement of both customer and laboratory. The specification of the analytical request should address the following issues:

- Analytical context
- Information required
- Criticality/acceptable risk
- Time constraints
- Cost constraints
- Sampling
- Traceability requirements
- Measurement uncertainty
- Method requirements, including sample preparation
- Identification/confirmation/fingerprinting
- Limit criteria
- QA/QC requirements
- Research plan requirements/approval

7.3.   The level of documentation should be commensurate with the scale and criticality of the task and include the output of any "information review" and "creative thought."

8. ANALYTICAL STRATEGY

8.1.   All analytical work should be adequately planned. Such a plan may, in its most basic form, be simply a notebook entry. More detailed plans will be appropriate for larger, more complicated tasks. For work carried out under GLP, there is a specific requirement that the work be performed to documented *study plans.*

8.2.   Plans will typically indicate the starting and intended finishing point of the particular task together with the strategy for achieving the desired aims. Where, during the course of the work, it is appropriate to change the strategy, the plan should be amended accordingly.

9. NONROUTINE ANALYSIS

9.1.   Nonroutine analysis can be considered as either tasks, but which are carried out infrequently, where reliable methodology is already established or tasks where every sample requires a different approach and methodology has to be established at the time. Guidance is given in Ref. A1.

9.2.   The costs of chemical measurement reflect the costs associated with the various stages of method development, validation, instrumentation, consumables, ongoing maintenance, staff input, calibration, quality control, etc. Many of these costs are independent of the number of samples subsequently analyzed using that method. Thus, where a single method can be used for a large throughput of samples, unit analytical costs will be comparatively low. Where a method has to be specially developed for just a few samples, the unit analytical costs can be very high. For such nonroutine analysis, some of the costs can be reduced by use of generic methods, i.e. methods which are very broadly applicable. In other instances, subcontracting the work to a laboratory that specializes in the particular type of work would be the most cost-effective solution. However, where work is subcontracted, appropriate QA procedures must be in place.

9.3.   In simple terms, a measurement can conveniently be described in terms of an isolation stage and a measurement stage. Rarely can an analyte be measured without first separating it from the sample matrix. Thus, the purpose of the isolation stage is to simplify the matrix in which the analyte is finally measured. Often the isolation procedure may vary very little for a wide variety of analytes in a range of sample matrices. A good example of a generic isolation procedure is the digestion technique to isolate trace metals in foods.

9.4.   Similarly, once analytes have been isolated from the sample matrix and are presented in a comparatively clean environment, such as a solvent, it may be possible to have a single generic method to cover the measurement of a wide variety of analytes. For example, gas chromatography, or UV–visible spectrophotometry.

9.5.   The documentation of such generic methods should be designed so that it can easily accommodate the small changes which relate to the extraction, clean-up, or measurement of different analytes, for example by the use of tables. The sort of parameters which might be varied are sample size, amount and type of extraction solvents, extraction conditions, chromatographic columns or separation conditions, or spectrometer wavelength settings.

9.6.   The value of such methods for nonroutine analysis is that where a new analyte/matrix combination is encountered, it is frequently possible to incorporate it within an existing generic method with appropriate additional validation, measurement uncertainty calculations, and documentation. Thus, the additional costs incurred are minimized in comparison to the development of a whole new method. The method should define the checks which will need to be carried out for the different analyte or sample type in order to check that the analysis is valid. Sufficient information will need to be recorded in order that the work can be repeated in precisely the same manner at a later date. Where a particular analysis subsequently becomes routine, a specific method may be validated and documented.

9.7.   It is possible to accredit nonroutine analysis and most accreditation bodies will have a policy for assessing such methods and describing them in the laboratory's accreditation scope or schedule. The onus will be on the laboratory to demonstrate to the assessors that in using these techniques, it is meeting all of the criteria of the relevant quality standard. In particular, the experience, expertise, and training of the staff involved will be a major factor in determining whether or not such analyses can be accredited.

10.   STAFF

10.1.   The laboratory management should normally define the minimum levels of qualification and experience necessary for the key posts within the laboratory. Chemical analysis must be carried out by, or under the supervision of a qualified, experienced, and competent analyst. Other senior laboratory staff will normally possess similar competencies. Lower formal qualifications may be acceptable when staff have extensive relevant experience and/or the scope of activities is limited. Staff qualified to degree level will normally have at least two years relevant work experience before being considered as experienced analysts. Staff undergoing training or with no relevant qualifications may undertake analyses provided that they have demonstrably received an adequate level of training and are adequately supervised.

10.2.   In certain circumstances, the minimum requirements for qualifications and experience for staff carrying out particular types of analysis may be specified in regulations.

10.3.   The laboratory must ensure that all staff receive training adequate to the competent performance of the tests and operation of equipment. Where appropriate, this will include training in the principles and theory behind particular techniques. Where possible, objective measures should be used to assess the attainment of competence during training. Only analysts who can demonstrate the necessary competence, or who are adequately supervised may perform tests on samples. Continued competence must be monitored, for example, using quality control techniques. The need to periodically retrain staff must be considered where a method or technique is not in regular use. Although the laboratory management is responsible for ensuring that adequate training is provided, it must be emphasized that a strong element of self-training takes place, particularly amongst more experienced analysts.

10.4. The laboratory shall maintain an up-to-date record of the training that each member of staff has received. The purpose of these records is to provide evidence that individual members of staff have been adequately trained and their competence to carry out particular tests has been assessed. In some cases, it may be pertinent to state any particular limitations to evidence about competence. The records should typically include:

i.   academic qualifications;
ii.   external and internal courses attended;
iii. relevant on-the-job training (and retraining as necessary).

Possibly also:

iv.   participation in QC and/or proficiency testing schemes, with associated data;
v.   technical papers published and presentations given at conferences.

10.5.   In some cases, it may be more appropriate to record competence in terms of particular techniques rather than methods.

10.6.   Access to these training records will be necessary in the course of everyday work. Access to other staff records, usually held centrally by the laboratory and listing personal details may be restricted by national legislation on data protection.

11. SAMPLING, SAMPLE HANDLING, AND PREPARATION

11.1.   Analytical tests may be required for a variety of reasons, including establishing an average analyte value across a material, establishing an analyte concentration profile across a material, or determining local contamination in a material. In some cases, for example forensic analysis, it may be appropriate to examine the entire material. In others, it is appropriate to take some sort of sample. Clearly, the way samples are taken will depend on the reason for the analysis.

11.2.   The importance of the sampling stage cannot be overemphasized. If the test portion is not representative of the original material, it will not be possible to relate the analytical result measured to that in the original material, no matter how good the analytical method is nor how carefully the analysis is performed. Sampling plans may be random, systematic, or sequential and they may be undertaken to obtain quantitative or qualitative information, or to determine conformance or nonconformance with a specification.

11.3. Sampling always contributes to the measurement uncertainty. As analytical methodology improves and methods allow or require the use of smaller test portions, the uncertainties associated with sampling become increasingly important and can increase the total uncertainty of the measurement process. The measurement uncertainty associated with subsampling, etc., should always be included in the test result measurement uncertainty, but the measurement uncertainty associated with the basic sampling process is commonly treated separately.

11.4. In many areas of chemical testing, the problems associated with sampling have been addressed and methods have been validated and published. Analysts should also refer to national or sectoral standards as appropriate. Where specific methods are not available, the analyst should rely on experience or adapt methods from similar applications. When in doubt, the material of interest and any samples taken from it, should always be treated as heterogeneous.

11.5. Selection of an appropriate sample or samples, from a larger amount of material, is a very important stage in chemical analysis. It is rarely straightforward. Ideally, if the final results produced are to be of any practical value, the sampling stages should be carried out by, or under the direction of, a skilled sampler with an understanding of the overall context of the analysis. Such a person is likely to be an experienced analyst or someone specifically trained in sampling. Where it is not practical to use such skilled people to take the samples, the laboratory is encouraged to liaise with the customer to provide advice and possibly practical assistance, in order to ensure the sampling is as appropriate as possible. It is a very common pitfall to underestimate the importance of the sampling procedure and delegate it to an unskilled and untrained employee.

11.6. The terminology used in sampling is complicated and can be confusing. Also the terms used may not be consistent from one application to another. It is important when documenting a sampling procedure to ensure that all of the terms used are clearly defined, so that the procedure will be clear to other users. Similarly, it is important to ensure when comparing two separate procedures that the terminology used is consistent. For example, care should be taken in the use of the word "bulk" since this can refer to either the combining of individual samples, or an undifferentiated mass.

11.7. One of the best treatments of sampling terminology is given in recommendations published by IUPAC (Ref. E7), which describes the terms used in the sampling of bulk goods or packaged goods. In this example, the sampling procedure reduces the original *consignment* through *lots* or *batches*, *increments*, *primary* or *gross samples*, *composite* or *aggregate samples*, *subsamples* or *secondary samples* to a *laboratory sample*. The *laboratory sample*, if heterogeneous, may be further prepared to produce the *test sample*. The *laboratory sample* or the *test sample* is deemed to be the end of the sampling procedure. Operations within this procedure are likely to be subject to sampling uncertainties.

11.8.   For the purposes of the guidance given below the following definitions as proposed by IUPAC have been used:

> *Sample*: A portion of material selected to represent a larger body of material.
> *Sample handling*: This refers to the manipulation to which samples are exposed during the sampling process, from the selection from the original material through to the disposal of all samples and test portions.
> *Subsample*: This refers to a portion of the sample obtained by selection or division; an individual unit of the lot taken as part of the sample or; the final unit of multistage sampling.
> *Laboratory sample*: Primary material delivered to the laboratory.
> *Test sample*: The sample prepared from the laboratory sample.
> *Sample preparation*: This describes the procedures followed to select the test portion from the sample (or subsample) and includes: in-laboratory processing; mixing; reducing; coning and quartering; riffling; and milling and grinding.
> *Test portion*: This refers to the actual material weighed or measured for the analysis.

11.9.   Once received into the laboratory, the *laboratory sample(s)* may require further treatment such as subdivision and or milling and grinding prior to analysis.

11.10. Unless otherwise specified, the test portion taken for analysis must be representative of the laboratory sample. To ensure that the test portion is homogeneous, it may be necessary to reduce the particle size by grinding or milling. If the laboratory sample is large, it may be necessary to subdivide it prior to grinding or milling. Care should be taken to ensure that segregation does not occur during subdivision. In some cases, it will be necessary to crush or coarsely grind the sample prior to subdivision into test samples. The sample may be subdivided by a variety of mechanisms, including coning and quartering, riffling, or by means of a rotating sample divider or a centrifugal divider. The particle size reduction step may be performed either manually (mortar and pestle) or mechanically using crushers or mills. Care must be taken to avoid cross-contamination of samples, to ensure that the equipment does not contaminate the sample (e.g. metals) and that the composition of the sample is not altered (e.g. loss of moisture) during milling or grinding. Many standard methods of analysis contain a section that details the preparation of the laboratory sample prior to the withdrawal of the test portion for analysis. In other instances, legislation deals with this aspect as a generic issue.

11.11. The analytical operations begin with the measuring out of a *test portion* from the laboratory sample or the test sample and proceeds through various operations to the final measurement.

11.12. There are important rules to be followed when designing, adapting, or following a sampling strategy:

11.12.1. The problem necessitating the taking of samples and subsequent analysis should be understood and the sampling procedure designed accordingly. The sampling strategy used will depend on the nature of the problem, e.g.:

    a. the average analyte concentration in the material is required;
    b. the analyte profile across the material is required;
    c. the material is suspected of contamination by a particular analyte;
    d. the contaminant is heterogeneously distributed (occurs in hot spots) in the material;
    e. there may be other, nonanalytical factors to consider, including the nature of the area under examination.

11.12.2. Care should be taken in assuming that a material is homogeneous, even when it appears to be. Where a material is clearly in two or more physical phases, the distribution of the analyte may vary within each phase. It may be appropriate to separate the phases and treat them as separate samples. Similarly, it may be appropriate to combine and homogenize the phases to form a single sample. In solids, there may be a considerable variation in analyte concentration if the particle size distribution of the main material varies significantly and over a period of time the material may settle. Before sampling, it may be appropriate, if practical, to mix the material to ensure a representative particle size distribution. Similarly, analyte concentration may vary across a solid where different parts of the material have been subjected to different stresses. For example, consider the measurement of vinyl chloride monomer (VCM) in the fabric of a PVC bottle. The concentration of VCM varies significantly depending on whether it is measured at the neck of the bottle, the shoulder, the sides, or the base.

11.12.3. The properties of the analyte(s) of interest should be taken into account. Volatility, sensitivity to light, thermal lability, and chemical reactivity may be important considerations in designing the sampling strategy and choosing equipment, packaging, and storage conditions. Equipment used for sampling, subsampling, sample handling, sample preparation, and sample extraction should be selected in order to avoid unintended changes to the nature of the sample which may influence the final results. The significance of gravimetric or volumetric errors during sampling should be considered and any critical equipment calibrated. It may be appropriate to add chemicals such as acids or antioxidants to the sample to stabilize it. This is of particular importance in trace analysis where there is a danger of adsorption of the analyte onto the storage vessel.

11.12.4.   It may be necessary to consider the use and value of the rest of the original material once a sample has been removed for analysis. Poorly considered sampling, especially if destructive, may render the whole consignment valueless or inoperative.

11.12.5.   Whatever strategy is used for the sampling, it is of vital importance that the sampler keeps a clear record of the procedures followed in order that the sampling process may be repeated exactly.

11.12.6.   Where more than one sample is taken from the original material, it may be useful to include a diagram as part of the documentation to indicate the pattern of sampling. This will make it easier to repeat the sampling at a later date and also may assist in drawing conclusions from the test results. A typical application where such a scheme would be useful is the sampling of soils over a wide area to monitor fall-out from stack emissions.

11.12.7.   Where the laboratory has not been responsible for the sampling stage, it should state in the report that the samples were analyzed as received. If the laboratory has conducted or directed the sampling stage, it should report on the procedures used and comment on any consequent limitations imposed on the results.

11.13.   Sample packaging and instruments used for sample manipulation should be selected so that all surfaces in contact with the sample are essentially inert. Particular attention should be paid to possible contamination of samples by metals or plasticizers leaching from the container or its stopper into the sample. The packaging should also ensure that the sample can be handled without causing a chemical, microbiological, or other hazard.

11.14.   The enclosure of the packaging should be adequate to ensure there is no leakage of sample from the container, and that the sample itself cannot be contaminated. In some circumstances, for example where samples have been taken for legal purposes, the sample may be sealed so that access to the sample is only possible by breaking the seal. Confirmation of the satisfactory condition of the seals will normally then form part of the analytical report.

11.15.   The sample label is an important aspect of documentation and should unambiguously identify the sample to related plans or notes. Labeling is particularly important, further into the analytical process, when the sample may have been divided, subsampled, or modified in some way. In such circumstances, additional information may be appropriate, such as references to the main sample, and to any processes used to extract or subsample the sample. Labeling must be firmly attached to the sample packaging and where appropriate, be resistant to fading, autoclaving, sample or reagent spillage, and reasonable changes in temperature and humidity.

11.16.   Some samples, those involved in litigation for example, may have special labeling and documentation requirements. Labels may be required to identify all those who have been involved with the sample, including the person taking the sample and the analysts involved in the testing. This may be supported by receipts, to testify that one signatory (as identified on the label) has handed the sample to the next signatory, thus proving that sample continuity has been maintained. This is commonly known as "chain of custody."

11.17.   Samples must be stored at an appropriate temperature and in such a manner that there is no hazard to laboratory staff and the integrity of the samples is preserved. Storage areas should be kept clean and organized so that there is no risk of contamination or cross-contamination, or of packaging and any related seals being damaged. Extremes of environmental conditions (e.g. temperature, humidity), which might change the composition of the sample should be avoided, as this can lead to loss of analyte through degradation or adsorption, or an increase in analyte concentration (mycotoxins). If necessary, environmental monitoring should be used. An appropriate level of security should be exercised to restrict unauthorized access to the samples.

11.18.   All staff concerned with administration of the sample handling system should be properly trained. The laboratory should have a documented policy for the retention and disposal of samples. The disposal procedure should take into account the guidelines set out above.

11.19.   To fully evaluate an analytical result for conformity assessment, or for other purposes, it is important to have knowledge of the sampling plan and its statistical basis. Sampling procedures for inspection by variables assumes that the characteristic being inspected is measurable and follows the normal distribution. Whereas sampling for inspection by attributes is a method whereby either the unit of product is classified as conforming or nonconforming, or the number of nonconformities in the unit of product is counted with respect to a given set of requirements. In inspection by attributes, the risk associated with acceptance/rejection of nonconformities is predetermined by the *acceptable quality level* (AQL) or the *limiting quality* (LQ).

12. ENVIRONMENT

12.1.   Samples, reagents, measurement standards, and reference materials must be stored so as to ensure their integrity. In particular, samples must be stored in such a way that cross-contamination is not possible. The laboratory should guard against their deterioration, contamination, and loss of identity.

12.2. The laboratory environment should be sufficiently uncrowded, clean, and tidy to ensure the quality of the work carried out is not compromised.

12.3. It may be necessary to restrict access to particular areas of a laboratory because of the nature of the work carried out there. Restrictions might be made because of security, safety, or sensitivity to contamination or interferences. Typical examples might be work involving explosives, radioactive materials, carcinogens, forensic examination, PCR techniques, and trace analysis. Where such restrictions are in force, staff should be made aware of:

  i.   the intended use of a particular area;
  ii.   the restrictions imposed on working within such areas;
  iii. the reasons for imposing such restrictions;
  iv.   the procedures to follow when such restrictions are breached.

12.4. When selecting designated areas for new work, account must be taken of the previous use of the area. Before use, checks should be made to ensure that the area is free of contamination.

12.5. The laboratory shall provide appropriate environmental conditions and controls necessary for particular tests or operation of particular equipment including temperature, humidity, freedom from vibration, freedom from airborne and dustborne microbiological contamination, special lighting, radiation screening, and particular services. Critical environmental conditions must be monitored and kept within predetermined limits.

12.6. A breakdown of critical environmental conditions may be indicated either by monitoring systems or by the analytical quality control within the particular tests. The impact of such failures may be assessed as part of ruggedness testing during method validation and where appropriate, emergency procedures established.

12.7. Decontamination procedures may be appropriate where environment or equipment is subject to change of use or where accidental contamination has occurred.

13. EQUIPMENT (See also Appendix B, p. 51)

13.1. Categories of Equipment

13.1.1. All equipment used in laboratories should be of a specification sufficient for the intended purpose, and kept in a state of maintenance and calibration consistent with its use. Equipment normally found in the chemical laboratory can be categorized as:

  i.   general service equipment not used for making measurements or with minimal influence on measurements (e.g. hotplates, stirrers, nonvolumetric glassware, and glassware used for rough volume measurements such as measuring cylinders) and laboratory heating or ventilation systems;

ii. volumetric equipment (e.g. flasks, pipettes, pyknometers, burettes, etc.) and measuring instruments (e.g. hydrometers, U-tube viscometers, thermometers, timers, spectrometers, chromatographs, electrochemical meters, balances, etc.).

iii. physical measurement standards (weights, reference thermometers);

iv. computers and data processors.

13.2. *General Service Equipment*

13.2.1. General service equipment will typically only be maintained by cleaning and safety checks as necessary. Calibrations or performance checks will be necessary where the setting can significantly affect the test or analytical result (e.g. the temperature of a muffle furnace or constant temperature bath). Such checks need to be documented.

13.3. *Volumetric Equipment and Measuring Instruments*

13.3.1. The correct use of this equipment is critical to analytical measurements and therefore it must be correctly used, maintained, and calibrated in line with environmental considerations (Sec. 12). The performance of some volumetric (and related) glassware is dependent on particular factors, which may be affected by cleaning methods, etc. As well as requiring strict procedures for maintenance, such apparatus may therefore require more regular calibration, depending on use. For example, the performance of pycnometers, U-tube viscometers, pipettes, and burettes is dependent on "wetting" and surface tension characteristics. Cleaning procedures must be chosen so as not to compromise these properties.

13.3.2. Attention should be paid to the possibility of contamination arising either from the fabric of the equipment itself, which may not be inert, or from cross-contamination from previous use. In the case of volumetric glassware, cleaning procedures, storage, and segregation of volumetric equipment may be critical, particularly for trace analyses where leaching and adsorption can be significant.

13.3.3. Correct use combined with periodic servicing, cleaning, and calibration will not necessarily ensure an instrument is performing adequately. Where appropriate, periodic performance checks should be carried out (e.g. to check the response, stability and linearity of sources, sensors and detectors, the separating efficiency of chromatographic systems, the resolution, alignment and wavelength accuracy of spectrometers, etc.) (see Appendix B).

13.3.4. The frequency of such performance checks may be specified in manuals or operating procedures. If not, then it will be determined by experience and based on need, type, and previous performance of the equipment. Intervals between checks should be shorter than the time the equipment has been found, in practice, to take to drift outside acceptable limits.

13.3.5. It is often possible to build performance checks—system suitability checks—into test methods (e.g. based on the levels of expected detector or sensor response to reference materials, the resolution of component mixtures by separation systems, the spectral characteristics of measurement standards, etc.). These checks must be satisfactorily completed before the equipment is used.

13.4. *Physical Measurement Standards*

13.4.1. Wherever physical parameters are critical to the correct performance of a particular test, the laboratory shall have or have access to the relevant measurement standard, as a means of calibration.

13.4.2. In some cases, a test and its performance is actually defined in terms of a particular piece of equipment and checks will be necessary to confirm that the equipment conforms to the relevant specification. For example, flashpoint values for a particular flammable sample are dependent of the dimensions and geometry of the apparatus used in the testing.

13.4.3. Measurement standard materials and any accompanying certificates should be stored and used in a manner consistent with preserving the calibration status. Particular consideration should be given to any storage advice given in the documentation supplied with the measurement standard.

13.5. *Computers and Data Processors*: Requirements for computers are given in Sec. 20.

14. REAGENTS

14.1. The quality of reagents and other consumable materials must be appropriate for their intended use. Consideration needs to be given to the selection, purchase, reception, and storage of reagents.

14.2. The grade of any critical reagent used (including water) should be stated in the method, together with guidance on any particular precautions which should be observed in its preparation, storage, and use. These precautions include toxicity, flammability, stability to heat, air, and light; reactivity to other chemicals; reactivity to particular containers; and other hazards. Reagents and reference materials prepared in the laboratory should be labeled to identify substance, strength, solvent (where not water), any special precautions or hazards, restrictions of use, and date of preparation and/or expiry. The person responsible for the preparation shall be identifiable either from the label or from records.

14.3. The correct disposal of reagents does not directly affect the quality of sample analysis, however, it is a matter of GLP and should comply with national environmental or health and safety regulations.

14.4. Where the quality of a reagent is critical to a test, the quality of a new batch should be verified against the outgoing batch before use, provided that the outgoing batch is known to be still serviceable.

15. TRACEABILITY

15.1. The formal definition of traceability is given in Sec. 3.10 and a CITAC policy statement has been prepared (Ref. A6). A guide on the traceability of chemical measurements is under development (Ref. A7). Traceability concerns the requirement to relate the results of measurements to the values of standards or references, the preferred reference points being the internationally recognized system of units, the SI. This is achieved through the use of primary standards (or other high level standards) which are used to establish secondary standards that can be used to calibrate working level standards and related measuring systems. Traceability is established at a stated level of measurement uncertainty, where every step in the traceability chain adds further uncertainty. Traceability is important because it provides the linkage that ensures that measurements made in different laboratories or at different times are comparable. It is a matter of choice, as indicated above, whether to claim traceability to local references, or to international references.

15.2. Chemical measurements are invariably made by calculating the value from a measurement equation that involves the measured values of other quantities, such as mass, volume, concentration of chemical standards, etc. For the measurement of interest to be traceable, all the measurements associated with the values used in the measurement equation used to calculate the result must also be traceable. Other quantities not present in the measurement equation, such as pH, temperature, etc., may also significantly affect the result. Where this is the case, the traceability of measurements used to control these quantities also need to be traceable to appropriate measurement standards.

15.3. Establishing the traceability of physical quantities such as mass, volume, etc., is readily achieved using transfer standards, at the level of uncertainty needed for chemical measurements. The problem areas for chemists are usually (chemical) method validation and calibration. Validation establishes that the method actually measures what it is intended to measure (e.g. methyl mercury in fish). Validation establishes that the measurement equation used to calculate the results is valid. Calibration is usually based on the use of gravimetrically prepared solutions of pure substance reference materials. The important issues here are identity and purity, the former being more of a problem in organic chemistry where much higher levels of structural detail are often required and confusion with similar components can readily occur.

The uncertainty of a measurement will in part depend on the uncertainty of the purity of the chemical standard used. However, only in the case of some organic materials, where purity and stability problems can be acute, or where high accuracy assay of major components is required, will purity be a major problem.

15.4. For many analyses, where extraction, digestion, derivatization, and saponification are commonly required, the main problem can be gaining good knowledge of the amount of analyte in the original sample relative to that in the sample presented to the end measurement process. This bias (sometimes called "recovery") can be due to processing losses, contamination, or interferences. Some of these effects are manifest within reproducibility uncertainties but others are systematic effects that need separate consideration. The strategies available to address method bias include:

- Use of primary or reference methods of known and small bias
- Comparisons with closely matched matrix CRMs
- Measurement of gravimetrically spiked samples and blanks
- Study of losses, contamination, interferences, and matrix effects

Establishing the traceability of this part of the measurement process requires relating the measurement bias to appropriate stated references, such as the values carried by matrix matched reference materials. It should be noted that the measurement of the recovery of spiked samples does not necessarily simulate the extraction of the native analyte from the samples. In practice, this is not normally a problem where the samples are liquid and/or totally digested. However, problems can occur with the extraction of solids. For example, a spiked analyte may be freely available on the surface of the sample particles, whereas the native analyte may be strongly adsorbed within the particles and therefore much less readily extracted.

15.5. Most chemical measurement can, in principle, be made traceable to the mole. When, however, the analyte is defined in functional terms, such as fat or protein based on a nitrogen determination, then specification of the measurement in terms of the mole is not feasible. In such cases, the quantity being measured is defined by the method. In these cases, traceability is for standards of the component quantities used to calculate the result, for example mass and volume, and the values produced by a standard method and/or the values carried by a reference material. Such methods are called empirical methods. In other case, the limitation in achieving traceability to SI derives from difficulty in evaluating bias and its uncertainty, such as the recovery of the analytes in complex matrices. The options here are to define the measurand by the method and establish traceability to stated references, including a reference method/reference material. Such measurements have a 'lower level' of traceability, but also have a smaller measurement uncertainty, relative to the stated references. Alternatively, the bias can be estimated and corrected for and the uncertainty due to the bias can also be estimated and included in the overall uncertainty evaluation. This would allow traceability to the SI to be claimed.

16. MEASUREMENT UNCERTAINTY

16.1. Measurement uncertainty is formally defined in Sec. 3.11. Good practice in the evaluation of measurement uncertainty is described in an ISO Guide (Ref. B7) and an interpretation for chemical measurement including a number of worked examples is given in a CITAC/EURACHEM Guide (Ref. A2). Measurement uncertainty characterizes the range of values within which the true value is asserted to lie, with a specified level of confidence. Every measurement has an uncertainty associated with it, resulting from errors arising in the various stages of sampling and analysis and from imperfect knowledge of factors affecting the result. For measurements to be of practical value it is necessary to have some knowledge of their reliability or uncertainty. A statement of the uncertainty associated with a result conveys to the customer the 'quality' of the result.

16.2. ISO/IEC 17025:1999 requires laboratories to evaluate their measurement uncertainty. There is also a requirement to report measurement uncertainty under specific circumstances, for example, where it is relevant to the interpretation of the test result (which is often the case). Thus, statement of measurement uncertainty in test reports should become common practice in the future (Ref. B18).

16.3. A statement of uncertainty is a quantitative estimate of the limits within which the value of a measurand (such as an analyte concentration) is expected to lie. Uncertainty may be expressed as a standard deviation or a calculated multiple of the standard deviation. In obtaining or estimating the uncertainty relating to a particular method and analyte, it is essential to ensure that the estimate explicitly considers all the possible sources of uncertainty and evaluates significant components. Repeatability or reproducibility, for example, are usually not full estimates of the uncertainty, since neither takes full account of any uncertainties associated with systematic effects inherent in a method.

16.4. A wide variety of factors make any analytical measurement result liable to deviate from the true value. For example, temperature effects on volumetric equipment, reflection and stray light in spectroscopic instruments, variations in electrical supply voltages, individual analysts' interpretation of specified methods, and incomplete extraction recoveries, all potentially influence the result. As far as reasonably possible, such errors must be minimized by external control or explicitly corrected for, for example by applying a suitable correction factor. The exact deviation of a single measurement result from the (unknown) true value is, however, impossible to obtain. This is because the different factors vary from experiment to experiment, and because the effect of each factor on the result is never known exactly. The likely range of deviation must therefore be estimated.

16.5.   The primary task in assigning a value to the uncertainty of a measurement is the identification of the relevant sources of uncertainty and the assignment of a value to each significant contribution. The separate contributions must then be combined (as shown in Sec. 16.13) in order to give an overall value. A record should be kept of the individual sources of uncertainty identified, the value of each contribution, and the source of the value (for example, repeat measurements, literature reference, CRM data, etc.).

16.6.   In identifying relevant sources of uncertainty, consideration must be given to the complete sequence of events necessary to achieve the purpose of the analysis. Typically, this sequence includes sampling and subsampling, sample preparation, extraction, clean-up, concentration or dilution, instrument calibration (including reference material preparation), instrumental analysis, raw data processing, and transcription of the output result.

16.7.   Each of the stages will have associated sources of uncertainty. The component uncertainties can be evaluated individually or in convenient groups. For example, the repeatability of a measurement may serve as an estimate of the total contribution of random variability, due to a number of steps in a measurement process. Similarly, an estimate of overall bias and its uncertainty may be derived from studies of matrix matched certified reference materials and spiking studies.

16.8.   The size of uncertainty contributions can be estimated in a variety of ways. The value of an uncertainty component associated with random variations in influence factors may be estimated by measuring the dispersion in results over a suitable number of determinations under a representative range of conditions. (In such an investigation, the number of measurements should not normally be less than 10.) Uncertainty components arising from imperfect knowledge, for example a bias or potential bias, can be estimated on the basis of a mathematical model, informed professional judgement, international laboratory intercomparisons, experiments on model systems, etc. These different methods of estimating individual uncertainty components can be valid.

16.9.   Where uncertainty contributions are estimated in groups, it is nonetheless important to record the sources of uncertainty which are considered to be included in each group, and to measure and record individual uncertainty component values where available as a check on the group contribution.

16.10.   If information from interlaboratory trials is used, it is essential to consider uncertainties arising outside the scope of such studies. For example, nominal values for reference materials are typically quoted as a range, and where several laboratories use the same reference material in a collaborative trial, the uncertainty in the reference material value is not included in the interlaboratory variation. Similarly, interlaboratory trials typically use a restricted range of test materials, usually carefully homogenized, so the possibility of inhomogeneity and differences in matrix between real samples and collaborative trial test materials should also be taken into account.

16.11. Typically, uncertainty contributions for analytical results might fall into four main groups:

    i. Contributions from short-term random variability, typically estimated from repeatability experiments.

    ii. Contributions such as operator effects, calibration uncertainty, scale graduation errors, equipment and laboratory effects, estimates from interlaboratory reproducibility trials, in-house intercomparisons, proficiency test results, or by professional judgement.

    iii. Contributions outside the scope of interlaboratory trials, such as reference material uncertainty.

    iv. Other sources of uncertainty, such as sampling variability (inhomogeneity), matrix effects, and uncertainty about underlying assumptions (such as assumptions about completeness of derivatization).

16.12. The uncertainty contributions for each source must all be expressed in the same way, ideally as standard deviations or relative standard deviations. In some cases, this will entail some conversion. For example, reference material limits are often presumed to have absolute limits. A rectangular distribution of width W has a standard deviation $W/(2\sqrt{3})$. Confidence intervals may be converted to standard deviations by dividing by the appropriate value of Student's t for large (statistical) samples (1.96 for 95% confidence limits).

16.13. Once a list of uncertainties is available, the individual components can be combined. Where individual sources of uncertainty are independent, the general expression for the combined standard uncertainty u is:

$$u = \sqrt{\sum (\partial R/\partial x_i)^2 \cdot u(x_i)^2}$$

where $\partial R/\partial x_i$ is the partial differential of the result R with respect to each intermediate value (or other 'influence quantity' such as a correction x_i and $u(x_i)$ is the uncertainty component associated with x_i.

16.14. This expression simplifies considerably for the two most common cases. Where the influence quantities or intermediate results are added or subtracted to give the result, the uncertainty u is equal to the square root of the sum of the squared contributing uncertainty components, all expressed as standard deviations. Where interim results are combined by multiplication or division, the combined *relative* standard deviation (RSD) is calculated by taking the square root of the sum of the squared RSDs for each interim result, and the combined standard uncertainty u calculated from the combined RSD and the result.

16.15. The overall uncertainty should be expressed as a multiple of the calculated standard deviation. The recommended multiplier is 2, that is, the uncertainty is equal to $2u$. Where the contributions arise from normally distributed errors, this value will correspond approximately to a 95% confidence interval.

16.16. It is not normally safe to extend this argument to higher levels of confidence without knowledge of the distributions concerned. In particular, it is commonly found that experimental uncertainty distributions are far wider at the 99% level of confidence than would be predicted by assumptions of normality.

16.17. It is often not necessary to evaluate uncertainties for every test and sample type. It will normally be sufficient to investigate the uncertainty once only, for a particular method and to use the information to estimate the measurement uncertainty for all tests carried out within the scope of that method.

17. METHODS/PROCEDURES FOR CALIBRATIONS AND TESTS

17.1. It is the laboratory's responsibility to use methods which are appropriate for the required application. The laboratory may use its own judgement, it may select a method in consultation with the customer, or the method may be specified in regulation or by the customer.

17.2. Quality standards often favor the use of standard or collaboratively tested methods wherever possible. Whilst this may be desirable in situations where a method is to be widely used, or defined in regulation, sometimes a laboratory may have a more suitable method of its own. The most important considerations are that the method should be suitable for the purpose intended, be adequately validated and documented and provide results that are traceable to stated references at an appropriate level of uncertainty.

17.3. The validation of a standard or collaboratively tested methods should not be taken for granted, no matter how impeccable the method's pedigree—the laboratory should satisfy itself that the degree of validation of a particular method is adequate for the required purpose, and that the laboratory is itself able to verify any stated performance criteria.

17.4. Methods developed in-house must be adequately validated, documented, and authorized before use. Where they are available, matrix matched reference materials should be used to determine any bias, or where this is not possible, results should be compared with other technique(s), preferably based on different principles of measurement. Measurement of the recovery of gravimetrically added spike analyte, measurement of blanks and the study of interferences and matrix effects can also be used to check for bias or imperfect recovery. Estimation of uncertainty must form part of this validation process and in addition to covering the above factors, should address issues such as sample homogeneity and sample stability. Advice on method validation is given in Sec. 18.

17.5.   Documentation of methods shall include validation data, limitations of applicability, procedures for quality control, calibration, and document control. A laboratory documenting methods may find it convenient to adopt a common format, such as ISO 78-2: (Ref. C10), which provides a useful model. In addition, advice on documentation of methods is available from other sources such as national standardization bodies and accreditation bodies.

17.6.   Developments in methodology and techniques will require methods to be changed from time to time and therefore method documentation must be subject to adequate document control. Each copy of the method should show issue number/date, issuing authority, and copy number. It must be possible to determine from records which is the most up-to-date version of each method is authorized for use.

17.7.   Obsolete methods should be withdrawn but must be retained for archive purposes and clearly labeled as obsolete. The difference in performance between revised and obsolete methods should be established so that it is possible to compare new and old data.

17.8.   When methods are revised then the validation also needs to be updated. The revision may be of a minor nature, involving different sample sizes, different reagents, etc. Alternatively, it may involve significant changes, such as the use of radically different technology or methodology. The level of revalidation required increases with the scale of the changes made to the method.

18. METHOD VALIDATION

18.1.   Checks need to be carried out to ensure that the performance characteristics of a method are understood and to demonstrate that the method is scientifically sound under the conditions in which it is to be applied. These checks are collectively known as validation. Validation of a method establishes, by systematic laboratory studies that the method is fit-for-purpose, i.e. its performance characteristics are capable of producing results in line with the needs of the analytical problem. The important performance characteristics include:

- Selectivity and specificity (description of the measurand),
- Measurement range,
- Calibration and traceability,
- Bias[*]
- Linearity,

- Limit of detection/limit of quantitation,
- Ruggedness,
- Precision.

*In some fields of chemical measurement, the term recovery is used to describe the total bias, in other fields, recovery is used in relation to certain elements of bias.

The above characteristics are interrelated, many of these contribute to the overall measurement uncertainty and the data generated may be used to evaluate the measurement uncertainty (see Sec. 16 and Ref. C13) during validation.

The Good practice in method validation is described in a EURACHEM Guide (Ref. A3). Note that there is no unanimous agreement on the interpretation of some of the above terms nor the conventions used in their determination. Thus, when stating validation data, it is advisable to state any conventions followed.

18.2.   The extent of validation must be clearly stated in the documented method so that the user can assess the suitability of the method for their particular needs.

18.3.   Standard methods will have been developed and validated collaboratively by a group of experts (Refs. C14–C19). This development should include consideration of all of the necessary aspects of validation and related uncertainty. However, the responsibility remains firmly with the user to ensure that the validation documented in the method is sufficiently complete to fully meet their needs. Even if the validation is complete, the user will still need to verify that the documented performance characteristics (e.g. trueness and precision) can be met in their own laboratory.

18.4.   As indicated above, there are different opinions concerning the terminology and the process of method validation. The following explanations supplement those in other parts of this guide and are intended as a guide rather than a definitive view.

18.5.   *Selectivity* of a method refers to the extent to which it can determine particular analyte(s) in a complex mixture without interference from the other components in the mixture. A method which is selective for an analyte or group of analytes is said to be *specific*. The applicability of the method should be studied using various samples, ranging from pure measurement standards to mixtures with complex matrices. In each case, the recovery of the analyte(s) of interest should be determined and the influences of suspected interferences duly stated. Any restrictions in the applicability of the technique should be documented in the method. This work will allow a clear description of the measurand to be made.

18.6.   *Range*: For quantitative analysis, the working range for a method is determined by examining samples with different analyte concentrations and determining the concentration range for which acceptable uncertainty can be achieved. The working range is generally more extensive than the linear range, which is determined by the analysis of a number of samples of varying analyte concentrations and calculating the regression from the results, usually using the method of least squares. The relationship of analyte response to concentration does not have to be perfectly linear for a method to be effective. For methods showing good linearity it is usually sufficient to plot a calibration curve using measurement standards at five different concentration levels (+ blank). More measurement standards will be required where linearity is poor. In qualitative analysis, it is common place to examine replicate samples and measurement standards over a range of concentrations to establish at what concentration a reliable cut-off point can be drawn between detection and nondetection (also see Sec. 18.8).

18.7. *Linearity* for quantitative methods is determined by the measurement of samples with analyte concentrations spanning the claimed range of the method. The results are used to calculate a regression line against analyte calculation using the least squares method. It is convenient if a method is linear over a particular range but it is not an absolute requirement. Where linearity is unattainable for a particular procedure, a suitable algorithm for calculations should be determined.

18.8. For qualitative methods, there is likely to be a concentration threshold below which positive identification becomes unreliable. The response range should be examined by testing a series of samples and measurement standards, consisting of sample blanks, and samples containing a range of analyte levels. At each concentration level, it will be necessary to measure approximately 10 replicates. A response curve of % positive (or negative) results vs. concentration should be constructed. From this it will be possible to determine the threshold concentration at which the test becomes unreliable. In the example shown below, positive identification of the analyte ceases to be 100% reliable below $100 \, \mu g \, g^{-1}$.

Example:

Concentration ($\mu g \, g^{-1}$)	No. replicates	Positive/negative
200	10	10/0
100	10	10/0
75	10	5/5
50	10	1/9
25	10	0/10
0	10	0/10

18.9. The *limit of detection* of an analyte is often determined by repeat analysis of a blank test portion and is the analyte concentration, the response of which is equivalent to the mean blank response plus 3 standard deviations. Its value is likely to be different for different types of sample.

18.10. The *limit of quantitation* is the lowest concentration of analyte that can be determined with an acceptable level of uncertainty. It should be established using an appropriate measurement standard or sample, i.e., it is usually the lowest point on the calibration curve (excluding the blank). It should not be determined by extrapolation. Various conventions take the limit to be 5, 6, or 10 standard deviations of the blank measurement.

18.11. *Ruggedness*: Sometimes also called *robustness*. Where different laboratories use the same method they inevitably introduce small variations in the procedure, which may or may not have a significant influence on the performance of the method. The ruggedness of a method is tested by deliberately introducing small changes to the method and examining the consequences. A large number of factors may need to be considered, but because most of these will have a negligible effect, it will normally be possible to vary several at once. Ruggedness is normally evaluated by the originating laboratory, before other laboratories collaborate.

18.12. The *bias (sometimes called recovery)* of a measuring system (method) is the systematic error of that measuring system. The issues associated with the estimation of bias and recovery are discussed in Sec. 15.4.

In addition to evaluating the bias, it is important to estimate the measurement uncertainty associated with the bias and to include this component in the overall estimate of measurement uncertainty.

18.13. The *precision* of a method is a statement of the closeness of agreement between mutually independent test results and is usually stated in terms of standard deviation. It is generally dependent on analyte concentration, and this dependence should be determined and documented. It may be stated in different ways depending on the conditions in which it is calculated. Repeatability is a type of precision relating to measurements made under repeatable conditions, i.e.: same method; same material; same operator; same laboratory; narrow time period. Reproducibility is a concept of precision relating to measurements made under reproducible conditions, i.e.: same method; different operator; different laboratories; different equipment; long time period. Precision is a component of measurement uncertainty (see Sec. 16).

18.14. Note that these statements of precision relate to quantitative analysis. Qualitative analysis can be treated in a slightly different way. Qualitative analysis effectively is a yes/no measurement at a given threshold analyte value. For qualitative methods, the precision cannot be expressed as a standard deviation or RSD, but may be expressed as true and false positive (and negative) rates. These rates should be determined at a number of concentrations, below, at and above the threshold level. Data from a confirmatory method comparison should be used if such an appropriate method is available. If such a method is not available, spiked and unspiked blank samples can be analyzed.

% false positive = falise positives $\times 100$/total known negatives

% false negatives = false negatives $\times 100$/total known positives

18.15. Confirmation is sometimes confused with repeatability. Whereas repeatability requires the measurement to be performed several times by one technique, confirmation requires the measurement to be performed by more than one technique. Confirmation increases confidence in the technique under examination and is especially useful where the additional techniques operate on significantly different principles. In some applications, for example, the analysis of unknown organics by gas chromatography, the use of confirmatory techniques is essential.

19. CALIBRATION

19.1. Calibration is a set of operations that establish, under specified conditions, the relationship between values of quantities indicated by a measuring instrument or measuring system, or values represented by a material measure, and the corresponding values realized by standards (Ref. VIM-B6). The usual way to perform calibration is to subject known amounts of the quantity (e.g. using a measurement standard or reference material) to the measurement process and monitor the measurement response. More detailed information on reference materials is given in the next chapter.

19.2. The overall program for calibration in the chemical laboratory shall be designed to ensure that all measurements that have a significant effect on test or calibration results are traceable to a measurement standard, preferably a national or international measurement standard such as a reference material. Where appropriate and where feasible, certified reference materials should be used. Where formally designated measurement standards are not available, a material with suitable properties and stability should be selected or prepared by the laboratory and used as a laboratory measurement standard. The required properties of this material should be characterized by repeat testing, preferably by more than one laboratory and using a variety of validated methods (see ISO Guide 35: Ref. C6).

19.3. Analytical tests may be subdivided into general classes depending on the type of calibration required.:

19.3.1. Some analytical tests depend critically on the measurement of physical properties, such as weight measurement in gravimetry and volume measurement in titrimetry. Since these measurements have a significant effect on the results of the test, a suitable calibration program for these quantities is essential. In addition, the calibration of measuring devises used to establish the purity or amount concentration of the chemical standards need to be considered.

19.3.2. Where a test is used to measure an empirical property of a sample, such as flashpoint, equipment is often defined in a national or international standard method and traceable reference materials should be used for calibration purposes where available. New or newly acquired equipment must be checked by the laboratory before use to ensure conformity with specified design, performance, and dimension requirements.

19.3.3. Instruments such as chromatographs and spectrometers, which require calibration as part of their normal operation, should be calibrated using reference materials of known composition (probably solutions of pure chemicals).

19.3.4. In some cases, calibration of the whole analytical process can be carried out by comparing the measurement output from a sample with the output produced by a suitable reference material that has been subjected to the same full analytical process as the sample. The reference material may be either a synthetic mixture prepared in the laboratory from materials of known (and preferably certified) purity, or a purchased certified matrix reference material. However, in such cases, a close match between the test sample and the matrix reference material, in terms of the nature of the matrix, and the concentration of the analyte has to be assured.

19.4. However, in many cases, calibration is only performed on the final measurement stage. For example, calibration of a gas chromatography method may be carried out using a series of measurement standards which are synthetic solutions of the analyte of interest at various concentrations. Such calibration does not take into account factors such as contamination or losses that occur during the sample preparation and extraction or derivatization stages. It is therefore essential during the method validation process to explore the potential problems of contamination and losses by taking matrix reference materials or spiked samples through the whole measurement process, and design the day-to-day calibration procedure and quality control checks accordingly (also see Sec. 15.4).

19.5. Individual calibration programs shall be established depending on the specific requirements of the analysis. Also, it may be necessary to check instrument calibration after any shutdown, whether deliberate or otherwise, and following service or other substantial maintenance. The level and frequency of calibration should be based on previous experience and should be at least that recommended by the manufacturer. Guidance on calibration is given in Appendix B and includes typical calibration intervals for various types of simple instruments and indicates the parameters which may require calibration in more complex analytical instruments. The frequency of calibration required will depend on the stability of the measurement system, the level of uncertainty required, and the criticality of the work.

19.6. Procedures for performing calibrations shall be adequately documented, either as part of specific analytical methods or as a general calibration document. The documentation should indicate how to perform the calibration, how often calibration is necessary action to be taken in the event of calibration failure. Frequency intervals for recalibration of physical measurement standards should also be indicated.

19.7.   The calibration of volumetric glassware normally relates to a particular solvent at a particular temperature. The calibration is rarely valid when the glassware is used with other solvents because of different densities, wetting characteristics, surface tension, etc. This is particularly pertinent for volumetric glassware calibrated to deliver a certain volume. Other volumetric equipment may be affected when using solvents with high rates of thermal expansion. In such situations, the glassware should be recalibrated using the relevant solvent, at the correct temperature. Alternatively, for the highest accuracy, measurements can often be made by mass rather than by volume.

19.8.   Figure 1 is a typical analytical process and illustrates the role of calibration in relation to method validation and quality control.

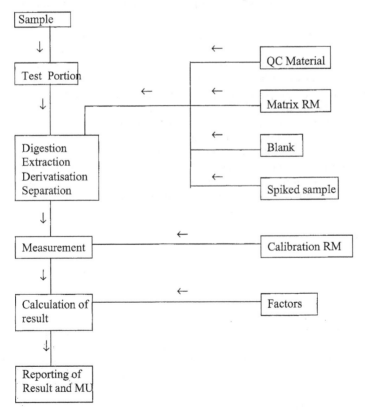

Figure   1

20. REFERENCE MATERIALS

20.1.   A series of ISO Guides relating to reference materials are available (Refs. C1–C6).

20.2.   *Reference materials and certified reference materials* are defined in Sec. 3. They are used for calibration, method validation, measurement verification, evaluating measurement uncertainty, and for training purposes.

20.3.   Reference materials may take a variety of forms, including pure substance RMs, matrix RMs, and solutions or mixtures. The following are all examples of reference materials:

- 95% pure sodium chloride;
- an aqueous solution containing 1% (w/v) copper (II) sulfate and 2% (w/v) magnesium chloride;
- a powdered polymer with a particular weight distribution range;
- a crystalline solid melting in the range 150–151°C;
- a dried milk powder containing a known amount of vitamin C.

20.4.   For many types of analysis, calibration may be carried out using reference materials prepared within the laboratory from chemicals of known purity and composition. Some chemicals may be purchased with a manufacturers certificate stating purity. Alternatively, chemicals of a stated but uncertified purity may be purchased from reputable suppliers. Whatever the source, it is the users' responsibility to establish that the quality of such materials is satisfactory. Sometimes additional tests will need to be carried out by the laboratory. Normally, a new batch of a chemical should be checked against the previous batch. Ideally, all chemicals to be used for reference material purposes should be purchased from producers with demonstrated QA systems. However, a QA system does not automatically guarantee the quality of the producer's products and laboratories should take all reasonable steps to confirm the quality of critical materials. The control of impurities is important, especially for trace analysis, where they may cause interferences. Due regard should be paid to the manufacturers recommendations on storage and shelf life. In addition, caution is needed, as suppliers do not always provide information about all impurities.

20.5.   The use of appropriate reference materials can provide essential traceability and enable analysts to demonstrate the accuracy of results, calibrate equipment and methods, monitor laboratory performance and validate methods, and enable comparison of methods by use as transfer (measurement) standards. Their use is strongly encouraged wherever appropriate.

20.6.   The uncertainty of purity of a pure substance reference material needs to be considered in relation to the uncertainty associated with other aspects of the method. Ideally, the uncertainty associated with a reference material, used for calibration purposes, should not contribute more than one-third of the overall measurement uncertainty.

20.7.   The composition of the certified reference material should be as close as possible to that of the samples. Where matrix interferences exist, ideally a method should be validated using a matched matrix reference material certified in a reliable manner. If such a material is not available, it may be acceptable to use a sample spiked with a reference material.

20.8.   It is important that any *certified* reference material used has been produced and characterized in a technically valid manner. Users of CRMs should be aware that not all materials are validated with the same degree of rigor. Details of homogeneity trials, stability trials, the methods used in certification, and the uncertainties and variations in the stated analyte values are usually available from the producer and should be used to judge the pedigree. The material must be accompanied by a certificate, which includes an estimate of uncertainty of the certified value (see Sec. 16). ISO Guide 34 (Ref. C5) and an ILAC Guide (Ref. B15) deal with criteria for the competence of reference material producers. These guides may provide the basis for future assessment of reference material producers.

20.9.   Reference materials and certified reference materials should be clearly labeled so that they are unambiguously identified and referenced against accompanying certificates or other documentation. Information should be available indicating shelf life, storage conditions, applicability, and restrictions of use. Reference materials made up within the laboratory, e.g. as solutions should be treated as reagents for the purposes of labeling (see Sec. 14.2).

20.10.   Reference materials and measurement standards should be handled in order to safeguard against contamination or degradation. Staff training procedures should reflect these requirements.

21.   QUALITY CONTROL AND PROFICIENCY TESTING

21.1.   The meaning of the terms 'quality control' and 'quality assurance (QA)' often vary according to the context. In practical terms, QA relates to the overall measures taken by the laboratory to regulate quality, whereas quality control describes the individual measures which relate to the quality of individual samples or batches of samples.

21.2.   As part of their quality systems, and to monitor day-to-day and batch-to-batch analytical performance, laboratories must operate an appropriate level of internal quality control (QC) checks and participate wherever possible in appropriate proficiency testing schemes (external QC). The level and type of QC will depend on criticality, nature of the analysis, frequency of analysis, batch size, degree of automation, and test difficulty and reliability.

21.3. *Internal QC*: This may take a variety of forms including the use of: blanks; measurement standards; spiked samples; blind samples; replicate analysis; and QC samples. The use of control charts is recommended, particularly for monitoring QC control samples (Ref. C20–22).

21.3.1. The level of QC adopted must be demonstrably sufficient to ensure the validity of the results. Different types of quality control may be used to monitor different types of variation within the process. QC samples analyzed at intervals in the sample batch will indicate drift in the system; use of various types of blank will indicate what are the contributions to the instrument besides those from the analyte; duplicate analyses give a check of repeatability, as do blind samples.

21.3.2. QC samples are typical samples which are sufficiently stable and available in sufficient quantities as to be available for analysis over an extended period of time. Over this period, the random variation in performance of the analytical process can be monitored by monitoring the analyzed value of the QC sample, usually by plotting it on a control chart. As long as the QC sample value is acceptable, it is likely that results from samples in the same batch as the QC sample can be taken as reliable. The acceptability of the value obtained with the QC sample should be verified as early as practicable in the analytical process so that in the event of system failure as little effort as possible has been wasted on unreliable sample analysis.

21.3.3. It is the responsibility of the analyst to set and justify an appropriate level of quality control, based on a risk assessment taking into account the reliability of the method, and the criticality of the work. It is widely accepted that for routine analysis, a level of internal QC of 5% has been identified as reasonable, i.e. 1 in every 20 samples analyzed should be a QC sample. However, for robust routine methods with high sample throughput, a lower level of QC may be reasonable. For more complex procedures, a level of 20% is not unusual and on occasions even 50% may be required. For analyses performed infrequently, a full system validation should be performed on each occasion. This may typically involve the use of a reference material containing a certified or known concentration of analyte, followed by replicate analyses of the sample and spiked sample (a sample to which a known amount of the analyte has been deliberately added). Those analyses undertaken more frequently should be subject to systematic QC procedures incorporating the use of control charts and check samples.

21.4. *Proficiency Testing (External QC)*: One of the best ways for an analytical laboratory to monitor its performance against both its own requirements and the norm of other laboratories is to participate regularly in proficiency testing schemes (Ref. C7). Proficiency testing helps to highlight not only repeatability and reproducibility performance between laboratories, but also systematic errors, i.e. bias. Proficiency testing and other types of intercomparisons are accepted as being an important means of monitoring quality at national and international levels.

21.5. Accreditation bodies also recognize the benefit of these schemes as objective evidence of competence of the laboratory and of the effectiveness of the assessment process itself. Where possible, laboratories should select proficiency testing schemes which operate according to good international practice (Ref. C7) and have transparent evidence of quality, e.g. by accreditation or other peer review (Ref. B16). Accredited laboratories are normally required to participate in proficiency testing (where suitable schemes exist) as an integral part of their QA protocols. It is important to monitor proficiency testing results as a means of checking performance and to take corrective action as necessary.

22. COMPUTERS AND COMPUTER CONTROLLED SYSTEMS

22.1. In chemical testing laboratories, computers have a wide variety of uses including:

- control of critical environmental conditions;
- monitoring and control of inventories;
- calibration and maintenance schedules;
- stock control of reagents and measurement standards;
- design and performance of statistical experiments;
- scheduling of samples and monitoring of work throughput;
- control chart generation;
- monitoring of test procedures;
- control of automated instrumentation;
- capture, storage, retrieval, processing of data, manually or automatically;
- matching of sample and library data;
- generation of test reports;
- word processing;
- communication.

22.2. Interfaces and cables provide physical connections between different parts of the computer or between different computers. It is important that interfaces and cables are chosen to suit the particular application since they can seriously affect speed and quality of data transfer.

22.3. The chemical testing environment creates particular hazards for the operation of computers and storage of computer media. Advice can usually be found in the operating manuals, however, particular care should be taken to avoid damage due to chemical, microbiological or dust contamination, heat, damp, and magnetic fields.

22.4. Initial validation should verify as many aspects of a computer's operation as possible. Similar checks should be carried out if the computer's use is changed, or after maintenance, or revision of software. Where a computer is used to gather and process data associated with chemical testing, for validation of that function, it is usually sufficient to assume correct operation if the computer produces expected answers when input with known parameters. Computer programs performing calculations can be validated by comparison with manually generated results. It should be noted that some faults will occur only when a particular set of parameters is input. In chemical testing, suitable checks on the data gathering and handling functions could be made using a certified reference material for the initial validation, with a secondary measurement standard such as a quality control material used for regular repeat checks. Any recommendations made by the manufacturer should be taken into consideration. The validation procedure used for a particular system and any data recorded during validation should be documented. It may be difficult to validate these systems in isolation from the analytical instrument producing the original signal. Usually, the whole system is validated in one go, by using chemical measurement standards or reference materials. Such validation is normally acceptable. It is convenient to illustrate validation using examples of typical applications:

22.4.1. *Word-Processing Packages* are widely used in laboratories to generate a wide variety of documentation. The laboratory should ensure that the use of word processing packages is controlled sufficiently to prevent the production of unauthorized reports or other documents. In the most simple cases, where the computer acts as little more than an electronic typewriter, validation is achieved by manually checking hard copies. More sophisticated systems read and process data to automatically produce reports in predetermined formats. Such systems will require additional checks.

22.4.2. *Microprocessor Controlled Instruments* will normally have a self-checking routine which is activated when the instrument is switched-on, and will include the recognition and checking of all peripheral equipment. Often the software is not accessible. Under most circumstances, validation can be performed by testing the various aspects of instrument function using known parameters, e.g. by testing reference materials, physical or chemical measurement standards or quality control samples.

22.4.3. *Data Handling or Processing Systems, Integration Systems.* Before it can be processed, the output from the analytical instrument will usually need to be converted to a digital signal using an analogue/digital converter. The digitized data is then translated into a recognizable signal (numbers, peaks, spectra according to the system) by the software algorithm. The algorithm makes various decisions (such as deciding where peaks start and finish, or whether a number should be rounded up or down) according to programed instructions. The algorithm is a common source of unexpected performance and validation should test the logic behind the decisions made by the algorithm.

22.4.4. *Computer Controlled Automated System.* This may embrace one or more of the foregoing examples, operated either simultaneously or in controlled time sequence. Such systems will normally be validated by checking for satisfactory operation (including performance under extreme circumstances) and establishing the reliability of the system before it is allowed to run unattended. The validation should consist of a validation of individual components, plus an overall check on the dialogue between individual components and the controlling computer. An assessment should be made of the likely causes of system malfunction. One important consideration is that the computer, interfaces, and connecting cabling have sufficient capacity for the required tasks. If any part of the system is overloaded, its operation will slow down and possibly data may be lost. This could have serious consequences where the operations include time-sequenced routines. Where possible the controlling software should be tailored to identify and highlight any such malfunctions and tag associated data. The use of quality control samples and standards run at intervals in the sample batches should then be sufficient to monitor correct performance on a day-to-day basis. Calculation routines can be checked by testing with known parameter values. Electronic transfer of data should be checked to ensure that no corruption has occurred during transmission. This can be achieved on the computer by the use of 'verification files' but, wherever practical, the transmission should be backed-up by a hard copy of the data.

22.4.5. *Laboratory Information Management Systems (LIMS)*. LIMS systems are increasingly popular as a way of managing laboratory activities. A LIMS is a computer-based system with software which allows the electronic collation, calculation, and dissemination of data, often received directly from analytical instruments. It incorporates word-processing, database, spreadsheet, and data processing capabilities and can perform a variety of functions, including sample registration and tracking; test assignment and allocation; worksheet generation; processing captured data; quality control; financial control; and report generation. The operation of the LIMS may be confined to the laboratory itself or it may form part of a company wide computer system. Information may be input manually or downloaded directly from analytical instrumentation or other electronic devices such as barcode readers. Information can be output either electronically or as hardcopies. Electronic outputs could consist of raw or processed data written to other computers either within the organization, or remote, perhaps transmitted via a modem or electronic mail. Similarly, the information could be downloaded to a disk. Where data crosses from one system to another there may be a risk of data corruption through system incompatibility or the need to reformat the information. A well-designed system enables high levels of QA to be achieved, right from the point of sample entry to the production of the final report. Particular validation requirements include management of access to the various functions, and audit trails to catalogue alterations and file management. Where data are transmitted electronically, it will be necessary to build in safety checks to guard against data corruption and unauthorized access.

23. LABORATORY AUDIT AND REVIEW

23.1.   See Sec. 3.6 for terminology.

23.2.   An important aspect of quality management is the periodic re-exami-
nation of the quality system by the laboratory's own management. In general,
all aspects of the quality system should be examined at least once a year. The
system should be examined in two ways. Firstly, it should be examined to
ensure that it is sufficiently well documented to enable adequate and consistent implementation, and that staff are actually following the system
described. This examination is commonly known as auditing (as opposed
to the external auditing or assessment carried out by accreditation or certification bodies). Secondly, the system should be examined to see whether it
meets the requirements of the laboratory, its customers and, if appropriate,
the quality management standard. Over a period of time, the needs of the
laboratory and its customers will change and the quality system should
evolve to continue to fulfil its purpose. This second type of examination is
commonly known as review and should be carried out at least annually. It
is carried out by the laboratory management and draws on information from
a number of sources, including results from internal audits, external assessments, proficiency testing scheme participation, internal quality control
studies, market trends, customer complaints and compliments, etc.

23.3.   The program of audits and review is normally co-ordinated by the
laboratory quality manager, who is responsible for ensuring that auditors
have the correct training, guidance, and authority necessary for their work.
Audits are normally carried out by laboratory staff who work outside of the
area they are examining. This is of course not always possible where staff
numbers are small.

23.4.   Audits may be carried out in two basic ways. In the horizontal audit,
the auditor will examine in detail single aspects of the quality system, for
example calibration or reports. In the vertical audit, the auditor will select
a sample and follow its progress from receipt to disposal, examining all
aspects of the quality system relating to its testing.

23.5.   A check list, detailing the aspects of a chemical laboratory which
should be examined during a quality audit is listed in Appendix A of this
Guide.

23.6.   The management review should be carried out at regular intervals.
Once a year is normally sufficient, although, for laboratories with extensive
scopes of accreditation it may be necessary to split the review into discrete
modules that can be examined during the course of the year. Issues which
should be covered at the annual review include the quality system and issues
which affect analytical quality, internal audits, corrective and preventative
action, client feedback and complaints.

REFERENCES AND BIBLIOGRAPHY

The following section provides useful References (Subsections A, B, C—these are referred to in the text—Website addresses (D), a Bibliography (E)).

A. CITAC and EURACHEM GUIDES (Available on CITAC www.citac.ws and EURACHEM www.eurachem.org)

1. Quality Assurance for Research and Development and Non-Routine Analysis: 1998 (CITAC/EURACHEM).
2. Quantifying Uncertainty in Analytical Measurement: 2000 (CITAC/ EURACHEM) (see also website—Ref. D12).
3. The Fitness for Purpose of Analytical Methods: A Laboratory Guide to Method Validation and Related Topics: 1998 (EURACHEM).
4. Harmonised Guidelines for the Use of Recovery Information in Analytical Measurement: 1998 (EURACHEM/IUPAC/ISO/AOACI).
5. Selection, Use and Interpretation of Proficiency Testing (PT) Schemes by Laboratories: 2000 (EURACHEM).
6. CITAC Policy Statement on Traceability in Chemical Measurement: 2000.
7. CITAC/EURACHEM Guide on Traceability in Chemical Measurements: 2002 (under preparation).

B. Key References

1. ISO/IEC 17025:1999 General Requirements for the Competence of Testing and Calibration Laboratories.
2. ISO 9000:2000 Quality Management Systems—Fundamentals and Vocabulary.
3. OECD Principles of Good Laboratory Practice: 1998 (Code: ENV/MC/ CHEM (98) 17 download: http://www1.oecd.org/ehs/ehsmono/01E88455.pdf).
4. ISO/IEC Guide 2:1996 Standardization and Related Activities—General Vocabulary (Currently Under Revision as ISO 17000).
5. ISO 9001:2000 Quality Management Systems—Requirements.
6. International Vocabulary of Basic and General Terms in Metrology (VIM)— 2nd ed. 1993 (ISO/BIPM/IEC/IFCC/IUPAC/IUPAP/OIML).
7. Guide to the Expression of Uncertainty in Measurement (GUM), ISO Geneva Switzerland, 1995.
8. Meeting the Measurement Uncertainty and Traceability Requirements of ISO/IEC 17025 in Chemical Analysis—B. King, Fresenius Journal, 2001.
9. The Selection and Use of Reference Materials—A Basic Guide for Laboratories and Accreditation Bodies—Draft EEEE/RM 2002—prepared by B. King 2000.
10. Position of Third Party Quality Assessment of Reference Materials and Their Production EEEE/RM/069 rev 1: Draft 2001.
11. APLAC Policy and Guidance on the Estimation of Uncertainty of Measurement in Testing—Draft April 2002.

12. ILAC P10: 2002 ILAC Policy on Traceability of Measurements Results.
13. ILAC G8: 1996 Guidelines on Assessment and Reporting of Compliance with Specification.
14. ILAC G9: 1996 Guidelines for the Selection and Use of Certified Reference Materials.
15. ILAC G12: 2000 Guidelines for the Requirements for the Competence of Reference Material Producers.
16. ILAC G13: 2000 Guidelines for the Requirements for the Competence of Providers of Proficiency Testing Schemes.
17. ILAC G15: 2001 Guidance for Accreditation to ISO/IEC 17025.
18. ILAC G17: 2002 Guidance for Introducing the Concept of Uncertainty of Measurement in Testing in Association with the Application of the Standard ISO/IEC 17025

    Note: Other Guidelines produced by Regional Accreditation Bodies are also relevant here (see website addresses in Sec. D, nos. 7, 8, and 9 below). In addition, most national accreditation bodies issue guidance in support of their requirements (usually based on ISO standards).

C. Other References (ISO Guides and Standards)

1. ISO Guide 30:1992 Terms and definitions used in connection with reference materials.
2. ISO Guide 31:2000 Reference materials—contents of certificates and labels.
3. ISO Guide 32:1997 Calibration in analytical chemistry and use of certified reference materials.
4. ISO Guide 33:2000 Uses of certified reference materials.
5. ISO Guide 34:2000 General requirements for the competence of reference material producers.
6. ISO Guide 35:1989 (under revision) Certification of reference materials—general and statistical principles.
7. ISO/IEC Guide 43:1997 Proficiency testing by interlaboratory comparisons—Part 1: Development and operation of proficiency testing schemes and Part 2: Selection and use of proficiency testing schemes by laboratory accreditation bodies.
8. ISO/IEC Guide 58: 1993 Calibration and testing laboratory accreditation systems—general requirements for operation and recognition. (To be replaced by ISO/IEC 17011 General requirements for bodies providing assessment and accreditation).
9. ISO/IEC Guide 62:1996 General requirements for bodies operating assessment and certification/registration of quality systems.
10. ISO 78-2:1999 Chemistry—layouts for standards—Part 2: Methods of chemical analysis.
11. ISO/DIS 10576-1:2001 Statistical methods—guidelines for the evaluation with specified requirements Pt 1. General principles.
12. ISO 3534 Statistics—vocabulary and symbols—Parts 1, 2, and 3 (1999).
13. ISO/DTS 21748–2002 (under preparation) Guide to the use of repeatability and reproducibility and trueness estimates in measurement uncertainty estimation.

14. ISO 5725-1:1994 Accuracy (trueness and precision) of measurement methods and results—Part 1: General principles and definitions ISO 5725-1:1994/Cor 1:1998.
15. ISO 5725-2:1994 Accuracy (trueness and precision) of measurement methods and results—Part 2: Basic method for the determination of repeatability and reproducibility of a standard measurement method.
16. ISO 5725-3:1994 Accuracy (trueness and precision) of measurement methods and results—Part 3: Intermediate measures of the precision of a standard measurement method.
17. ISO 5725-4:1994 Accuracy (trueness and precision) of measurement methods and results—Part 4: Basic methods for the determination of the trueness of a standard measurement method.
18. ISO 5725-5:1998 Accuracy (trueness and precision) of measurement methods and results—Part 5: Alternative methods for the determination of the precision of a standard measurement method.
19. ISO 5725-6:1994 Accuracy (trueness and precision) of measurement methods and results—Part 6: Use in practice of accuracy values.
20. ISO 7870:1993 Control charts—general guide and introduction.
21. ISO 7966:1993 Acceptance control charts.
22. ISO 8258:1991 Shewhart control charts.

D. Useful Website Addresses

1. CITAC: www.citac.ws.
2. EURACHEM: www.eurachem.org.
3. ISO: www.iso.ch.
4. (ISO) REMCO: www.iso.org/remco.
5. COMAR (Reference Material Data Base): www.comar.bam.de.
6. AOAC: www.aoac.org.
7. ILAC: www.ilac.org.
8. APLAC: www.ianz.govt.nz/aplac.
9. EA: www.european-accreditation.org.
10. BIPM: www.bipm.fr.
11. OECD: www.oecd.org.
12. www.mutraining.com (web based training on measurement uncertainty and accreditation).
13. www.measurementuncertainty.org (MU forum/search engine—linked to Ref. A2).

E. Bibliography

1. AOAC International—ISO 17025 and the Laboratory—An Introduction to Laboratory Accreditation: 2000.
2. Garfield FM, Klesten E, Husch J. AOAC International—Quality Assurance Principles for Analytical Laboratories—3rd ed. 2000: ISBN-0-935584-70-6.
3. Crosby, Neil T, Patel, Indu. General Principles of Good Sampling Practice. Cambridge: Royal Society of Chemistry, 1995.

4. Enell JW. "Which sampling plan should I choose?." J Quality Technol 1984; 16(3):168–171.
5. Garfield FM. "Sampling in the analytical scheme." J Assoc Off Anal Chem 1989; 72(3):405–411.
6. Gy, Pierre. Sampling for Analytical Purposes. Chichester: Wiley, 1998.
7. Horwitz W. "Nomenclature for sampling in Analytical Chemistry". IUPAC. Pure Appl Chem 1990; 62(6):1193–1208.
8. Horwitz W. "Problems of samplings and analytical methods." IUPAC, J Assoc Off Anal Chem 1976; 59(6):1197–1203.
9. Horwitz W. "Design, conduct and interpretation of method performance studies." IUPAC, Protocol 1994.
10. Kateman G, Buydens L. Quality Control in Analytical Chemistry. 2nd ed. New York: Wiley, 1993.
11. Keith LH. Environmental Sampling and Analysis. A Practical Guide. Chelsea. MI: Lewis Publishers, 1991.
12. Keith LH. Principles of Environmental Sampling. Washington DC: ACS, 1988.
13. Keith LH, ed. Principles of Environmental Sampling. 2nd ed. Washington, DC: American Chemical Society cl996.
14. Kratochvil B, Wallace D, Taylor JK. "Sampling for chemical analysis." Anal Chem 1984; 56(5):113R–129R.
15. Miller JC, Miller JN. Statistics for Analytical Chemistry 4th ed. Ellis Horwood, 1998.
16. Prichard E. Analytical Measurement Terminology (UK's Valid Analytical Measurement Program, LGC Ltd) ISBN 0-85404-443-4, 2000.
17. Prichard E. Quality in the Analytical Chemistry Laboratory. ACOL, Wiley, 1997.
18. Stoeppler M, ed. Sampling and Sample Preparation: Practical Guide for Analytical Chemists. Berlin: Springer Verlag, 1997.
19. Taylor BN, Kuyatt CE. Guidelines for evaluating and expressing uncertainty in NIST measurement results, NIST technical note 1297, 1994, National Institute of Standards and Technology.
20. Taylor JK. Quality assurance of chemical measurements. Michigan: Lewis Publishers, 1987.
21. UK DTI VAM Programme—General Guidelines for Use with a Protocol for QA of Trace Analysis 1998.
22. Youden WJ, Steiner EH. Statistical Manual of the Association of Official Analytical Chemists. Statistical Techniques for Collaborative Tests. Planning and Analysis of Results of Collaborative Tests. Washington DC: AOAC, 1975.

Acronyms

Some common acronyms follow:

AOAC: Association of Official Analytical Chemists (USA)
APLAC: Asia-Pacific Laboratory Accreditation Cooperation
BIPM: International Bureau of Weights and Measures
CCQM: Consultative Committee for Amount of Substance
CITAC: Cooperation on International Traceability in Analytical Chemistry
EA: European Cooperation for Accreditation
IEC: International Electrotechnical Commission
ILAC: International Laboratory Accreditation Cooperation
ISO: International Organization for Standardisation
ISO/REMCO: International Organization for Standardisation, Committee on Reference Materials
IUPAC: International Union of Pure and Applied Chemistry
JCTLM: Joint Committee on Traceability in Laboratory Medicine
OECD: Organization for Economic Cooperation and Development
OIML: International Organization on Legal Metrology

Appendix A

Quality Audit—Areas of particular importance to a chemistry laboratory.

1. Staff

    i. Staff have the appropriate blend of background, academic or vocational qualifications, experience and on-the-job training for the work they do.
    ii. On-the-job training is carried out against established criteria, which wherever possible are objective. Up-to-date records of the training are maintained.
    iii. Tests are only carried out by authorized analysts.
    iv. The performance of staff carrying out analyses is observed by the auditor.

2. Environment

    i. The laboratory environment is suitable for the work carried out.
    ii. The laboratory services and facilities are adequate for the work carried out.
    iii. There is adequate separation of potentially conflicting work.
    iv. The laboratory areas are sufficiently clean and tidy to ensure the quality of the work carried out is not compromised.
    v. There is adequate separation of sample reception, preparation, clean-up, and measurement areas, to ensure the quality of the work carried out is not compromised.
    vi. Adherence to safety regulations is consistent with the requirements of the quality management standard.

3. Equipment

    i. The equipment in use is suited to its purpose.

    ii. Major instruments are correctly maintained and records of this maintenance are kept.

    iii. Appropriate instructions for use of equipment are available.

    iv. Critical equipment, e.g. balances, thermometers, glassware, timepieces, pipettes, etc., are uniquely identified, appropriately calibrated (with suitable traceability), and the corresponding certificates or other records demonstrating traceability to national measurement standards are available.

    v. Calibrated equipment is appropriately labeled or otherwise identified to ensure that it is not confused with uncalibrated equipment and to ensure that its calibration status is clear to the user.

    vi. Instrument calibration procedures and performance checks are documented and available to users.

    vii. Instrument performance checks and calibration procedures are carried out at appropriate intervals and show that calibration is maintained and day-to-day performance is acceptable. Appropriate corrective action is taken where necessary.

    viii. Records of calibration, performance checks, and corrective action are maintained.

4. Methods and Procedures

    i. In-house methods are fully documented, appropriately validated, and authorized for use.

    ii. Alterations to methods are appropriately authorized.

    iii. Copies of published and official methods are available.

    iv. The most up-to-date version of the method is available to the analyst.

    v. Analyses are (observed to be) following the methods specified.

    vi. Methods have an appropriate level of advice on calibration and quality control.

    vii. Uncertainty has been estimated.

5. Chemical and Physical Measurement Standards, Certified Reference Materials and eagents.

    i. The measurement standards required for the tests are readily available.

    ii. The measurement standards are certified or are the "best" available.

    iii. The preparation of working measurement standards and reagents is documented.

    iv. Measurement standards, reference materials, and reagents are properly labeled and correctly stored. Where appropriate "opening" and "use-by" dates are used.

v. New batches of measurement standards, and reagents critical to the performance of the method are compared against old batches before use.

vi. The correct grade of materials is being used in the tests.

vii. Where measurement standards, or reference materials are certified, copies of the certificate are available for inspection.

6. Quality Control

i. There is an appropriate level of quality control for each test.

ii. Where control charts are used, performance has been maintained within acceptable criteria.

iii. QC check samples are being tested by the defined procedures, at the required frequency and there is an up-to-date record of the results and actions taken where results have exceeded action limits.

iv. Results from the random reanalysis of samples show an acceptable measure of agreement with the original analyses.

v. Where appropriate, performance in proficiency testing schemes and/or interlaboratory comparisons is satisfactory and has not highlighted any problems or potential problems.

vi. There is an effective system for linking proficiency testing performance into day-to-day quality control.

7. Sample Management

i. There is an effective documented system for receiving samples, identifying samples against requests for analysis, showing progress of analysis, issue of report, and fate of sample.

ii. Samples are properly labeled and stored.

8. Records

i. Notebooks/worksheets or other records show the date of test, analyst, analyte(s), sample details, test observations, quality control, all rough calculations, any relevant instrument traces, rough data, and relevant calibration data.

ii. Notebooks/worksheets are indelible, mistakes are crossed out rather than erased or obliterated, and the records are signed by the analysts.

iii. Where a mistake is corrected, the alteration is traceable to the person making the correction.

iv. The laboratory has procedures for checking data transfers and calculations and is using them.

9. Test Reports

i. The information given in reports is consistent with the requirements of the standard, the customer and reflects any provisions made in the documented method.

10. Miscellaneous

    i. Documented procedures are in operation to handle queries and complaints and system failures.

    ii. There is adequate evidence of corrective action (in the case of system failures) and preventive action. Effectiveness is evaluated in both cases.

    iii. The Laboratory Quality Manual is up-to-date and is accessible to all relevant staff.

    iv. There are documented procedures for subcontracting work, including verification of suitability.

    v. Vertical audits on random samples (i.e. checks made on a sample, examining all procedures associated with its testing from receipt through to the issue of a report) have not highlighted any problems.

Appendix B

Calibration Intervals and Performance Checks.

B1. Guidance is given in Table Appendix B-1 on the calibration of equipment in common use in analytical laboratories and on which the calibration of other instruments may be dependent. More comprehensive advice is available in the literature (see bibliography #32) and also in equipment manuals.

Table Appendix B-1

	Type of instrument	Frequency of check	Parameters to be checked
(a)	Balances	Depends on use	Linearity, zero point, accuracy (using calibrated weights)
(b)	Volumetric glassware	Depends on use	Accuracy, precision (pipettes/burettes)
(c)	Hydrometers (working)	Annually	One point calibration versus reference hydrometer
(d)	Hydrometers (reference)	5 years	One point calibration using measurement standard of known specific gravity
(e)	Barometers*	5 years	One point
(f)	Timers (see note)	2 years or less depending on use	Accuracy
(g)	Thermometers (reference)	5 years Annually	Critical points on scale, fixed points, e.g. ice point
(h)	Thermometers	Annually depending on use	Check specific points against reference thermometer

Note: Those instruments marked * will normally be calibrated in an accredited calibration laboratory, but should at least show traceability to national measurement standards.

National radio-time signals, or telephone time signals, provide a suitable source of traceable calibration of both absolute time and time difference. Timers with quartz/electronic movements are generally more accurate and stable than conventional mechanical timers and will need to be calibrated less often.

B2. The following aspects of the instruments listed below may need to be checked depending on the method:

B2.1. *Chromatographs (general)*:

i. Overall system checks, precision of repeat sample injections, carry-over.
ii. Column performance (capacity, resolution, retention).
iii. Detector performance (output, response, noise, drift, selectivity, linearity).
iv. System heating/thermostatting (accuracy, precision, stability, ramping characteristics).
v. Autosampler (accuracy and precision of time routines).

B2.2. *Liquid and Ion Chromatography*:

i. Composition of mobile phase.
ii. Mobile phase delivery system (precision, accuracy, pulse-free).

B2.3. *Electrode/meter systems, including conductivity, pH and ion-selective*:

i. Electrode drift or reduced response.
ii. Fixed point and slope checks using chemical measurement standards.

B2.4. *Heating/cooling apparatus, including freeze dryers, freezers, furnaces, hot air sterilizers, incubators, melting and boiling point apparatus, oil baths, ovens, steam sterilizers and water baths*:

i. Periodic calibration of temperature sensing system using the appropriate calibrated thermometer or pyroprobe.
ii. Thermal stability, reproducibility.
iii. Heating/cooling rates and cycles.
iv. Ability to achieve and sustain pressure or vacuum.

B2.5. *Spectrometers and spectrophotometers, including atomic absorption, fluorimetric, inductively coupled plasma—optical emission, infrared, luminescence, mass, nuclear magnetic resonance, ultra-violet/visible, visible, and X-ray fluorescence*:

i. Selected wavelength accuracy, precision, stability.

ii. Source stability.

iii. Detector performance (resolution, selectivity, stability, linearity, accuracy, precision).

iv. Signal to noise ratio.

v.   Detector calibration (mass, ppm, wavelength, frequency, absorbance, transmittance, bandwidth, intensity, etc.).

vi. Internal temperature controllers and indicators where applicable.

B2.6 *Microscopes:*

i.   Resolving power.

ii. Performance under various lighting conditions (fluorescence, polarization, etc.).

iii. Graticule calibration (for length measurement).

B2.7 *Autosamplers:*

i.   Accuracy and precision of timing systems.

ii. Reliability of sequencing programs.

iii. Accuracy and precision of sample delivery systems.

Appendix C

Comparison between ISO/IEC 17025:1999 and ISO/IEC Guide: 1990 (This table is reproduced from ILAC G15:2001, Guidelines for Accreditation to ISO/IEC 17025.)

Item from contents list of ISO/IEC 17025	ISO/IEC 17025 Clause	ISO/IEC Guide 25
Scope	1.1	1.1
	1.2	—
	1.3	—
	1.4	1.3
	1.5	*7.6 Note*
	1.6	*Intro*
Normative references	2	2
Terms and definitions	3	3
Management requirements		
Organization	4.1.1	4.1
	4.1.2	1.2
	4.1.3	4.1
	4.1.4	—

Continued

Item from contents list of ISO/IEC 17025	ISO/IEC 17025 Clause	ISO/IEC Guide 25
	4.1.5 (a)	4.2 (a)
	4.1.5 (b)	4.2 (b)
	4.1.5 (c)	4.2 (i)
	4.1.5 (d)	4.2 (c)
	4.1.5 (e)	5.2 (b), 5.2 (c)
	4.1.5 (f)	4.2 (d)
	4.1.5 (g)	4.2 (e)
	4.1.5 (h)	4.2 (f)
	4.1.5 (i)	4.2 (g)
	4.1.5 (j)	4.2 (h)
Quality system	4.2.1	5.1
	4.2.2	5.1, 5.2 (a)
	4.2.2 (a)	5.1
	4.2.2 (b)	5.2 (a)
	4.2.2 (c)	5.2 (a)
	4.2.2 (d)	5.1
	4.2.2 (e)	5.2
	4.2.3	5.2 (m)
	4.2.4	5.2 (e)
Document control	4.3.1	5.2 (d)
	4.3.2.1	5.2 (d)
	4.3.2.2 (a)	5.1, 5.2 (d)
	4.3.2.2 (b)	5.2 (d)
	4.3.2.2 (c)	5.2 (d)
	4.3.2.2 (d)	5.2 (d)
	4.3.2.3	5.2 (d)
	4.3.3.1	5.2 (d)
	4.3.3.2	5.2 (d)
	4.3.3.3	5.2 (d)
	4.3.3.4	5.2 (d)
Review of requests, tenders and contracts	4.4.1	5.2 (i)
	4.4.1 (a)	5.2 (i)
	4.4.1 (b)	5.2 (i)
	4.4.1 (c)	5.2 (i)
	4.4.2	5.2 (i)
	4.4.3	5.2 (i)
	4.4.4	5.2 (i)
	4.4.5	5.2 (i)
Subcontracting of tests and calibrations	4.5.1	14.1
	4.5.2	14.1
	4.5.3	—
	4.5.4	14.2
Purchasing services and supplies	4.6.1	10.8, 15.2
	4.6.2	15.1

Continued

Item from contents list of ISO/IEC 17025	ISO/IEC 17025 Clause	ISO/IEC Guide 25
	4.6.3	—
	4.6.4	15.3
Service to the client	4.7	—
Complaints	4.8	16.1
Control of	4.9.1	5.2 (o)
nonconforming	4.9.1 (a)	5.2 (o)
work	4.9.1 (b)	5.2 (o)
	4.9.1 (c)	5.2 (o)
	4.9.1 (d)	5.2 (o), 13.6
	4.9.1 (e)	5.2 (o)
	4.9.2	16.2
Corrective action	4.10.1	5.2 (o)
	4.10.2	5.2 (o)
	4.10.3	5.2 (o)
	4.10.4	5.2 (o)
	4.10.5	16.2
Preventive action	4.11.1	—
	4.11.2	—
Control of records	4.12.1.1	12.1
	4.12.1.2	12.2
	4.12.1.3	12.2
	4.12.1.4	10.7 (e)
	4.12.2.1	12.1
	4.12.2.2	—
	4.12.2.3	—
Internal audits	4.13.1	5.3
	4.13.2	5.3
	4.13.3	5.5
	4.13.4	—
Management reviews	4.14.1	5.4
	4.14.2	5.5
Technical requirements		
General	5.1.1	—
	5.1.2	—
Personnel	5.2.1	6.1
	5.2.2	6.2
	5.2.3	—
	5.2.4	5.2 (e)
	5.2.5	6.3
Accommodation and	5.3.1	7.1, 7.2
environmental	5.3.2	7.3
conditions	5.3.3	7.4
	5.3.4	7.5
	5.3.5	7.6

Continued

Item from contents list of ISO/IEC 17025	ISO/IEC 17025 Clause	ISO/IEC Guide 25
Test and calibration	5.4.1	10.2, 10.1, 10.5
methods and	5.4.2	10.3
method validation	5.4.3	—
	5.4.4	10.4
	5.4.5.1	—
	5.4.5.2	10.4
	5.4.5.3	—
	5.4.6.1	10.2
	5.4.6.2	10.2
	5.4.6.3	—
	5.4.7.1	10.6
	5.4.7.2	10.7
	5.4.7.2 (a)	10.7 (b)
	5.4.7.2 (b)	10.7 (c)
	5.4.7.2 (c)	10.7 (d)
Equipment	5.5.1	8.1
	5.5.2	9.1
	5.5.3	10.1
	5.5.4	—
	5.5.5 (a)	8.4 (a)
	5.5.5 (b)	8.4 (b)
	5.5.5 (c)	—
	5.5.5 (d)	8.4 (d)
	5.5.5 (e)	8.4 (f)
	5.5.5 (f)	8.4 (g)
	5.5.5 (g)	8.4 (h)
	5.5.5 (h)	8.4 (i)
	5.5.6	8.2
	5.5.7	8.2
	5.5.8	8.3
	5.5.9	—
	5.5.10	—
	5.5.11	—
	5.5.12	—
Measurement	5.6.1	9.1
traceability	5.6.2.1.1	9.2
	5.6.2.1.2	9.3
	5.6.2.2.1	9.2
	5.6.2.2.2	9.3
Measurement	5.6.3.1	9.4, 9.5
traceability (cont.)	5.6.3.2	9.7
	5.6.3.3	9.6
	5.6.3.4	
Sampling	5.7.1	10.5

Continued

Item from contents list of ISO/IEC 17025	ISO/IEC 17025 Clause	ISO/IEC Guide 25
	5.7.2	—
	5.7.3	—
Handling of test and calibration items	5.8.1	11.4
	5.8.2	11.1
	5.8.3	11.2
	5.8.4	11.3
Assuring the quality of test and calibration results	5.9	5.6, 5.6 (a)
	5.9 (a)	5.6 (c)
	5.9 (b)	5.6 (b)
	5.9 (c)	5.6 (d)
	5.9 (d)	5.6 (e)
	5.9 (e)	5.6 (f)
Reporting the results	5.10.1	13.1
	5.10.2 (a)	13.2 (a)
	5.10.2 (b)	13.2 (b)
	5.10.2 (c)	13.2 (c)
	5.10.2 (d)	13.2 (d)
	5.10.2 (e)	13.2 (h)
	5.10.2 (f)	13.2 (e), 13.2 (f)
	5.10.2 (g)	13.2 (g)
	5.10.2 (h)	13.2 (i)
	5.10.2 (i)	13.2 (k)
	5.10.2 (j)	13.2 (m)
	5.10.2 (k)	13.2 (n)
	5.10.3.1 (a)	13.2 (j)
	5.10.3.1 (b)	—
	5.10.3.1 (c)	13.2 (l)
	5.10.3.1 (d)	—
	5.10.3.1 (e)	—
	5.10.3.2 (a)	—
	5.10.3.2 (b)	—
	5.10.3.2 (c)	—
	5.10.3.2 (d)	—
	5.10.3.2 (e)	—
	5.10.3.2 (f)	—
	5.10.4.1 (a)	13.2 (j)
	5.10.4.1 (b)	13.2 (1)
	5.10.4.1 (c)	—
	5.10.4.2	—
	5.10.4.3	
	5.10.4.4	—
	5.10.5	—
	5.10.6	13.3

Continued

Item from contents list of ISO/IEC 17025	ISO/IEC 17025 Clause	ISO/IEC Guide 25
	5.10.7	13.7
	5.10.8	13.4
	5.10.9	13.5

Appendix D

QUALITY ASSURANCE FOR RESEARCH AND DEVELOPMENT AND NONROUTINE ANALYSIS* (EURACHEM/CITAC)

1. AIMS AND OBJECTIVES

1.1. Who This Guide Is For

1.1.1. This guide is intended to be used by managers and analytical staff, both in industry and the academic world, involved in the planning, performance, and management of nonroutine measurements in analytical science and associated research and development (R&D). Those responsible for the evaluation of the quality of such work will also find the guide useful. It provides principles from which assessing organizations such as accreditation or certification bodies could specify assessment criteria.

1.2. Using This Guide

1.2.1. This guide aims to state and promote quality assurance (QA) *good practice* or at least practice that meets the professional standards of the peer group. Many of these practices have already been stated in an earlier Co-operation on International Traceability in Analytical Chemistry (CITAC) guide (CG1) (1), which provides advice for mainly routine analysis, and an earlier EURACHEM-/WELAC guide (2), which advises on the interpretation of EN 45001 and ISO Guide 25 for chemistry laboratories. Predictably, there is likely to be a high degree of overlap between what is good practice in a routine situation and what

---

* This document has been produced primarily by a joint EURACHEM/CITAC Working Group, the membership of which is listed in Annexure A.

is good practice in a nonroutine situation. To avoid duplication, those practices are only repeated in the following, where it has been considered appropriate that further clarification is necessary for nonroutine purposes. Where the guidance has *not* been restated, reference to the relevant part of the CITAC guide has been stated instead. Thus, this guide should be used in conjunction with CG1.

1.3. Emphasis of Guidance

1.3.1.   There is still much discussion as to how applicable the various established quality standards/protocols such as ISO Guide 25(3), EN 45001 (4), ISO 9000 (5), and OECD Principles of Good Laboratory Practice (GLP) (6) are to nonroutine work. The GLP is study based, and the studies often involve nonroutine or developmental work. Research and Development is compatible with the design element of ISO 9001. However, it is widely argued that nonroutine work does not fit easily into a highly documented and formalized quality system. For this reason, the guidance is directed towards good practice rather than compliance with formal standards. The two approaches are not necessarily at odds with one another, but compliance may occasionally place requirements that are considered to be over and above what is considered to be best practice. Conversely, no single quality standard necessarily covers all the elements of activity that might be considered relevant as best practice. The aim is to produce guidelines for analysts, their customers, and their managers, and not a quality manual template for an organization. In addition, note that external verification, such as can be provided against a formal quality standard, is not mandatory, even though it may be desirable in some cases.

1.3.2.   It is anticipated that once this guide is published, it may be possible for accreditation bodies and other authoritative organizations to adapt the text for compliance purposes, for example, to the published standards/ protocols mentioned in Section 1.3.1.

1.4. Customers

1.4.1.   Nonroutine work regulated by this guidance may be performed for a number of different types of customers such as:

- other departments within the same organization, which lack the specialist skills the work demands;
- external customers who commission specific tasks;
- regulatory bodies that commission the work to help enforce law, regulatory, or licensing requirements;
- funding bodies that commission large work programmes, within which specific tasks lie.

2. INTRODUCTION

2.1. What Is Research and Development?

2.1.1.   *Research* is a scientific investigation aimed at discovering and applying new facts, techniques, and natural laws (7). At its heart is inquiry into the unknown, addressing questions not previously asked. Research is done by a wide range of organizations: universities and colleges; government agencies; industry and contract organizations. Research projects vary widely in content and also in style, from open-ended exploration of concepts to working towards specific targets.

Development in an industrial context is the work done to finalize the specification of a new project or new manufacturing process. It uses many of the methods of scientific inquiry and may generate much new knowledge, but its aim is to create practicable economic solutions.

The combined term *Research and Development* (R&D) can be seen as the work in an industrial or government context concentrating on finding new or improved processes, products, etc. and also on ways of introducing such innovations.

The use of the term R&D may not wholly encompass the activities intended to be covered by the guidelines, but has been adopted by the authors as the most appropriate and convenient single term.

2.1.2. These guidelines are intended to cover analytical testing or measurements where for various reasons, the work is nonroutine or necessary procedures are not already in place; for example:

- Methods already exists for the analytical problem, but have not previously been applied to the particular type of sample now encountered. The existing methods need to be evaluated and extended or adapted as necessary.
- The analytical problem is entirely new, but may be tackled by applying existing methods or techniques.
- The analytical problem is entirely new, there is no established method, and something has to be developed from the beginning.

Annexure E provides some additional ideas for those carrying out R&D to develop analytical instrumentation.

2.2. Importance of QA

2.2.1. The importance of QA is well established and accepted for routine analysis. It is less well established for R&D.

2.3. What Needs to Be Controlled in R&D?

2.3.1. Figure 1 shows a hierarchical approach to QA within an organization. The outer layer represents the elements of QA that apply to all levels of activity within the organization—so-called *organizational quality elements*. These are described in Chapter 5. Examples at this level include a quality management structure with a defined role within the organization, a quality system, documented procedures for key activities, a recruitment and training policy for all staff, etc. The next layer, *technical quality elements,* described in Chapter 6, forms a subset and comprises specific QA elements that apply to the technical activities of the organization, such as policy and procedures

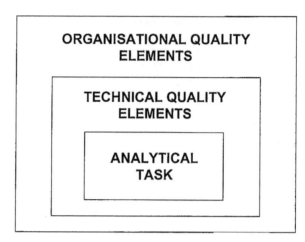

Figure 1 Nested structure of activities.

for instrument calibration and performance checks, use of calibrants and reference materials, and use of statistical procedures. The inner layer, *analytical task quality elements,* described in Chapter 7, represents the activities carried out for particular projects or individual analytical tasks. It includes the planning, control, and reporting practices recommended at the start of, during, and at the completion of R&D work.

3.   DEFINITIONS

3.1.   Accreditation

Procedure by which an authoritative body gives formal recognition that a body or person is competent to carry out specific tasks (ISO/CASCO 193 (Rev. 2), 1.11 (8) and ISO Guide 2: 1996, 12.11) (9).

3.2.   Certification

Procedure by which a third party gives written assurance that a product, process, or service conforms to specified requirements (ISO/CASCO 193 (Rev. 2), 4.1.2 (8) and ISO Guide 2: 1996, 15.1.2) (9).

3.3.   Contract

An agreement made between two or more parties on specified terms. Typically as applied to analytical work, it refers to an agreement to a laboratory (the contractor) to do work for the customer, at a specified price and within a specified timescale, with perhaps other conditions specified.

3.4.   Customer

A purchaser of goods or services.

3.5. Project

A research or study assignment, a plan, scheme, or proposal (10). In the analytical context, a project refers to a discrete job starting with a particular problem and involving one or more tasks undertaken to solve the problem (see also Study).

3.6. Quality Assurance (QA)

All the planned and systematic actions implemented within the quality system and demonstrated as needed, to provide adequate confidence that an entity will fulfill requirements for quality (ISO 8402: 1994, 3.5) (11).

3.7. Quality Control (QC)

Operational techniques and activities that are used to fulfill requirements for quality (ISO 8402: 1994, 3.4) (11).

3.8. Registration

Procedure by which a body indicates relevant characteristics of a product, process, or service, or particulars of a body or person, in an appropriate, publicly available list [ISO/CASCO 193 (Rev. 2), 1.10 (18) and ISO Guide 2: 1996, 12.10).

3.9. Routine Analysis

The analytical problem will have been encountered before. A suitable validated method for solving the problem will exist and may be in regular use. The degree of associated staff training, calibration, and QC used with the method will depend on sample throughput.

3.10. Study

An attentive or detailed examination (10).

N.B: Use of the terms "Project" and "Study" in this guide does not mean that the guide is applicable only to GLP work.

3.11. System (Quality)

The organizational structure, procedures, processes, and resources needed to implement quality management (ISO 8402: 1994, 3.6) (11).

3.11.1. *System* has been used in this guide to refer more generally to the infrastructure within which a laboratory undertakes analytical work and in this context does not necessarily constitute a quality system. This is entirely consistent with the ISO definition.

3.12. Task

No formal definition. The use of task in this guide denotes a small discrete piece of work, several tasks making up a project or study.

3.13. Validation

Confirmation by examination and provision of objective evidence that the particular requirement for a specified end use is fulfilled (ISO 8402: 1994, 2.18) (11).

3.14. Verification

Confirmation by examination and provision of objective evidence that specified requirements have been fulfilled (ISO 8402: 1994, 2.17) (11).

4. PRINCIPLES FOR MAKING VALID ANALYTICAL R&D MEASUREMENTS

Six basic principles have been identified as important for laboratories making measurements to follow (12).

4.1. Analytical Measurements Should Be Made to Satisfy an Agreed Requirement

In routine work, it is usually a straightforward process to define the problem for which the analytical work is being carried out. In R&D, specification of the problem is usually done as the part of project definition. The customer may only have a vague idea of what the problem is and how chemical analysis can solve it, and will rely on the laboratory's technical expertise to design a suitable technical work programme. Cost and time constraints will have to be considered as part of the programme design. The programme will define how results will be reported and the importance of only using results in the appropriate context. Results can be badly misunderstood or misused if extrapolated outside the boundary conditions of the programme.

4.2. Analytical Measurements Should Be Made Using Methods and Equipment which Have Been Tested to Ensure They Are Fit for Purpose

Whatever type of measurements are made, suitable, well maintained, and calibrated equipment is vital to ensure success. It is of the utmost importance that performance characteristics of methods should be evaluated to the extent, necessary to show that they are suitable for the measurements for which they are being used.

4.3. Staff Making Analytical Measurements Should Be Both Qualified and Competent to Undertake the Task

In R&D work, it may not be possible to guarantee that the staffs are totally competent as the full extent of the expertise required. The needs may not be fully appreciated when the work is started. It is possible that the analyst will not have much previous experience of the problem, but should have at least a basic knowledge of the underlying concepts involved in the work.

4.4. There Should Be Regular Independent Assessment of the Technical Performance of a Laboratory

A laboratory's internal QC may indicate consistency in the measurements made within that laboratory. Independent assessment of the measurement capability by participation in proficiency testing schemes or measurement of well-characterized reference materials gives an idea of how well the laboratory's performance would compare with that of its peers. However, it is recognized that the options for such independent assessment may be limited in an R&D environment.

4.5. Analytical Measurements Made in One Location Should Be Consistent with Those Made Elsewhere

Use of reference materials (where available) and assessment of measurement uncertainty of the methods in use will help ensure traceability and compatibility with others making similar measurements.

4.6. Organizations Making Analytical Measurements Should Have Well Defined QC and QA Procedures

All of the various measures taken to ensure quality of measurements within a laboratory should be incorporated into a quality system to ensure transparent and consistent implementation. If possible, some sort of external audit is desirable to verify the working of this quality system.

5. ORGANIZATIONAL QUALITY ELEMENTS

5.1. Administrative and Technical Planning of the Work

See also CITAC guide CG1, Section 11 (1).

5.1.1. Laboratories that carry out analytical R&D need to have staff with suitable managerial and technical abilities to plan, control, deliver, and report each project. This is considered in more detail in Section 7.1.3.

5.1.2. Where a laboratory is carrying out a number of projects simultaneously, co-ordination of the project management related to use of facilities is advised. Management needs to be aware of the different projects in progress in the laboratory at a given time and the corresponding risks of one project affecting another, both from a resource point of view, but also from cross contamination. Similarly, where projects are spread across several departments within a laboratory or involve input from external laboratories, suitable co-ordination is necessary to ensure coherent delivery of the work without any adverse effect on quality.

5.2. Quality Management, Corporate and Local

5.2.1. Regardless of whether the laboratory is formally recognized as compliant with a published quality management standard, it is recommended that it has a quality management system, whether formal or informal, through which its declared quality policy can be implemented. Typically, this will involve staff with specific responsibilities for quality, who act as the focus and co-ordinators for quality matters within the laboratory. Quality also needs to be managed at various lower levels, for example, group, team, or section. This may involve individuals having particular quality-related responsibilities as part of their duties, and each member of staff should be aware of what role they have in the delivery of quality within the laboratory.

5.2.2. The management of quality in an R&D environment can be a delicate issue. A balance needs to be struck between maintaining a suitable level of control while at the same time not inhibiting creativity.

5.3. Record Keeping and Document Control

5.3.1. The purpose of keeping records is that information and data held or gathered by the laboratory can be used to compile reports, make comparisons with other data (whether contemporary or historical), repeat work, and develop new or similar processes. Record keeping and document control are sufficiently important to justify a laboratory having a centralized policy, including relevant training for staff and competence assessment. The policy might typically cover:

- use of various types of media for record keeping;
- external considerations (such as recording requirements for patent applications);
- minimum levels of information for particular operations;
- use of forms and other approved formats;
- legibility, clarity, layout of information, and ease of data retrieval;
- traceability of records to time, date, analyst, sample, equipment, and project;
- use of audit trails;
- authorization of records by the use of signatures and other methods;
- methods for ensuring completion of a record;
- cross-referencing copying restrictions;

- rules for amending and authorizing amendments to records;
- rules for minimum retention of data, reports, and other useful information.

5.3.2. Useful information should be recorded at the time or immediately after the work is completed.

5.3.3. Document control should be extended to all formal documents used in the analytical work, i.e., those documents whose use is recognized within the quality system (as defined in the quality manual) and whose format, content, and use has to be reviewed and authorized. It is not unusual for a laboratory to use a hierarchical approach for its quality system documentation.

This ensures a maximum of flexibility, as work patterns change. The following table shows four levels of formal document.

Level	Documentation	Subject/examples
1 (Highest)	Corporate quality policy	Quality manual
2	Formalized internal procedures operable across the laboratory	Standard operating procedures (SOPs)
	Other (external) normative documents	Relevant laws, regulations, standards (ISO/CEN, etc.), official methods (e.g., AOACI), codes of practice
3	Technical work instructions (specific applications)	In-house methods
4 (Lowest)	Records	Instrument logbooks, calibration records, laboratory notebooks and other raw data, correspondence, reports

5.3.4. Clear responsibilities for document control should be assigned to staff. To maximize flexibility, authorization should be devolved as far down the management chain as possible, bearing in mind the need for those authorized to have sufficient expertise to make sound judgments.

5.3.5. For all controlled documents, there should be a system for recalling and archiving versions of documents when they are upgraded or replaced. Suitable facilities for archiving information should be available and their use laid down within the document control policy. The use of computer-based systems is recommended to facilitate the control of documents, but care is advised to ensure that access to the system is only available to the authorized staff.

5.4. Staff Qualifications, Training, and Supervision of Staff

See also CITAC CG1, Section 10 (1).

5.4.1. Analytical R&D must be carried out by staff having appropriate experience, knowledge, and competence, consistent with the particular role they have in the work. Suitable qualifications may be academic, professional, or technical, preferably with a specialization in analytical chemistry and may also feature on-the-job training. For R&D leaders, a high level of qualifications and relevant experience is necessary. Published guidance is available (13). The balance between academic qualifications and experience required for particular types of analytical work may vary from country to country.

Staff should receive relevant on-the-job training. The training programme should be assessed regularly and adjusted, as necessary to ensure that it continues to be relevant to the type of work carried out.

5.4.2. It is the responsibility of management to establish appropriate levels of supervision for each task, depending on the difficulty of the work and the capability of the analyst. It is recognized that analysts may be given unfamiliar tasks as part of their training; in such cases, management should take extra care to ensure that the level of supervision is appropriate.

5.4.3. Analysts involved with R&D will need to have or develop particular skills. For example, they will have to exercise high levels of judgment about how to approach the analysis, about the selection of best methods, and about interpretation of results. They will occasionally encounter problems which are beyond their own experience and possibly also that of the laboratory, and so should have experience of literature searching and other information gathering techniques. They should maintain and develop their expertise by reading scientific literature, attending seminars and courses, participate in professional activities, and be aware of colleagues who are experts in the various analytical subjects who might be able to give advice. They should also maintain an up-to-date awareness of QA. Management is responsible for ensuring that staffs have the resources to maintain these professional skills.

5.4.4. Staff records are an important aspect of establishing the suitability of staff to undertake the analytical work. As a minimum, they should include:

- education, leading to formal qualification, for example, academic, professional, technical/vocational*;
- methodological/technical expertise;

---

* Vocational training is practical training related to a particular job, accompanied by study of the relevant theoretical knowledge. Part of the training may be provided within the laboratory, but the competence may be assessed independently and recognized via a formal qualification (14–16).

- external and/or internal training courses attended;
- relevant on the job training;
- previous R&D experience, in terms of subject areas covered;
- list of scientific papers published, posters presented or lectures given.

5.5. Equipment

See CITAC CG1, Section 12. For computer-controlled equipment, see CITAC CG1 Section 17 and App. C (1) and GLP guidance (17).

5.5.1. Equipment should be purchased against technical specifications derived from anticipated use and required performance capability. Where an instrument is sold on such a basis, there is an obligation on the agent or manufacturer to demonstrate to the purchaser, if required, that the instrument can meet that specification. Newly acquired items of equipment should be formally commissioned before being put into routine laboratory use, so that correct functioning and compliance with the appropriate specifications can be verified (18).

5.5.2. A list of equipment should be kept, indicating the equipment name, identification, records of commissioning, and related operating procedures, where appropriate. Records of calibration and maintenance should be kept.

5.5.3 It is not uncommon in R&D for a piece of equipment to be used by different persons, for a number of applications, perhaps in different projects, within a brief timescale. Where this is the case, special precautions for instrument cleaning and maintenance are advised, together with records detailing what the equipment has been used for, when, and by whom. This may help reduce unexpected observations that might have been caused by cross-contamination.

5.5.4. Research and Development may actually involve the modification of existing equipment or design of new equipment. Accepted engineering and scientific practices should be applied to design and construction. Method validation procedures and use of blanks, standards, old samples reference material can be used as part of the commissioning process.

5.6. Monitoring Quality

See CITAC CG1 Section 18 (1).

5.6.1. Regular and systematic monitoring of quality is necessary to ensure that it is appropriate to the laboratory's needs, and all aspects of it are functioning properly. Monitoring may be carried out by external bodies (different types of external assessments are described in more detail in Section 8) or internally, using laboratory staff. Where there is a formal quality system, internal assessment is conducted to formal procedures and known variously as audit or review (19–22).

5.6.2.   One approach to internal assessment is for a laboratory to train some of its own staff to act as internal auditors. The laboratory will benefit by involving its staff in monitoring the quality system. Assessors can be the staffs at any level in the organization and should be independent of the work they are assessing, but have sufficient technical expertise and experience to be able to examine it critically.

5.6.3.   All areas of the laboratory whose operations affect quality should be assessed in a systematic manner, typically at least once a year. Assessments should examine adequacy of procedures and ensure that these procedures are being followed, that suitable records are kept, and that appropriate actions are taken. Ideally, a preplanned timetable should be followed, and over an agreed period should cover the whole quality system. It is unnecessary to examine the entire output of the laboratory—the assessment should be done on a sampling basis. In the case of research, it will be appropriate to select and examine entire projects or studies.

5.6.4.   Even if a research laboratory's quality system is not fully documented to the requirements specified in quality standards, provided some form of work-plan is available, an appropriate assessment can be made against this. For example, some of the questions that could be asked in assessment of the work's plan could include:

- Is the analytical task clearly described and understood?
- Is there an analytical working plan or study plan, and is there evidence of adequate experimental design?
- Are the task leader and other technical staff sufficiently competent?
- Are the applied procedures and equipment fit for purpose?
- Are calibration levels adequate and traceability suitable?
- What measures are taken to confirm the reliability of results, and are the results plausible (e.g., duplicate analysis, use of RM/CRM (certified reference material), spiked samples, cross-checking by other personnel, other internal and external QC)?
- Has the work been completed and does the test report contain sufficient information (analytical results, interpretation, reference to customer requirements)?
- Is the level of record keeping sufficient for its purpose?
- Are scheduled milestones and deliverables being met?
- Are any relevant regulatory requirements being met?

5.6.5.   Where changes to procedures are required, staff should be identified to carry out them over an agreed timescale. Subsequent completion of the changes should be confirmed.

5.6.6.   In R&D, it is not unusual to make ad hoc deviations from procedures. These may adversely influence software or hardware performance, data collection, calculations, and interpretation of results. A simple system recording deviations, as they occur and confirming that consequences have been evaluated and where appropriate corrective action has been taken, should ensure that there is no inadvertant loss of quality arising from the deviations.

5.7. Subcontracting

5.7.1. The laboratory should consult with the customer before placing any part of a contract with subcontractors.

5.7.2. Where one laboratory (A) subcontracts work to a second laboratory (B), B should operate to at least equivalent levels of quality as A. Laboratory A should put in place whatever procedures are appropriate to assure itself of the quality of the capabilities of B and the quality of the work it is producing. This might include:

- assessing the quality of subcontractors;
- establishing a list of laboratories approved to act as subcontractors;
- reviewing data and reports of subcontractors for scientific content;
- limiting the scope for the subcontractor to work independently on the subcontract;
- checking the subcontractor work against the initial specification and defining corrective action if necessary.

Note that the subcontractor and the laboratory placing the subcontract could be two different laboratories within the same organisation, i.e. the arrangement could be purely internal.

6. TECHNICAL QUALITY ELEMENTS

6.1. Unit Operations

6.1.1. The R&D projects can be considered as a collection of discrete tasks or workpackages, each consisting of a number of unit processes, themselves composed of modules containing routine unit operations. The unit processes are characterized as being separated by natural dividing lines at which work can be interrupted and the test portion or extract can be stored without detriment before the next step. This is illustrated in Figure 2.

6.1.2. The benefit of this modular approach to defining R&D projects is that new R&D work is likely to contain at least some components that are familiar to the laboratory and may even be performed routinely. This approach offers benefits in terms of establishing staff competence and also in documentation of procedures.

6.2. Technical Capability of the Laboratory

6.2.1. It is a common practice to allow the general acceptance of laboratory performances by a *type of test* approach. This means, if the laboratory has

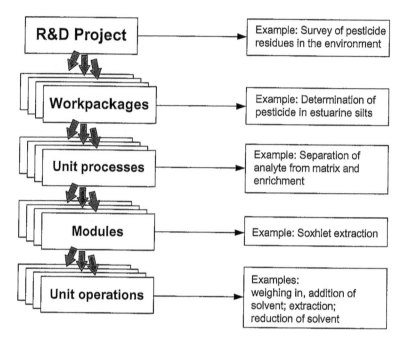

Figure 2   Illustration of the breakdown of R&D projects into unit operations.

demonstrated its ability to perform a particular method, it is also accepted as fit to perform similar closely related methods. This logical, but knowledge- and experience-oriented approach, enables the demonstration of valid analytical measurements to external experts without the need for elaborate validation of every single unit operation or module or process.

6.3. Methodology

6.3.1.   It is likely that procedures for carrying out *unit operations* and perhaps even *modules* (Figure 2) will be sufficiently routine and/or com- mon to other work to warrant full documentation as a written SOP. Using this principle, any new test procedure can be described by the appropriate combination of the SOPs of the relevant unit processes or modules, keeping new documentation to a minimum. Representation of new test methods by recombination of existing SOPs has a number of advantages in terms of using existing validation information and uncertainty contribution estimations. Validation of the whole workpackage or task will often be necessary, but can be achieved using reference materials, etc. In practice, SOPs might even cover individual workpackages, but care should be exercised if this reduces the flexibility of operations.

6.3.2.   Standard operating procedures provide a source of information to which analysts, carrying out a particular operation, can refer in order to ensure a consistent approach. A closely followed, well written SOP can improve the consistency of data produced for a particular process, between analysts, between laboratories, and over time intervals. Thus, an SOP should contain whatever level of information is necessary to avoid ambiguity. A well-written SOP also helps auditors to follow the course of the work done and therefore assess the validity of the data. In an R&D environment, it is expected that as the science improves, SOPs can be reviewed and changed to reflect the improvements (e.g., in speed, in material and money savings, in waste production, etc.) as long as the results are convincingly demonstrated to be comparable or better than those obtained with existing versions. Changes must be authorized, prior to use, in line with document control policy.

6.3.3. Where SOPs do not already exist or are inappropriate, contemporaneous notes should be made to describe the procedures used in the work. Sufficient detail should be recorded, so that at some later time, the procedures used can be reconstructed, if necessary. Where a number of procedures were attempted before one was found that was satisfactory, records should be kept of the failures, so that they can be avoided in future.

6.4. Reagents, Reference Materials, and Calibrants

See CITAC CG1, Sections 13 and 16 (1).

6.4.1. Special attention should be given to chemical and physical properties of reagents, reference materials, and calibrants (chemical and physical measurement standards). Careless preparation or poor storage may result in inadvertant degradation. This is particularly important where chemical metabolites or chemicals about which little is known are involved. Sometimes, the use of added preservatives or storage under inert atmospheres (e.g., Ar or N_2) may be appropriate.

6.4.2. Reagents, calibrants, and reference materials prepared for specific R&D applications should be appropriately labeled, and if appropriate, their use restricted, to prevent contamination through widespread use. Details of preparation, etc. should be recorded in SOPs.

6.5. Calibration and Traceability

See CITAC CG1, Section 15 (1).

6.5.1. Calibration establishes, for specified conditions, how the response of the measurement system relates to the parameter being measured. Calibration is usually performed using a reference material of established composition, or calibrant in which the property of interest (e.g., the chemical purity) is well characterized.

6.5.2. In R&D, one is more likely to encounter the situation where calibrants are absent or, if available, are poorly characterized. Where the stoichiometry of the calibrant is not known, an approximate amount should be weighed and the exact amount of calibrant constituent determined with an absolute method (coulometry, volumetry, gravimetry). Where no suitable calibrant is available, the method for determining the response for the property analyte should be demonstrated.

6.5.3. Validation of the unit processes, together with appropriate traceability, is important to ensure that data produced is comparable with data for similar measurements made at different times, or by different analysts or laboratories, or using different methods and different samples. Traceability can be achieved by calibration using various calibrants, reference materials, or even standardized procedures. Caution is advised that when using standardized procedures as frequently, they contain bias that may be poorly controlled.

6.5.4. Traceability to (the) SI is often possible in chemical analysis at some level of uncertainty. Traceability can be to a standard/calibrant, whether national or international, which has been accepted as the point of reference by the analytical community concerned and which all interested parties have access to, either directly or indirectly, through a chain of subsidiary standards. Similarly, traceability can also be established to a reference method.

6.5.5. Traceability is not to be confused with the traceability from the sample via the test procedure to the final test result. This has been tentatively termed "trackability" (from tracking back).

6.6. Instrument Performance

6.6.1. For instrumentation, design, installation, operational, and performance qualifications are of equal importance in R&D, as they are in routine work. Design and operational qualifications are briefly dealt with in Section 5.5.1. This section deals with operational and performance qualifications— Does the instrument/system work in the specific application and what could be the interferences? Does the instrument continue to work in the manner intended (continuing fitness for purpose)?

6.6.2. In R&D, it is not sufficient to adapt existing work without demonstrating that the instrumentation works properly with the new application. Care is also needed with novel or modified instrumentation, where the performance claims of the manufacturer may no longer be true because of the modification.

6.6.3. The ultimate performance test for any calibrated analytical instrument is to analyze a CRM and obtain a result within the uncertainty range stated for the CRM. If the matrix of the CRM is similar to that of the samples, and the CRM is subjected to the whole analytical process, then this serves to validate the entire procedure (23–25).

6.6.4. Often in R&D, no CRM is available and it is not possible to relate a property to an existing national or international standard or calibrant. Instead, in-house reference materials can be used. It is advisable to specify one or two materials with characterized property values appropriate to the scope of the procedure that can be used for instrument performance checks, calibration, or QC. Specific mixtures of analytes can be contrived to test certain performance parameters, for example, the resolution of two compounds in a separation process.

6.6.5. In critical instances, the use of a different analytical procedure and/or technique, susceptible to different interferences, is advised to check results. This check is more valuable than, for example, interlaboratory comparisons involving only a limited number of laboratories using exactly the same overall procedure and measurement technique. However, interlaboratory comparisons involving larger number of laboratories and different techniques are more useful still.

6.6.6. Where R&D involves testing a large number of similar samples using a particular procedure, control samples and charts can be used to monitor the continuing stability of instrument performance.

6.7. Use of Statistics

6.7.1. Statistical techniques are an invaluable tool in the design or use of analytical methods. During the lifetime of an R&D method, statistics can be used in four basic areas:

    I. experimental design of the method;
   II. characterization of method performance, ruggedness, and determination of uncertainty;
  III. QC of the method (once the method is in use);
  IV. interpretation of populations of results.

6.7.2. In each of these areas, a variety of statistical techniques may be applied or indeed are necessary, depending on the different parameters to be studied, and such chemometric approaches can also reduce time and costs. A detailed study of this area is beyond the scope of this guide; references to a number of suitable texts are provided in Section 9.

6.7.3.1. *Experimental design*: In any analytical procedure, the performance can be influenced by a number of different variables, such as matrix interferences in the samples, reagent concentrations, temperature, derivatization time, etc. Experimental design is usually used to describe the stages of identifying the different factors that affect the result of an experiment, designing the experiment so that the effect of these factors is minimized, and using statistical analysis to separate the effects of the factors involved. For example, a ruggedness test will indicate firstly whether a particular method will stand up to everyday use and will indicate which parts of the method are vulnerable to change and need to be subject to QC. As part of the design process, regression or multiple regression analysis may be used, together with ANOVA (ANalysis Of VAriance) determinations and MANOVA (Multiple ANalysis Of Variance) (26,27).

6.7.3.2. Statistical methods are very important in the design of sampling schemes. If used properly, they can enable the desired results to be obtained with the minimum of samples and subsequent analysis. Internationally available standards have been published for the use of statistics in certain types of sampling (28). However, a broad knowledge of the history of the sample substantially helps to design a more intelligent sampling plan and reduces sampling time and costs.

6.7.3.3. SIMPLEX optimization can be used for rapid method development where a number of factors affect method performance and to investigate all possible combinations would involve vast amount of work (29). Other specialized techniques that may be used in a similar way include full factorial designs, fractions of factorial designs, and Taguchi designs.

6.7.3.4. Where a large number of samples need to be processed and only a few are expected to yield "positive" results, screening techniques may be used for eliminating the large numbers of negative samples to leave the positive samples which can then be examined in more detail.

6.7.4. *Characterization of method performance and determination of uncertainty:* This involves the evaluation of various parameters associated with the performance of the method, such as precision, trueness, etc., followed by a judgment as to whether these performance capabilities are sufficient to meet the needs of the method. The process is generally referred to as method validation (see Section 6.8.5). Determination of measurement uncertainty use similar measures to those, determined during method validation and involves identification, determination, and final recombination of all the sources of uncertainty arising at all stages of the analytical procedure to give an overall measure (see Section 6.8.6). Both method validation and measurement uncertainty make use of simple statistical measures such as means, standard deviation, variance, etc.

6.7.5. *Development of QC:* The QC procedures developed for a new method should concentrate on those parameters that have been identified as critically influencing the method. However, for R&D work, there may be problems in finding suitable samples for QC purposes, and control-charting techniques are less relevant in nonroutine situations. Control charts can still be applied, for example, to monitor instrument calibration, and the main thrust of QC in the R&D situation is probably best directed towards ensuring that instrumentation is working properly and calibrated, monitoring values from reference materials where available, and replicate analysis (consecutive and random, to monitor short- and long-term variation, respectively).

6.7.6. *Interpretation of results*: The problems associated with validation of methods in R&D and the subsequent design of adequate QC should be borne in mind when interpreting sets of data produced in R&D. Techniques used for the detection of outliers and measures of distribution of result populations, such as standard deviation, are particularly relevant in this case.

6.8. Technical Requirements Related to Particular Unit Processes

6.8.1. In most analytical R&D situations, the following unit processes (which may or may not have subsidiary modules and unit operations) may be encountered: sampling, sample preparation, separation of the analyte from the matrix and enrichment, measurement, calculation, and presentation and interpretation of the result. Guidance is generally limited to information specific or more relevant to R&D.

6.8.2. Sampling

See also CITAC CG1, Section 19 (1).

6.8.2.1. Extensive guidance on sampling exists in the scientific literature (28). There is actually little advice on sampling in R&D that is not also applicable to routine measurements.

6.8.2.2. Where R&D involves the development of new test procedures for subsequent use on real samples, method development needs to consider practical sample sizes that will typically be available for testing. During the development stages, it may be useful to have large quantities of real sample available for method validation, etc.

6.8.2.3. Research and Development may involve taking types of samples that have never been encountered before, with unknown or unfamiliar analyte contents or matrix types. The samples may present unknown hazards or problems with stability, handling, and storage. The sampling strategy should try to anticipate potential problems and if possible make suitable allowances. Customer's declarations of the expected contents of samples should be treated with caution. Sampling plans should be detailed even if some of the information recorded is subsequently not needed. The analytical staff involved with the R&D should use their scientific expertise to help ensure that the sampling procedure is as appropriate as possible. Where appropriate, procedures should be recorded.

6.8.2.4. Similarly, for unfamiliar samples, storage conditions should err on the side of caution. In critical cases, it is strongly advised that samples are retained after analysis at least until the validity of the tests results have been confirmed by suitable review.

6.8.2.5. With samples taken for R&D purposes, little may be known about their homogeneity. It is particularly important to investigate this before any subsampling is carried out to reduce the effective bulk of the sample. Any means used to homogenize the sample must not compromise its integrity. It may be appropriate to separate phases in inhomogeneous samples and treat the separate phases as different samples. Conversely, it may be appropriate to homogenize the samples. The uncertainty of subsampling, which is determined by the level of homogeneity, may be estimated by setting up a specific study and taking more subsamples and determining the uncertainty statistically.

6.8.2.6. It may be convenient to have a single SOP describing the variety of sample treatment methods (solvation, dissolution, digestion, extraction, surface cleaning, melting, combustion, etc.) used by the laboratory and containing detail on the special precautions to be taken for the different analyte groups. It should also describe how the methods are applied to blanks (spiked and unspiked), reference materials and other calibrants, and other materials used for QC purposes.

6.8.3. Isolation of the Analyte(s) Using Separation and Enrichment

6.8.3.1. Diverse techniques are available for separation and enrichment. The experience of the analyst will be an important factor in choosing the most appropriate for a particular application. For future reference, records should indicate the logic behind a particular choice.

6.8.4. Measurements

6.8.4.1. The measurement process consists of using a calibrated instrument to determine the net instrument signals of the test portions and various different blanks. Within run and between run changes in instrument response can be monitored using QC samples and calibration standards.

6.8.4.2. Depending on the circumstances, this determination step may be repeated several times to allow a statistical data treatment of this single step. The determination of more than one test portion from the same sample can be used to determine (at least an estimate of) the overall repeatability of the analytical method. Where there is a suspicion that interferences are present, results obtained from test portions using external standard calibration (using a calibration curve) can be checked by spiking test portions with known amount of the analyte of interest.

6.8.4.3. Blank corrections for measurements should be made by calculating actual concentrations of sample and blank, as indicated by the respective instrument signals and then subtracting one from the other. The practice of subtracting the blank signal from the sample signal and then calculating the result using the net signal is not recommended.

6.8.5. Validation

See also CITAC CGI, Section 22 (1).

6.8.5.1. There is a clear responsibility on the part of the test laboratory and its staff to justify the trust of the customer or data user by providing reliable data that can be used to solve the analytical problem. An implication of this is that methods developed in-house must be adequately validated, documented, and authorized before use. Validation is normally quite straight forward for routine work, but can be expensive and time consuming. For methods used or developed during the course of R&D, validation is equally important, but less straight forward. General guidance has been produced by EURA-CHEM (31).

6.8.5.2. Various options exist for the characterization of method performance. The trueness of a new method could be assessed against that of an established method, repeatability could be assessed using reference materials, and reproducibility through interlaboratory comparisons. In R&D, many of these options may not be available. Validation tools may be limited to the use of in-house reference materials, and uncertainty estimations based on error propagation principles relying on a solid understanding of the theoretical principles of the method and the practical experience of the research workers.

6.8.5.3. A suitable unit process for data treatment should include validation of the overall procedure. This means the evaluation of various performance parameters of the method and the consideration of their adequacy relative to the analytical requirement. Parameters such as: limit of detection, limit of quantification, dynamic measuring range, sensitivity, repeatability (same analyst, same instrument, same laboratory, same day), reproducibility (different analyst, different instrument, different laboratory, different day), accuracy (difference from the true value), and other terms (e.g., robustness or ruggedness) will need to be considered.

6.8.5.4. The extent to which validation is needed, as well as the effort given to this task, depends on the use that will be made of the method or technique. At one limit, where new methods or techniques (or ones seldom applied) are being used, a customer requirement for durable methodology will justify extensive work on validation. In many situations, however, less than full validation is necessary or possible. Here, the professional judgment of the analysts will be introduced to decide those unit operations of the analysis that need to be investigated and those whose performances can be estimated from comparable systems. The extent of validation and the consequences in time and cost are one of the key issues to be agreed between analyst and customer when commissioning method development.

6.8.5.5. It is generally assumed that R&D requires an increasing effort for validation, as seldom applied or totally new techniques or methods are being used. The unit operation approach, described earlier, enables the possibility of recombination of the units into a large variety of testing methods: If these units can be individually validated, it may be possible to estimate the overall performance capability of subsequent combinations of the modules, which then require the minimum of further validation for verification. It is not necessary to define all unit operations for each possible analyte, but it might be sufficient for a group of analytes with a nearly similar matrix.

6.8.5.6.   Ideally, individual recovery studies should be performed for each analyte. This can be done using a synthetic matrix similar to the sample matrix or by analyte addition (spiking) to subsample aliquots and determination of the increase of the measured concentration. Often, the recovery factor depends strongly on the sample matrix. Guidance on acceptable recovery ranges for similar analyte/matrix combinations may be available in the literature. Whether results should be corrected for nonquantitative recoveries is the cause of much debate (32), and the client may have a preference. Reports should indicate clearly whether or not data has been changed to allow for nonquantitative recoveries.

6.8.5.7.   Ideally, the procedure should try to identify such a matrix effect, so that any blank correction procedures can be performed properly. In analytical R&D, the search for systematic errors is of greater importance, as per se less is known in those fields. Wherever possible, these systematic errors should be identified and, if possible, eliminated.

6.8.5.8.   It should be noted that methods could be validated at different levels. Analysis of CRMs with similar matrices to the test materials gives the highest confidence level for in-house validation. If the obtained results lie within the stated confidence range, then the total analytical process is under control and all involved unit processes are automatically included in this validation. This means there is no need for any further method or instrument validation and no need for other more formal demands. Other mechanisms for validation are described in what follows in order of decreasing confidence:

- taking part in interlaboratory comparison tests;
- performing a limited number of control analyses of the sample at a different test laboratory;
- employing several methods with different interferences possibility and obtaining only one and the same result;
- reanalysis of an in-house sample of known content.

6.8.5.   Measurement Uncertainty

See also CITAC CG1, Section 21 (1).

6.8.6.1.   Uncertainty should be estimated and quoted in a way that is widely accepted, internally consistent, and easy to interpret. More detailed guidance has been published by EURACHEM (32). Where appropriate, uncertainty should be quoted with the analytical result, so that the user can be assured of the degree of confidence that can be placed on the result.

6.8.6.2. The most significant contributions to the overall uncertainty of a measurement are usually due to the sampling processes and the accuracy of the determination of recovery factors. Contributions due to instrument performance are generally less significant.

7. ANALYTICAL TASK QUALITY ELEMENTS

7.1. Preparation and Planning Before Starting Work

7.1.1. Definition of Task and Project Design

7.1.1.1. Planning and preparation is a critical part of analytical R&D, especially where new analytical methods are generated or extensive validation of generic methods is required. The effort put into planning depends on the complexity and requirements of the work, the previous experience, the extent to which the work is unfamiliar or novel in its character, the number of persons or organizations involved, the expenditure for new equipment, the consequences of wrong results, the duration of the work, deadlines, etc. A flowchart such as the one shown in Annexure B may assist planning. As a rule of thumb, proportionally more planning is needed for high-risk work. When costing project work, it is important to correctly estimate the resources needed in the planning or subsequent management stages. The structure of the project should be flexible enough to allow creative problem solving. The project management team is responsible for planning activities within the project and allocating resources to cover these activities. The sort of activities involved includes:

- scoping;
- milestone planning;
- objective/goal setting;
- resource allocation and costing;
- contract control;
- financial control;
- change of management;
- liaison with customers.

7.1.1.2. Task definition is the first stage of planning and should provide sufficient information to allow more detailed planning or indicate viability of proceeding. Go/no-go decision criteria should be incorporated in the project structure at the earliest opportunity. It is vital to establish a good link with the client to ensure work is defined adequately and thus maximize the chances of a productive outcome to the project. The sort of areas covered in task definition may include the following:

- nature of the problem that the work is intended to address, seeking clarifying from the client as necessary;

- objective, goals and expected information, purpose of results/data, intended use of information;
- type of material/product/matrix to be analyzed/amount available-/safety considerations;
- sampling procedures/sampling plans, statistical methods;
- element/species/determinant/property to be analyzed/determined;
- methodology, generic methods to be used, destructive/nondestructive methods;
- required accuracy (or precision, bias, etc. as appropriate) and related equipment performance requirements;
- validation procedures and use of reference materials, standards, reference methods;
- required date of completion;
- available resources (personnel, equipment);
- expected use of subcontraction;
- success/failure criteria where appropriate;
- expected/permissible costs and expenditures;
- reference to exploratory work and review of literature required for definition and execution of the task;
- degree of confidentiality necessary;
- requirements and arrangements for archiving;
- ownership of intellectual property;
- possible strategy for dissemination and exploitation.

7.1.1.3. A questionnaire can be used to help define work. The example shown in Annexure C is adapted from one used for routine work. Note that it is not exhaustive but illustrates some of the issues that should be addressed.

7.1.1.4. Where limited amounts of sample are available, it is particularly critical to have a clear strategy in place before beginning the work. Use of nondestructive methods should be considered.

7.1.2. Project Design and Research Plan

7.1.2.1. Once task definition is complete, the research plan(s) can be drawn up. The laboratory management should involve the client, and the laboratory staff from the very beginning, in order to ensure that the finalized project as far as possible meets the clients' requirements, is technically possible, and suitable resources are available within the specified timescale. The project should be structured by a logical sequence of tasks or workpackages, points of decision where the work can change direction if necessary, and points of achievement (milestones, target dates) that enable progress to be monitored. All contractual or technical issues should be resolved, before the analytical work is begun. Particularly, where operations may be complex, use of flow-chart, such as that shown in Annexure B, a decision tree or other diagrams may help to clarify the procedure.

7.1.2.2. The *research plan* defines the following:

- Goals: Set clear final (and if appropriate, intermediate) goals (measurable objectives including go/no-go decision points/acceptance criteria). Establish what questions need to be answered at each stage and the corresponding results/data required to answer them.
- Tactics: Outline the strategy to be used at each stage. If necessary, subdivide the tasks into manageable, defined workpackages (unit operations) with discrete goals.
- Resources: Define the resources (personnel, equipment, facilities, consumables) needed at each stage.
- Time schedule: Define start and end of project, dead lines for intermediate goals, and minimum critical path for completing the work.

7.1.2.3. Research plans should contain as much detail as is necessary to define the tasks involved. For isolated tasks, the plan may simply be an entry in a notebook or a form. A more detailed plan will be necessary for larger, more complex tasks or when time and cost constraints are to be closely controlled, or when high risk or significant investments depend on the outcome of the work. If there is significant doubt as to whether the work can be completed successfully by a single route, then alternative plans should be defined.

7.1.2.4. A workpackage typically consists of a discrete piece of work with defined starting and finishing times/dates; necessary starting conditions (particularly, if the workpackage is one in a sequence); a goal (achievement of which indicates successful completion of the workpackage); a budget indicating financial, time, and other resource restrictions; a note of any particular resource requirements; a statement of the roles and responsibilities of the various staff involved with delivery at all levels from management to technician; and a specification for reporting the progress and the final goal.

7.1.2.5. Milestones are points of appraisal (usually) at the end of a workpackage. Their timing is normally fixed within the overall project timetable. They are points at which decisions can be made either to proceed with the project, to stop, or to select a particular path in the work plan for further action. Where appropriate, the client should be involved in any important decisions.

7.1.2.6. A number of tools are available to assist project design and control (33). They include:

- Bar chart (Gantt chart);
- PERT (program evaluation and review technique) chart;
- CPM (critical path method).

7.1.3. Resource Management of Task

7.1.3.1. Large or multitask projects may involve scientists from several departments of the laboratory and perhaps outside specialist subcontractors.

The role of project management is particularly important in order to ensure the project team functions smoothly, with all members co-operating and aware of their roles and responsibilities. Particular attention should be given to:

- definition of the project management hierarchy, with leaders in particular areas, and defined authority and responsibility for all team members;
- involvement of all personnel pertinent to the project (including the client) in defining the task and assignments and in planning the project;
- setting clear tasks and goals, which are challenging but achievable;
- early consultation with the management of specialists in other departments or organizations, involved in the project; unresolved questions concerning priorities and workload, and budget contributions often disrupt good team work;
- communication: hold meetings at appropriate intervals for exchange of information, problem solving, consultation, reporting, co-ordination, and decision making.

For small, simple projects the same principles can and should be applied in a cut-down form.

7.1.3.2. Resource management at the planning stage may include the following:

- Evaluation of the skills and facilities required for the project, comparing those against what is available, and plans to cover any shortfall. This includes special considerations such as environmental controls, special equipment and reagents, protective clothing, decontamination procedures.
- Costing the planned deployment of personnel and facilities and set budgets for the various parts of the work (time and finance budget).
- Establishment of a timetable for the work consistent with client requirements and the availability of personnel and facilities at each stage.
- Availability and allocation of resources to defined tasks and/or appointed dates/decision points (e.g., milestones) and including resource distribution in the project plans.
- Definition of a system for monitoring time and resource expenditure in the project.
- Identification of potential problems with disposal of samples, reagents, and contaminated equipment, arising as a result of the work.

7.2. While the Work Is in Progress

7.2.1. Progress Review Monitoring Analysis

7.2.1.1. Progress of work and status of expenditure should be controlled by comparing achievements and use of resources against the planned budgets at convenient points within the work, typically at regular intervals or completion of milestones. Informal reviewing should be carried out individually by the laboratory staff, as the work progresses. Unexpected difficulties or results, or major deviations from goals, may call for extraordinary reviews and interim reports with replanning of the work and reallocation of resources as necessary.

7.2.1.2. Progress should be reported to laboratory management or the client, in the format and at the time intervals agreed at the planning stage. Typically, reports might cover a review of the project plans, information on whether the work is running on schedule and will achieve its objectives—on-time/late/at all—an account of technical progress with achievements and failures/setbacks, and information on resources.

7.2.1.3. Effective project management requires records of laboratory data, observations, and reported progress against milestones or goals to be clear and comprehensive, so that decisions made during the project and the underlying reasons are easily understood, and laboratory work and results can be repeated if required. Records should include laboratory note books, computer print-outs, instrument charts indicating all activities, working conditions, and instrument setting, observations during experimental work, as well as justification for tactics and/or changing plans.

7.2.1.4. Ultimately, the level of data recorded should comply with customer requirements, or those laid down for scientific papers, published standard methods, or other requirements such as patents or licenses. It should be sufficient to enable other scientists to repeat the experiments and obtain data compatible with the original work. Thus

- all experimental details, observations, and data necessary for possible replication of the work must be recorded;
- records should be made at the time and kept as up-to-date as possible;
- records should be traceable to particular samples, tasks or projects, people, and time;
- details of unsuccessful work should be recorded—in R&D, it is worthwhile reporting failures, as well as successes.

7.2.2. Data Verification

7.2.2.1. Data verification should show that a new or adapted method gives consistent results with a particular sample. If results are not consistent with established data, the analytical procedure may need to be improved until the required consistency is achieved. Management should be aware that data and method validation costs form a significant part of the total costs of R&D.

7.2.2.2. The unit operations, as listed in Section 6.8.1, may influence one another, but contribute individually to variations in results. A step-by-step verification may often be impractical, although it may be feasible and useful to study particular performance characteristics of particular stages of the sequence of operations. In R&D, the plausibility of data may be checked either using literature data and theoretical considerations or using specially prepared reference materials and model substances.

7.2.3. Changing Direction

7.2.3.1.   Where a review of progress shows that a particular line of investigation is likely to be unsuccessful, goals and/or chosen tactics and tasks may have to be changed. Such a change may already have been anticipated during planning. Changes should be made in consultation with the client where appropriate and justified in reports.

7.3. When the Work Is Complete

7.3.1. Achievement Review

7.3.1.1.   The completed work should be reviewed by the management to evaluate achievements. Experiences gained at all stages of the project may provide lessons for planning and carrying out similar work in the future. The review might typically cover:

- aspects of technical achievement such as differences between goals and results, problems encountered and how they were solved, and usefulness of the results;
- compliance with budgeted costs and timescales, with explanations for any deviations, correlation of expenditures, and technical results;
- quality of work of individual contributors;
- consequences of project and results to the laboratory (organization, personnel, equipment, methods and procedures, and possibility of dissemination or exploitation);
- satisfaction of client.

7.3.1.2.   The achievement review may be supplemented by an external peer review, for example, when data is published in scientific journals, or by a third party review (audit).

7.3.2. Reporting, Technology Transfer, and Publication

7.3.2.1.   Research and Development may be reported in various ways. Primarily, a report should be made to the client in the format previously agreed and be written in a language that the client can readily understand. The report should provide sufficient information to enable the client, any subsequent user, or assessor of the report to be able to follow any arguments, and if required, repeat any or all stages of the experimental work and obtain compatible results such as:

- the meaning of the test results should not be distorted by the reporting process;
- appropriate use should be made of conventions for rounding of numbers and expression of decimal places and significant figures;
- where appropriate, results should include an estimate of the associated uncertainty with its corresponding confidence level.

7.3.2.2.   Compared to scientific publications, project reports typically contain project-oriented information (technical, financial statements, etc.), conclusions, and recommendations, and usually present the findings in a less technical way.

7.3.2.3.   If the work has yielded data, observations, new methods, techniques, or new knowledge of interest to the wider community, then dissemination or exploitation of the work is an important issue. Dissemination or exploitation can take a number of forms such as lectures, publications in journals, patents, licenses, standards, and training material. Permission for dissemination or exploitation must be sought from the laboratory, the client, or whoever else owns the intellectual property. Where it is hoped that new methods can be adopted more widely, further performance evaluation may be required, perhaps using collaborative study. Methodology must be described unambiguously and in sufficient detail to allow others to be able to follow the arguments and replicate the work; otherwise, its credibility may be adversely affected.

7.3.3.   Archiving

7.3.3.1.   Archiving primarily involves the secure storage of samples, analytical records, results, methods, and other information for later retrieval and use. The method of archiving and the time for which material is kept depend on what is archived and why. It may be done for a number of reasons such as:

- legal or regulatory requirement;
- requirement of customer or some other external agency (e.g., accreditation body);
- verification of previous work and procedure at later stages of the project;
- validation of methods and results after completion of laboratory work and reporting/publication;
- proficiency testing or collaborative studies with samples;
- postreport questioning by client or peer review;
- problems associated with duplication of work/results, technology transfer;
- keeping the information, which benefits the laboratory.

7.3.3.2.   Samples should normally be stored until the likelihood of their requiring retest has been ruled out or they have deteriorated to an extent where retest would be meaningless (unless study of their deterioration is a part of the work).

7.3.3.3. An important feature of an effective archive system is knowing what it contains and being able to find things quickly. Use of a searchable, database is recommended and offers some protection against illness, death, or transfer of expert staff and also helps to save time and money by providing a means of preventing the inadvertant duplication of earlier work.

7.3.3.4. Where space is important, text-based material can usually be archived in electronic or photographic form. Back-up copies should be kept in remote, flameproof storage. The use of different media may be preferred in different sectors, and use of others prohibited.

7.3.3.5. Retention of data, reports, and other useful information should be consistent with regulatory and customer requirements.

8. EXTERNAL VERIFICATION

8.1.1. Although the laboratory may monitor the quality of its work by internal assessment, independent external assessment may be useful, in order to:

- demonstrate its quality to customers, regulatory bodies, funding bodies, or other external parties;
- compare its level of quality with others in order to make improvements.

8.1.2. Although it is a straightforward process for a laboratory carrying out routine work to apply a structured QA system and use it to regulate laboratory performance, the ever-changing nature of work in an R&D laboratory demands a more flexible and less bureaucratic approach. It is a widely held opinion that the rigidity of conventional formal QA systems and their associated means of external assessment restrict the creativity of thought and practice required in an R&D environment. A number of options are available for externally assessing R&D:

- formal assessment against conventional QA standards (ISO Guide 25, ISO 9000, and GLP);
- benchmarking;
- visiting groups and peer review of publications;
- ranking of laboratories;
- external quality assessment.

8.2. Formal Assessment Against Published QA Standards

8.2.1. ISO Guide 25 (3)

8.2.1.1. Traditionally, the preferred route for routine laboratory environments, formal accreditation against standards derived from ISO Guide 25, provides an independent assessment against an objective criterion that a laboratory is competent to perform specific calibration or testing measurements. The assessment is carried out by peers, i.e., specific measurement methods are assessed by colleagues from other organizations with expertise in those measurements, who can judge whether the procedures in use are technically valid. Accreditation is granted on the basis of the laboratory's ability to perform tests and does not cover peripheral issues, such as administrative procedures not related to the measurements, and perhaps more important, expert but subjective interpretation of the measurement data. Accreditation cannot guarantee the reliability of a measurement result. However, it does provide recognition that the conditions under which the measurement was made maximizes the probability of the measurement being verifiable. Even where there is no formal verification of compliance against ISO Guide 25, it remains a very useful technical QA model for laboratories to refer to in order to regulate the quality of R&D.

8.2.1.2. Although accreditation is granted against a specified schedule of measurements, it is currently difficult and expensive to apply it to R&D. The 1998 revision of ISO Guide 25 now incorporates much of ISO 9001 (34). However, the definition of R&D used in ISO Guide 25 may not necessarily correlate with its use in this document. In theory, R&D consisting of objective nonroutine measurements, which could be fully documented and validated, could be accredited, provided the laboratory considered it to be cost effective to do so.

8.2.1.3. It is sometimes possible for accreditation to be formally granted for groups of tests rather than specific tests, particularly where the laboratory in question has a proven quality system and has a high degree of established expertise in the technique relevant to the group of tests. It should be possible to extend this accreditation to whole types of test (see Annexure D). Whether or not accreditation could be granted for the unit operations described in Section 6 earlier is a matter for conjecture. Although there is a logical development of the principle of granting accreditation for test types, accreditation bodies currently only accredit the whole test. Some ideas of how accreditation of R&D might be achieved by type of test are given in Annexure D.

8.2.2. ISO 9001 (5)

8.2.2.1. ISO 9000 is unspecific about how technical work should be performed. The certification assessment is primarily aimed at the management of procedures, and assessors are not normally from a relevant technical background. ISO 9000 requires no specific assessment of the validity of work and enables the laboratory to set its own level of quality. Certification thus has merits for assessment of how the overall work is managed, but on its own does not assure its validity.

8.2.2.2. The main merit of applying ISO 9001 to an R&D environment lies in its use for controlling the organization and project management aspects of work. There should be no reason why a laboratory cannot have certification to ISO 9001 to organize, manage, and perform R&D work, using the more technically exacting requirements of ISO Guide 25 as a basis for the technical side of its work.

8.2.3.   Good Laboratory Practice (6)

8.2.3.1.   A laboratory operating to GLP (OECD Principles of Good Laboratory. Practice) will have demonstrated that it has a management system and laboratory procedures that would enable a third party to reconstruct any GLP compliant study. The GLP is concerned with traceability of the materials used, especially samples, and good descriptions of analytical methods. It is not, per se concerned with technical quality elements such as accuracy or precision, although many of the laboratory system elements required by GLP considerably assist in the delivery of technical quality. The GLP traces its origins to testing in support of toxicological assessments carried out in support of product registration, but in theory there is no reason why it cannot be applied to all areas of measurement. Eligibility of work for formal registration of compliance depends on the policy of the national bodies that administer GLP principles in each country.

8.3.   Benchmarking

8.3.1.   Benchmarking is a continuous, systematic process in which a laboratory/organization compares its practices and procedures with comparable activities in other organizations in order to make improvements. It can be carried out at various levels with various partners (who need not be laboratories): internal, external, competitive, noncompetitive, and *best practice* (the acknowledged leaders of the process being benchmarked). When benchmarking with other organizations, an agreed Code of Conduct is vital to ensure an effective, efficient, and ethical process, while protecting both parties. A typical benchmarking process is shown in Figure 3.

8.3.2.   Examples

1.   External: A laboratory can assess its purchasing procedures by benchmarking with another organization known to have very good purchasing procedures.
2.   Internal: Group A in a laboratory wins only 10% of possible contracts while group B in the same laboratory wins 50%. By benchmarking its bidding procedures against those of group B, group A ought to be able to improve its success rate at winning contracts.

8.4.   Visiting Groups and Peer Review

8.4.1.   These types of review involve the use of groups of senior level experts, probably from a wide range of sources, to evaluate a laboratory.

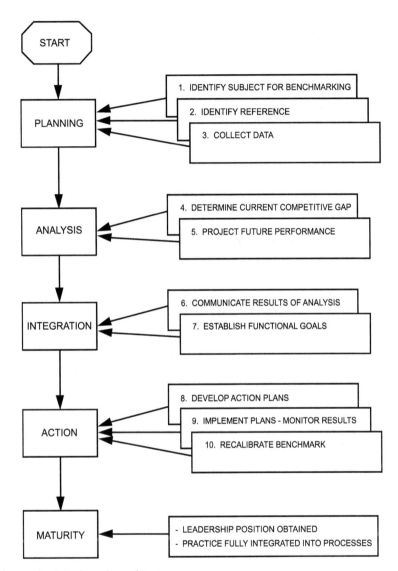

Figure 3 The benchmarking process.

The evaluation can be directed either at the laboratory itself or at the laboratory's scientific output.

8.4.2. In the former case, the evaluation is likely to be against the laboratory's stated objectives, with a strong emphasis on the excellence of the science, staff, and facilities. Such groups typically act on behalf of R&D funding bodies and are a popular form of assessment in the academic world.

The terms of reference of such groups may vary from group to group and there are no universally recognized criteria against which assessments are carried out. The sort of areas covered might include:

- whether staff have appropriate training and qualifications, and are fully conversant with the aims and objectives of their work;
- awareness of staff to published work in their subject areas;
- quality and availability of scientific support services;
- adequacy of resources;
- degree of scientific collaboration;
- effectiveness of technology transfer;
- management of the R&D programme;
- whether the organization of projects effectively meets the customer requirements.

8.4.3. The strength of the visiting groups' approach is that it concentrates on the quality of the science. However, the way it is used at present makes weak in several other respects such as:

- it lacks harmonized and transparent criteria;
- it tends to look at work retrospectively;
- it is subjective and susceptible to bias.

8.4.4. Assessment visits for accreditation/certification/registration purposes (see the earlier section) and visits by customers are a special subset of visiting groups/peer review. In the case of customers, those visiting may lack technical expertise in the areas concerned.

8.4.5. Peer review of publications, also known as citation analysis, involves the following:

- assessment of the number and quality of publications the laboratory under examination has published in the scientific press;
- assessment of how much those publications are being cited by colleagues within the same research field.

Citation analysis traces its origins to law but is now a widely used, significant research tool, adopted from the field of information science to a range of subject areas. The Science Citation Index (SCI) was first published in 1961. Four particular applications have been reported (35,36):

1. to assess the impact of individuals, institutions, and journals;
2. to investigate hypotheses about the history and sociology of science;
3. to study performance characteristics of information search and retrieval;
4. to evaluate the tool.

Increasingly, it is used in the analysis of departmental output or as a measure of the value of the work of a department (37,38).

8.4.6. Some journals will only accept papers for publication that have been the subject of satisfactory peer review (this is the most common type of peer-review mechanism in use today). As a consequence, it is more difficult to publish in these journals. From the citation analysis point of view, publication in a respected journal will score better than one in a less respected journal—the so-called impact factor. Criteria, ranking journals in order of merit, are published annually by the Institute for Scientific Information. This system has some merit, as published work often reflects the competence and expertise of the publishing laboratory. A laboratory can deliberately raise the profile of its work by publishing as often as possible in the most highly regarded journals. However, publication is not always an option and laboratories that do not publish are not necessarily producing poor-quality work. One should also be aware that the status of journals sometimes changes with time. Citation analysis has a number of other limitations, making it a dangerous technique to use in isolation:

- method papers are cited more often than empirical or theoretical papers and tend to be referenced due to utility rather than innovation or novelty;
- work ahead of its time is not cited, because there are no other scientists interested in the same field of work;
- citations are proned to discrepancies, for example, misspellings;
- citations are rarely complete or comprehensive. Citation counts need to be seen mainly as indicators, and comparisons can only be made if identical citable and citing pools are used;
- negative or contradictory citations tend to indicate a lack of value to the work.

8.4.7. Patents and licenses are other forms of dissemination and exploitation that can be used as a measure of a laboratory's output.

8.5. Ranking of Organizations

8.5.1. This involves comparing laboratories against a set of common criteria and ranking them on the basis of the comparison.

8.6. External Quality Assessment Procedures (Proficiency Testing)

8.6.1. Participation in external quality assessment schemes provides an external measure of performance. In nonroutine work or R&D, relevant schemes may be difficult to identify or may give an unrealistic impression of performance. Other types of interlaboratory comparison are perhaps more relevant to R&D, such as co-operative studies, but these do not give the same measure of laboratory performance. It should also be recognized that the proficiency testing schemes, which give the most reliable measure of performance, are those in which the participating laboratories receive the test samples blind.

8.7. Conclusions

8.7.1.   No single method of assessment stands out as being the most suitable for monitoring the quality of nonroutine and R&D work. It is recommended that where some kind of external assessment is required, a combination of approaches should be taken and formal assessment should be confined wherever possible to those parts of the quality system that remains stable from project to project, for example, the management levels and the technical infrastructure. Typically, this could be established for the three-tier quality system approach as follows:

	Verification	
Quality elements	Format	Informal
Organizational	Certification to ISO 9000	Follow ISO Guide 25 Benchmarking Self assessment
Technical	Accreditation to ISO Guide 25/EN 45001	Follow ISO Guide 25 Visiting groups Benchmarking Peer review
Analytical task	Registration to GLP Proficiency testing	Follow GLP principles

8.7.2.   The informal verification principles mentioned earlier could be made more formal if required, and the declared compliance with particular standards, guides, or protocol could be independently assessed by a suitable outside body, for example, a visiting group, consultant, or examining inputs, such as:

- existence of project plans where no elaborated methods are available;
- maintenance and calibration schedules;
- record keeping;

and outputs, such as:

- reports and publications;
- satisfactory participation in relevant proficiency testing, external quality assessment, or other intercomparisons.

8.7.3.   A well functioning quality system need not stifle creativity in R&D and is vital for ensuring the smooth transfer of technology from research to diagnostic or commercial environments. Research workers must have an appreciation of the quality requirements of clients, and quality must be designed into every process.

ACKNOWLEDGMENTS

The secretary (D.H.) would also like to thank all of those individuals and organizations who have contributed comments, advice, and background documentation. Production of this Guide was in part supported under contract with the UK Department of Trade and Industry as part of the National Measurement System Valid Analytical Measurement (VAM) Programme.

BIBLIOGRAPHY AND REFERENCES

References Cited in the Text

1. "International Guide to Quality in Analytical Chemistry—An Aid to Accreditation", CITAC CG1, 1st ed., 12/95. ISBN: 0948926090.
2. EAL-G4: "Accreditation for Chemical Laboratories—Guidance on the interpretation of the EN 45000 Series of Standards and ISO/IEC Guide 25", 1st ed., 4/93 (originally EURACHEM/WELAC GD1/WGD2).
3. ISO/IEC Guide 25: 1990, 3rd ed., "General requirements for the competence of calibration and testing laboratories".
4. EN 45001: 1989, "General criteria for the operating of test laboratories" (Note: CEN standard complying with [3]).
5. ISO 9000 series of standards (primarily ISO 9001: 1994, Quality systems—Model for quality assurance in design/development, production installation and servicing", ISO 9002:1994, "Quality systems—Model for quality assurance in production and installation", ISO 9003:1994, "Quality systems—Model for quality assurance in production and installation", and others).
6. "The OECD Principles of Good Laboratory Practice", Environmental Monograph No. 45, OCDE/GD(92)32, Organisation for Economic Co-operation and Development, 1992.
7. "McGraw-Hill Dictionary of Scientific and Technical Terms", 4th edition, Parker, New York, NY, 1989.
8. ISO/CASCO 193 (Rev. 2)—ISO/CASCO discussion paper providing up-to-date references on terminology between full updates of ISO Guide 2.
9. ISO/IEC Guide 2: 1996, "Standardisation and related activities—General vocabulary".
10. "Chambers 21st Century Dictionary", Chambers, 1996. ISBN: 0550105883.
11. ISO 8402: 1994, "Quality—Vocabulary".
12. "The Managers Guide to VAM", Department of Trade and Industry, UK, 9/96 (available through LGC).
13. Kellner R. "The WPAC-EUROCURRICULUM on analytical chemistry". J. Anal. Chem 1993; 347:1–2.
14. "The Monitor (The Journal for NVQ Professionals)", National Council for Vocational Qualifications, 222 Euston Rd, London, NW1 2BZ.
15. "Data News", National Council for Vocational Qualifications, 222 Euston Rd, London, NW1 2BZ.
16. "Development of Evidence Requirements and Assessment Guidelines for an NVQ and SVQ in Analytical Chemistry", Report to the Royal Society of Chemistry, Sept. UK, 1996.

17. "The Application of GLP Principles to Computer Systems, Advisory Leaflet No. 1", UK GLP Compliance Programme, London, 1989.
18. Bedson PJ. "Guidance on equipment qualification for analytical instruments". J. Accred. QA 1996; 1:265–274.
19. ISO 10011-1: 1990, "Guidelines for auditing quality systems. Part 1: Auditing".
20. ISO 10011-2: 1991, "Guidelines for auditing quality systems. Part 2: Qualification criteria for quality system auditors".
21. ISO 10011-3: 1991, "Guidelines for auditing quality systems. Part 3: Management of audit programmes".
22. EAL-G3. "Internal Quality Audits and Reviews", European Accreditation of Laboratories.
23. ASTM D 3864–96: "Standard Guide for Continual On-line Monitoring Systems for Water Analysis".
24. prEN 50228–1, "Performance requirements and test methods for control and monitoring pH analysers for use in the water industry. Part 1: Specification and test methods for assessing performance under laboratory conditions".
25. prEN 50228–2, "Performance requirements and test methods for control and monitoring pH analysers for use in the water industry. Part 2: Procedures and test methods for assessing performance under field conditions".
26. Aruajo PW, Brereton RG. "Experimental design II. Optimisation". Trend Anal Chem 1996; 15(2):63–70.
27. Caulcutt R. Statistics in Research and Development. Chapman and Hall: London, 1983.
28. Thomas, Paul CL, Schoefield H. "Sampling Source Book—An indexed biography of the literature of sampling", Butterworth Heinemann, 1st ed., 1995. ISBN: 0–7506–1947–3 (>2300 sampling references).
29. Brereton RG., "Chemometrics—Applications of mathematics and statistics to laboratory systems", Ellis Horwood, Chichester, 1990. ISBN: 0–13–131350–9.
30. "Harmonised guidelines for the use of recovery information in analytical measurement", IUPAC. From Technical Report, Pure Appl. Chem. 1999; 71(2):337–348.
31. "The Fitness for Purpose of Analytical Methods—A Laboratory Guide to Method Validation and Related Topics", EURACHEM, 1st ed. 1998. ISBN: 0–948926–12–0.
32. "Quantifying Uncertainty in Analytical Measurement", EURACHEM, 1st ed., 1995. ISBN: 0–948926–08–2.
33. Lock D. (ed.), "Gower Handbook of Project Management", Gower, 2nd ed., 1994. ISBN: 0–566–07391–9.
34. ISO/IEC Guide 25, revision of 3rd edition, 6th draft 1996, circulated for public comment, ISO CASCO WG10, International Standards Organisation, Geneva.
35. Peritz BC. "On the objectives of citation analysis: problems of theory and method". J Am Soc Inform Sci 1992; 43(6):448–451.
36. Synder H, Cronin B, Davenport E. "What's the use of citation? Citation analysis as a literature topic in selected disciplines of the social sciences". J Inform Sci 1995; 21(2):75–85.

37. Bradley SJ, Willett P, Wood FE. "A publication and citation analysis of the department of information studies, University of Sheffield, 1980–1990". J Inform Sci 1992; 18:225–232.
38. Sandison A. "Thinking about citation analysis". J Document 1989; 45(1): 59–64.

Selected Bibliography

Reference	Subject
Anderson, E. S., Grude, K. V., Hang, Y., "Goal Directed Project Management", Kogan Page, 2nd ed., 1995. ISBN: 0 7494 1389 1.	Project management
ANSI Z1-13–19XX "Quality System Guidelines for Research (Draft)".	R&D
"AOAC Peer Verified Methods Program—Manual on policies and procedures", AOAC International, Arlington VA 22201–3301, USA.	Food analysis
Association of Public Analysts, "A Protocol for Analytical Quality Assurance in Public Analysts' Laboratories", APA, London, 1986.	General QA
ASTM D 4210–89: "Standard Practice for Intra laboratory Quality Control Procedures and a Discussion on Reporting Low-Level Data."	QC
ASTM D 4697–87: "Standard Guide for Maintaining Test Methods in the User's Laboratory".	
ASTM E 1212–92: "Standard Practice for Establishment and Maintenance of Quality Control Systems for Nondestructive Testing Agencies".	General QA, QC
ASTM E 882–87: "Standard Guide for Accountability and Quality Control in the Chemical Analysis Laboratory".	QC
Bamfield, P., "Research and Development Management in the Chemical Industry", VCH Verlagsgesellschaft mbH, 1996. ISBN: 3–527–28778–7.	Management of R&D
BS2846 Part 4: 1976 (ISO 2854). "Techniques of estimation relating to means and variances".	Statistics

(*Continued*)

Reference	Subject
BS2846 part 7: 1984. "Tests for departure from normality".	Statistics
BS3534 part 1: 1993. "Statistics—Vocabulary and symbols—Probability and general statistical terms".	Statistics
BS 4778:Part 2:1991, "Quality vocabulary, part 2, quality concepts and related definitions"	Terminology
BS 5700: 1984 + AMD 5480: 1987, "Guide to process control using quality control chart methods and cusum techniques" (complements BS 5701: 1980 and BS 5703: 1980).	Statistics, QC
Carr-Brion, K. G., Clarke, J. R. P., "Sampling Systems for Process Analysis" Butterworth Heinemann, 2nd ed., 1996. ISBN: 0–7506–1247–9.	Sampling
Caulcutt, R., Boddy, R., "Statistics for Analytical Chemists", Chapman and Hall, London, 1991.	Statistics
Chatfield, C., "Statistics for technology", Chapman and Hall, London 3rd ed. (revised) 1995, ISBN: 0412 25340 2.	
"Criteria for evaluating acceptable methods of analysis for Codex purposes", Codex Alimentarius Commission on Methods of Analysis and Sampling (CX/MAS 95/3; 1995).	Food analysis
Cook S., "Practical Benchmarking—A Manager's Guide to Creating a Competitive Advantage" Kogan Page, London. ISBN: 0–7494–1551–7.	Benchmarking
DOE standard DOE-ER-STD-6001–92 "Implementation Guide for Quality Assurance Programs for Basic and Applied Research", US Department of Energy.	R&D
Dietrich, C. F., "Uncertainty, Calibration and Probability", Adam Hilger, London, 2nd ed., 1991.	Measurement uncertainty
Dux, J. P., "Handbook of Quality Assurance for the Analytical Chemistry Laboratory", Van Nostrand Reinhold, New York, 1986.	General QA

(*Continued*)

Reference	Subject
"Estimating Uncertainties in Testing—A Guide to Estimating and Reporting Uncertainties of Measurement in Testing", British Measurement and Testing Association (BMTA), 1st ed. October 1994.	Measurement uncertainty
Funk, W., Dammann, V., Donnevert, G., "Quality Assurance in Analytical Chemistry", VCH Weinheim, 1995. ISBN: 3-527-28668-3.	
Garfield, F. M., "Quality Assurance Principles for Analytical Laboratories", Association of Official Analytical Chemists, Arlington, VA, 1992.	General QA
"Güidelines for Collaborative Study—Procedures to Validate Characteristics of a Method of Analysis", AOAC International, revised May 1994, J. Assoc. Off. Anal. Chem., 1989, 72, 694–704 (originally published).	Collaborative study
Gunzler, H., "Accreditation and Quality Assurance in Analytical Chemistry", Springer Verlag, 1994. ISBN: 3-540-60103-1.	General QA
"Harmonised guidelines for internal quality control in analytical chemistry laboratories", (IUPAC). Pure Appl. Chem., 1995, 67(4), 649–666.	QC
"Harmonised protocols for the adoption of standardised analytical methods and for the presentation of their performance characteristics" (IUPAC). Pure Appl. Chem., 1990, 62(1), 149–162.	Collaborative study
ISO 3534-1: 1993, "Statistics—Vocabulary and symbols—Part 1: Probability and general statistical terms".	Statistics, terminology
ISO 3534-2: 1993, "Statistics—Vocabulary and symbols—Part 2: Statistical quality control".	Statistics, terminology
ISO 3534-3: 1985, "Statistics—Vocabulary and symbols—Part 3: Design of experiments".	Statistics, terminology

(*Continued*)

Reference	Subject
ISO 5725-1: 1994, "Accuracy (trueness and precision) of measurement methods and results—Part 1: General principles and definitions".	Terminology, accuracy
ISO 5725-2: 1994, "Accuracy (trueness and precision) of measurement methods and results—Part 2: Basic methods for the determination of repeatability and reproducibility of a standard measurement method".	Accuracy
ISO 5725-3: 1994, "Accuracy (trueness and precision) of measurement methods and results—Part 3: Intermediate measures of the precision of a standard measurement method".	Accuracy
ISO 5725-4: 1994, "Accuracy (trueness and precision) of measurement methods and results—Part 4: Basic methods for the determination of the trueness of a standard measurement method".	Accuracy
ISO 5725-6: 1994, "Accuracy (trueness and precision) of measurement methods and results—Use in practice of accuracy values".	Accuracy
ISO 78-2: 1982, "Layouts for standards —Part 2: Standard for chemical analysis" 1999.	Documentation of methods
ISO 7870: 1993, "Control charts— General guide and introduction".	Statistics, QC
ISO 7873: 1993, "Control charts for arithmetic average with warning limits".	Statistics, QC
ISO 7966: 1993, "Acceptance control charts".	Statistics, QC
ISO 8258: 1991, "Shewhart control charts".	Statistics, QC
"ISO VIM International vocabulary of basic and general terms in metrology", 2nd ed., 1993.	Terminology
ISO/CD 5725-5, "Accuracy (trueness and precision) of measurement methods and results—Part 5: Alternative	Accuracy

(*Continued*)

Reference	Subject
methods for the determination of the precision of a standard measurement method".	
ISO/DIS 11095 (April 1993), "Linear calibration using reference materials".	RMs, CRMs
ISO/IEC (with BIPM, IFCC, IUPAC, IUPAP, OIML): Guide to the expression of uncertainty in measurement", 1994. ISBN: 92 67 10188 9.	Measurement uncertainty
ISO/IEC Guide 30: 1992, "Terms and definitions used in conjunction with reference materials".	Terminology
ISO/IEC Guide 33: 1989, "Uses of certified reference materials".	CRMs
ISO/TAG4/WG3: 1993, "Guide to the Expression of Uncertainty in Measurement".	Measurement Uncertainty
Karlöf, B., Östblom, S., "Das Benchmarking Konzept-Wegweiser zur Spitzenleistung in Qualität und Productivität", Verlag Vahlen, ISBN: 3 8006 1831 1.	Benchmarking
Kateman, G., Buydens, L., "Quality Control in Analytical Chemistry", John Wiley & Sons, 1991.	General QA, QC
Litke, H. D., "Project management— Methoden, Techniken, Verhaltensweisen", Carl Hanser Verlag, Munchen, 1991. ISBN: 3–446–15836–7.	Project management
Lock, D., "The Essentials of Project Management", Gower, 1996. ISBN: 0–566–0774500.	Project management
McNair, C. J., Leibfreid, K. H. J., "Benchmarking: A Tool for Continuous Improvement", Harper Business, London, 1992.	Benchmarking
Mesley, R. J., Pocklington, W. D., Walker R. F., "Analytical Quality Assurance— A Review". Analyst, 1991, 116(10), 975–1092.	General QA
Miller, J. G., Miller, J. N., "Statistics for Analytical Chemistry", Ellis Horwood and Prentice Hall, London, 3rd ed., 1993. ISBN: 0–13–030990–7.	Statistics

(*Continued*)

Reference	Subject
Murdoch, J., "Control Charts", Macmillan, 1979.	Statistics, QC
"Nomenclature for the Presentation of Results of chemical analysis (IUPAC Recommendations, 1994)". Pure Appl. Chem., 1994, 66(3), 595–608.	Terminology, statistics
Parkany, M., "Quality Assurance for Analytical Laboratories", Special-publication 130, Royal Society of Chemistry, UK. ISBN: 0–85186–705–7.	General QA
Prichard, F. E. "Quality in the Analytical Chemistry Laboratory", ACOL, Wiley, 1995. ISBN: 0–471–95541–8.	General QA
Prichard, E., Mackay, G. M., Points, J., "Trace Analysis—A Structured Approach to Obtaining Reliable Results", Royal Society of Chemistry, 1996. ISBN: 0–85404–417–5.	QA specifically for trace analysis
"Protocol for the Design, Conduct and Interpretation of Method-Performance Studies (IUPAC)". Pure Appl. Chem., 1995, 67(2), 331–343.	Collaborative study
Safety Series No. 50-SG-QA16 "Quality Assurance for Research and Development", International Atomic Energy Authority.	R&D
Taylor, B. N., Kuyatt, C. E., "Guidelines for evaluating and expressing uncertainty in NIST measurement results2, NIST technical note 1297, National Institute of Standards and Technology, 1994.	Measurement uncertainty
Taylor, J. K., "Quality Assurance of Chemical Measurements", Lewis, Michigan, 1987. ISBN: 0–87371–097–5.	General QA
"Report on Minimum Criteria to Assure Data Quality", USA—Environment Protection Agency, July 1989.	General QA Environmental analysis
"Validation of test methods—General principles and consepts (sic)", Committee paper EAL/GA(96)58, European Accreditation of Laboratories, Oct. 1996.	Method validation

(Continued)

Reference	Subject
Youden, W. J., Steiner, E. H., "Statistical Manual of the AOAC", Association of Official Analytical Chemists, 1975.	Statistics
Zairi, DR M., Leonard, P., "Practical Benchmarking: A Complete Guide", Chapman and Hall, Cambridge, 1994.	Benchmarking

ANNEXURE A

Composition of EURACHEM/CITAC R&D/Nonroutine Analysis Working Group

Name	Affiliation	Country	Background	Liaison links
Prof C. Adams	Unilever	United Kingdom	Industry/ Academic	DIRAG, EURACHEM, U.K.
Prof K. Cammann	ICBFhM	Germany	Academic	EURACHEM, EUROLAB, ISO/IUPAC/ AOAC
ir H. A. Deckers	RvA	Netherlands	Industry	WOBAC, EUROLAB EAL
Prof Z. Dobkowski	Ind. Chem. Res. Inst.	Poland	Government	Polish Chem. Soc., AOAC, EURACHEM
Mr D. Holcombe	LGC	United Kingdom	Government	ISO/IUPAC/ AOAC
Dr P. D. LaFleur	Kodak	United States	Industry	CITAC
Dr P. Radvila	EMPA	Switzerland	Government	SAPUZ, EUROLAB-CH, EURACHEM
Dr C. Rohrer	Lenzing AG	Austria	Industry	
Dr W. Steck	BASF AG	Germany	Industry	CITAC, EURACHEM
ir P. Vermaercke	S.C.K.	Belgium	Industry	BELTEST

Other inputs to the guide were made by the following:

    a. CITAC Working Group with members from Australia, Austria, Belgium, China, Germany, Hong Kong, Japan, Korea, Mexico, the Netherlands, Russia, Switzerland, United Kingdom, and United States.

b. EURACHEM full, associate, and observer members from Austria, Belgium, Commission of the EC, Cyprus, Czech Rep., Denmark, Finland, France, Germany, Greece, Hungary, Iceland, Ireland, Italy, Luxembourg, Malta, the Netherlands, Norway, Poland, Portugal, Russia, Slovakia, Slovenia, Spain, Sweden, Switzerland, Turkey, United Kingdom, United States, AOACI, and FECS.

c. Miscellaneous inputs have been made by colleagues from Australia, Austria, Belgium, Canada, Cyprus, Czech Rep., Denmark, European Commission SMT Programme, Finland, Germany, Hungary, Iceland, Ireland, Israel, Italy, Malta, the Netherlands, Norway, Portugal, Spain, Sweden, Switzerland, United Kingdom, United States.

A.1. About CITAC and EURACHEM

A.1.1. CITAC: *Co-operation on International Traceability in Analytical Chemistry* arose from an international workshop held in association with the Pittsburgh Conference in Atlanta, March 1993. CITAC aims to foster collaboration between existing organizations to improve the international comparability of chemical measurement. A working group co-ordinates activities that include production of a directory of reference materials under development, preparation of quality system guidelines for the production of reference materials, preparation of a directory of chemical metrology activities, definition of criteria for establishing traceability to the mole, and preparation of an international guide to quality in analytical chemistry (1).

A.1.2. EURACHEM was established in 1989 to provide a focus for analytical chemistry and quality-related issues in Europe. It is a network of European national laboratories that have an interest or responsibility for chemical analysis. It provides a framework that facilitates collaboration between analysts throughout Europe to improve the quality of analytical measurements and provides a forum for the discussion of common problems and for developing an informed and considered approach to both technical and policy issues.

Up-to-date information on EURACHEM activities is available from its twice yearly newsletter or from its website http://www.vtt.fi/ket/eurachem.html.

Both CITAC and EURACHEM secretariats can be contacted via the drafting secretary at the address shown at the front of the guide.

ANNEXURE B

Flowchart Showing Typical Lifecycle of R&D Project

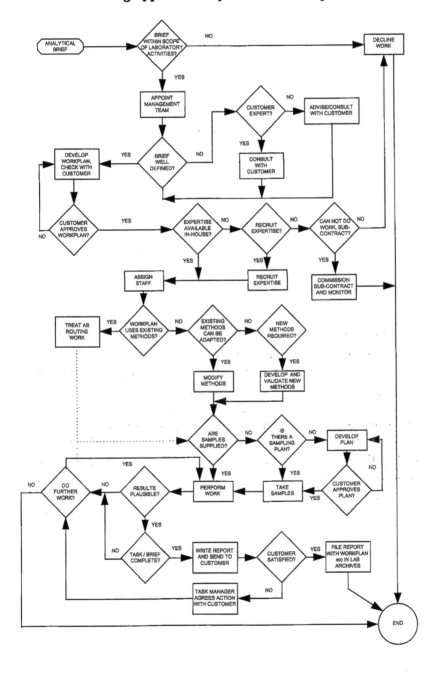

ANNEXURE C
Questionnaire for Analytical Work

A. Client

Contact person:
Tel:/Fax:
Address:

B. Objective/goals/required information

Requested analysis:
☐ Qualitative/semiquantitative, limit of detection:
☐ Quantitative, range of concentration:
Previous analysis/results:

C. Costs

Expected costs:
Cost limits:

D. Date of completion/schedule

Date of intermediate results/reports:
Date for final results/report:

E. Sampling   ☐ Client   ☐ Laboratory   ☐ Other

Date of sampling:
Source/producer:
Responsible person:
Number of samples:

F. Description of sample(s)

Identification:
Approx. composition:
Main component: Minor constituent:
Intended use:
Packaging/stability:
Special care for storage/transport/stabilisation:
Pretreatment/preconditioning:
Reference materials/reference sample:

G. Methodology

Description of methods used for sampling, sample preparation, measurement:
Standard method:
Generic method:

New/adapted method: R&D for new method:
Validation for adopted method:

ANNEXURE D

Proposals for the Accreditation of R&D Tests by Type

D.1. Purpose

The accreditation of types of tests serves to provide a flexible description for the scope of accreditation. This annexure sets out proposals for possible conditions under which accreditation might be granted for tests by type. Note the responsibility for defining such conditions strictly lies with accreditation bodies.

D.2. Area of Application

These proposals should be applicable to all testing laboratories aiming for flexibility in their scope of accreditation, especially with regard to R&D work.

D.3. Definitions

D.3.1. Type of Test

"Sector (of a testing field) with similar technical-methodological features, with comparable calibration, validation, and training principles." Types of test may be defined on a technology- or application-related basis. For example:

- gas chromatography (or perhaps more broadly "separation techniques");
- atomic spectroscopy;
- thermoanalysis;
- primary fire characteristics.

D.3.2. Testing Field

Testing fields are sizable sectors distinguished by common fundamentals of a technical-, methodological-, and training-related nature. For example:

- chemical and physicochemical analysis;
- biological investigations;
- medical laboratory diagnostics.

D.3.3. Flexibilization

Flexibilization of the scope of accreditation is understood to comprise all measures to be taken for accreditation not directed exclusively at the accreditation of individual test methods.

D.4. General

The accreditation of types of tests means that the testing laboratories are given the opportunity to introduce new test methods within the approved type of test or of modifying existing methods without having to obtain approval from the accreditation authority in each individual case in advance. It also allows confirmation of the competence of R&D analytical activities on the basis of general work.

Accreditation of a type of test is granted under certain conditions and within the limits governed by the experience that has already been demonstrated by the laboratory for that type of test. Making the scope of accreditation flexible with respect to the methods used does not necessarily imply making it flexible with respect to the sample types under test.

D.5. Recommended Conditions for the Accreditation of Types of Tests

For every type of test for which the laboratory requires accreditation, it should submit to the accrediting body:

- a sufficient number of different test methods, SOPs, or test reports;
- procedures for validation or verification as part of the type of test;
- corresponding records of validation and verification.

The methods submitted must reflect adequate operator competence (e.g., technical range) within the type of test applied for. For new or modified test methods, complete documentation and validation are required. For R&D, appropriate test reports and/or generic SOPs may be submitted instead of the test methods.

The laboratory should have available at all times a list of the methods currently covered by its accreditation. The list can be submitted to the accreditor as part of the monitoring procedure, with new or modified methods identified.

D.6. Assessment of the Scope of Accreditation

In the accreditation of types of tests, the assessment is directed, in particular, towards the following:

- the organizational prerequisites that the testing laboratory has to meet for it to validate or verify new or modified test methods;
- the qualifications and experience of the staff and management, and the policies on further training;
- the level of technical equipment;
- the procedures for testing;
- the quality management system;
- the records of validation and verification carried out.

The assessor has the responsibility for selecting and inspecting key test methods and equipment. The following criteria are amongst those that might be used as a basis for such selections:

- the technical complexity of the tests;
- the possible consequences of errors in performing the tests;
- the frequency of use of the test methods;
- the ratio of routine and nonroutine tests.

The extent of the checks should be sufficient to allow the accrediting body to be confident of the capability of the laboratory to introduce new methods or to modify existing methods or to carry out R&D. At the same time, the checks must not impose unreasonable costs on the laboratory. The assessors report should indicate to which test items the respective types of test relate.

D.7. Scope of Accreditation of Types of Tests

The scope of accreditation may be specified in terms of:

- testing field(s);
- type(s) of test(s);
- test method(s);
- item(s) under test.

ANNEXURE E

R&D to Develop Analytical Instrumentation

E.1. The following specific interpretation is recommended for R&D to develop analytical instrumentation.

E.2. Introduction

Instrumental R&D involves the improvement of existing analytical systems or development of entirely new systems. The basis for the R&D usually arises from the need for novel systems that are faster, more sensitive, more accurate, more precise, more discriminating, simpler (and easier to use), more economic, more environmentally friendly, or applicable to different particular analyte(s)/sample matrix combinations. Occasionally, it may be carried out on a purely speculative basis, i.e., with no particular end application in mind, for example, to investigate the practical potential of a particular measurement principle.

Instrumental R&D projects generally involve building and evaluating prototype instrumentation, making and evaluating changes until the prototype evolves either to a state where performance objectives have been met or further development is not viable. The prototype might be a whole new instrument or an accessory (such as a detector or a chromatography column) for an established instrument.

E.3. Planning

Instrumental R&D project planning involves objective setting, as with conventional analytical R&D. The research plan effectively involves setting out the strategy for the project and defining the criteria against which the performance of the prototype can be assessed.

E.4. Experimental Design

The project should include experiments to evaluate and validate instrument performance and to help define the behavior of the instrument under calibration. Long-term stability/acceptable performance should be monitored before the equipment is put into routine use. A means of controlling calibration should be established, either through external adjustment or fixed internally. Suitable standards, blanks, reference materials, or check samples of known content can be used in these experiments.

The criteria that cause deterioration of instrument performance should be identified, and wherever possible, routines established for controlling these criteria. Where instrument performance is particularly sensitive to operator skill, optimum-operating procedures should be established. Checking procedures, using standards, check samples, test mixtures, etc., should be established as part of the monitoring process.

Where the instrument under development involves the processing of raw data or signal through some form of algorithm, access to the raw data/signal is advised, so that the basic instrumental performance and signal processing can be checked independently.

A number of ways to evaluate and validate the novel instrumentation are possible. Where other techniques/procedures/instrumentation exists for the particular measurement application, these could be used for the parallel evaluation and validation of the novel instrumentation. Collaborative trial could be used, either involving several laboratories each evaluating the novel instrumentation, or the developing laboratory comparing results generated by its own use of the novel instrumentation against other laboratories using other techniques.

E.5. Data Recording

Data from instrument evaluation should include a record of conditions under which the instrument is, and is not, working satisfactorily. Typically, this will include information on analyte and matrix condition, presence of particular chemical, spectral, and physical interferents, temperature, humidity, electrical, and magnetic settings. Sufficient data should be recorded over extended time periods and differing conditions to establish the reliability of the technology.

E.6. Reports

Where new instrumentation is successfully developed, the reports from the prototype evaluation and validation stages will form the basis for the "instrumentation in more widespread use"; i.e., the report is effectively the operating manual. It should include user-friendly instructions for operation of the instrument, applicability, information on storage, calibration and maintenance, and performance checks. Where appropriate, there should be an explanation of how the raw signal is processed by the algorithm for zeroing purposes, so that in routine use, incorrect assumptions are not made in the subtraction of blanks. New instrumentation should be subject to equipment qualification procedures, before being put into use.

E.7. Evaluation

Where the novel instrumentation performance overlaps with the existing instrumentation, the success of the R&D can be evaluated by comparison of the two instruments against agreed performance criteria. Unless something is being developed for a particular end use, it is probably easier to test the instrumentation initially against simple problems and then more demanding problems, as familiarity with the technique and the behavior of the instrument improve. In general, if the instrument appears to function correctly with one analyte in a single matrix, this is not satisfactory evidence for the soundness of the technique per se. However, it may be acceptable where the particular analyte/matrix pair is the main reason for the R&D work.

Appendix E

ACCREDITATION FOR MICROBIOLOGICAL LABORATORIES (EURACHEM/EA)

1. INTRODUCTION AND SCOPE OF DOCUMENT

1.1. The general requirements for accreditation are laid down in the International Standard *General requirements for the competence of testing and calibration laboratories* (ISO/IEC 17025 1st Ed., 1999), hereafter referred to as ISO 17025. All of these requirements must be met by laboratories seeking accreditation.

1.2. This document supplements ISO 17025 by providing specific guidance for both assessors and laboratories carrying out microbiological testing. It gives detailed guidance on the interpretation of ISO 17025 for those undertaking the examination of materials, products, and substances. The guidance is applicable to the performance of all objective measurements, whether routine, nonroutine, or as part of research and development. Although it is written primarily for food and environmental microbiological testing, the general principles may be applied to other areas. ISO 17025 remains the authoritative document and, in cases of dispute, accreditation bodies will adjudicate on unresolved matters. The guidance given in this document may also be of use to those working towards registration under other quality standards such as GLP, GMP, and GCP.

1.3. This document can be considered as the "Application Document" for microbiological testing as set out in Annex B of ISO 17025. This document has been produced jointly by EURACHEM and EA as a means of promoting a consistent approach to laboratory accreditation amongst EA member bodies, particularly those participating in the EA Multilateral Agreement.

1.4.   Microbiological testing is taken to include sterility testing, detection, isolation, enumeration, and identification of micro-organisms (viruses, bacteria, fungi, and protozoa) and their metabolites in different materials and products, or any kind of assay using micro-organisms as part of a detection system as well as the use of micro-organisms for ecological testing. It follows that some of the guidance in this document, e.g. on laboratory environment, will need to be interpreted accordingly. This document can also provide guidance to laboratories using techniques in areas related to microbiology, such as biochemistry, molecular biology, and cell culture, although there may be additional requirements for such laboratories.

1.5.   This document is concerned with the quality of test results and is not specifically concerned with health and safety matters. However, laboratory practices should conform to national health and safety regulations. It is important to note that in some cases health and safety issues may have an effect on quality of testing and the laboratory will be required to take this into account.

1.6.   Definitions of the terms used are given in Appendix A.

2.   PERSONNEL

ISO 17025, paragraph 5.2

2.1.   Microbiological testing should be either performed or supervised by an experienced person, qualified to degree level in microbiology or equivalent. Alternative qualifications may meet requirements where staff have extensive relevant experience relating to the laboratory's scope of accreditation. Staff should have relevant practical work experience before being allowed to perform work covered by the scope of accreditation without supervision or before being considered as experienced for supervision of accredited work. Specific national regulations may override the guidance given in this document.

2.2.   If the laboratory includes opinions and interpretations of test results in reports, this shall be done by authorized personnel with suitable experience and relevant knowledge of the specific application, including, for example, legislative and technological requirements and acceptability criteria.

2.3.   The laboratory management shall ensure that all personnel have received adequate training for the competent performance of tests and operation of equipment. This should include training in basic techniques, e.g. plate pouring, counting of colonies, aseptic technique, etc., with acceptability determined using objective criteria. Personnel may only perform tests on samples if they are either recognized as competent to do so, or if they do so under adequate supervision. On-going competence should be monitored objectively with provision for retraining where necessary. Where a method or technique is not in regular use, verification of personnel performance before testing is undertaken may be necessary. The critical interval between performances of tests should be established and documented. The interpretation of test results for identification and verification of micro-organisms is strongly connected to the experience of the performing analyst and should be monitored for each analyst on a regular basis.

2.4. In some cases, it may be more appropriate to relate competence to a particular technique or instrument rather than to methods.

3. ENVIRONMENT

ISO 17025, paragraph 5.2

3.1. Premises

3.1.1. The typical laboratory is comprised of the testing facilities (where specific microbiological testing and associated activities are carried out) and ancillary facilities (entrances, corridors, administration blocks, cloak rooms and toilets, storage rooms, archives, etc.). In general there are specific environmental requirements for the testing facilities.

Depending on the type of testing being carried out, access to the micro-biological laboratory should be restricted to authorized personnel. Where such restrictions are in force, personnel should be made aware of:

- a. the intended use of a particular area;
- b. the restrictions imposed on working within such areas;
- c. the reasons for imposing such restrictions;
- d. the appropriate containment levels.

3.1.2. The laboratory should be arranged so as to minimize risks of cross-contamination, where these are significant to the type of test being performed. The ways to achieve these objectives are, for example

- a. to construct the laboratory to the 'no way back' layout principle;
- b. to carry out procedures in a sequential manner using appropriate precautions to ensure test and sample integrity (e.g. use of sealed containers);
- c. to segregate activities by time or space.

3.1.3. It is generally considered as good practice to have separate locations, or clearly designated areas, for the following:

- sample receipt and storage areas;
- sample preparation (e.g. a segregated location should be used for the preparation of powdery products likely to be highly contaminated);
- examination of samples, including incubation;
- maintenance of reference organisms;
- media and equipment preparation, including sterilization;
- sterility assessment;
- decontamination.

The area for washing (after decontamination) may be shared with other parts of the laboratory providing that the necessary precautions are taken to prevent transfer of traces of substances which could adversely affect microbial growth. The need for physical separation should be judged on the basis of the activities specific to the laboratory (e.g. number and type of tests carried out).

Laboratory equipment should not routinely be moved between areas to avoid accidental crosscontamination. In the molecular biology laboratory, dedicated pipettes, tips, centrifuges, tubes, etc. should be located in each work area (low–medium–high DNA working environments).

3.1.4. Space should be sufficient to allow work areas to be kept clean and tidy. The space required should be commensurate with the volume of analyses handled and the overall internal organization of the laboratory. The space should be as required according to the national regulations when available.

3.1.5. Workrooms should be appropriately ventilated at a suitable temperature. This may be done by natural or forced ventilation, or by the use of an air conditioner. Where, air conditioners are used, filters should be appropriate, inspected, maintained, and replaced according to the type of work being carried out.

3.1.6. Reduction of contamination may be achieved by having:

- smooth surfaces on walls, ceilings, floors, and benches (the smoothness of a surface is judged on how easily it may be cleaned). Tiles are not recommended as bench covering material;
- concave joints between the floor, walls, and ceiling;
- minimal opening of windows and doors while tests are being carried out;
- sun shades placed on the outside;
- easy access for cleaning of internal sun shades if it is impossible to fit them outside;
- fluid conveying pipes not passing above work surfaces unless placed in hermetically sealed casings;
- a dust-filtered air inlet for the ventilation system;
- separate hand-washing arrangements, preferably nonmanually controlled;
- cupboards up to the ceiling;
- no rough and bare wood;
- wooden surfaces of fixtures and fittings adequately sealed;
- stored items and equipment arranged to facilitate easy cleaning;
- no furniture, documents, or other items other than those strictly necessary for testing activities.

This list is not exhaustive, and not all examples will apply in every situation.

Ceilings, ideally, should have a smooth surface with flush lighting. When this is not possible (as with suspended ceilings and hanging lights), the laboratory should have documented evidence that they control any resulting risks to hygiene and have effective means of overcoming them, e.g. a surface-cleaning and inspection program.

3.1.7. Where laboratories are on manufacturing premises, personnel must be aware of the potential for contamination of production areas, and should demonstrate that they have taken appropriate measures to avoid any such occurrence.

3.2. Environmental Monitoring

3.2.1. An appropriate environmental monitoring program should be devised, including, for example, use of air settlement plates and surface swabbing. Acceptable background counts should be assigned and there should be a documented procedure for dealing with situations in which these limits are exceeded. Analysis of data should enable trends in levels of contamination to be determined.

3.3. Hygiene

3.3.1. There should be a documented cleaning program for laboratory fixtures, equipment, and surfaces. If should take into account the results of environmental monitoring and the possibility of crosscontamination. There should be a procedure for dealing with spillages.

3.3.2. Measures should be taken to avoid accumulation of dust, by the provision of sufficient storage space, by having minimal paperwork in the laboratory and by prohibiting plants and personal possessions from the laboratory work area.

3.3.3. Clothing appropriate to the type of testing being performed (including, if necessary, protection for hair, beard, hands, shoes, etc.) should be worn in the microbiological laboratory and removed before leaving the area. This is particularly important in the molecular biology laboratory, where for example, movement from an area of high DNA load to one of low DNA load may unwittingly introduce crosscontamination. In many laboratories a laboratory coat may suffice.

3.3.4. Adequate hand washing facilities should be available.

4. VALIDATION OF TEST METHODS

4.1. The validation of microbiological test methods should reflect actual test conditions. This may be achieved by using naturally contaminated products or products spiked with a predetermined level of contaminating organisms. The analyst should be aware that the addition of contaminating organisms to a matrix only mimics in a superficial way the presence of the naturally occurring contaminants. However, it is often the best and only solution available. The extent of validation necessary will depend on the method and application.

The laboratory shall validate standard methods applied to matrices not specified in the standard procedure.

4.2. Qualitative microbiological test methods, such as where the result is expressed in terms of detected/not detected and confirmation and identification procedures, should be validated by determining, if appropriate, the specificity, relative trueness, positive deviation, negative deviation, limit of detection, matrix effect, repeatability and reproducibility (see Appendix A for definitions).

4.3. For quantitative microbiological test methods, the specificity, sensitivity, relative trueness, positive deviation, negative deviation, repeatability, reproducibility and the limit of determination within a defined variability should be considered and, if necessary, quantitatively determined in assays. The differences due to the matrices must be taken into account when testing different types of samples. The results should be evaluated with appropriate statistical methods.

4.4. Laboratories shall retain validation data on commercial test systems (kits) used in the laboratory. These validation data may be obtained through collaborative testing and from validation data submitted by the manufacturers and subjected to third party evaluation (e.g. AOAC). If the validation data are not available or not wholly applicable, the laboratory shall be responsible for completing the validation of the method.

4.5. If a modified version of a method is required to meet the same specification as the original method, then comparisons should be carried out using replicates to ensure that this is the case. Experimental design and analysis of results must be statistically valid.

4.6. Even when validation is complete, the user will still need to verify on a regular basis that the documented performance can be met, e.g. by the use of spiked samples or reference materials incorporating relevant matrices.

5. UNCERTAINTY OF MEASUREMENT

5.1. The international definition for uncertainty of measurement is given in ISO International vocabulary of basic and general terms in metrology: 1993 (see Appendix B). The general approach to evaluating and expressing uncertainty in testing expected by European accreditation bodies is one based on the recommendations produced by the International Committee for Weights and Measures (CIPM), as described in the *Guide to the Expression of Uncertainty in Measurement*, 1995, ISO Geneva.

5.2. Microbiological tests generally come into the category of those that preclude the rigorous, metrologically and statistically valid calculation of uncertainty of measurement. It is generally appropriate to base the estimate of uncertainty on repeatability and reproducibility data alone, but ideally including bias (e.g. from proficiency testing scheme results). The individual components of uncertainty should be identified and demonstrated to be under control and their contribution to the variability of results evaluated. Some components (e.g. pipetting, weighing, and dilution effects) may be readily measured and easily evaluated to demonstrate a negligible contribution to overall uncertainty. Other components (e.g. sample stability and sample preparation) cannot be measured directly and their contribution cannot be evaluated in a statistical manner but their importance to the variability of results should be considered also.

5.3.   It is expected that accredited microbiological testing laboratories will have an understanding of the distributions of organisms within the matrices they test and take this into account when subsampling. However, it is not recommended that this component of uncertainty is included in estimates unless the client's needs dictate otherwise. The principal reasons for this are that uncertainty due to distribution of organisms within the product matrix is not a function of the laboratory's performance and may be unique to individual samples tested and because test methods should specify the sample size to be used taking into account poor homogeneity.

5.4.   The concept of uncertainty cannot be applied directly to qualitative test results such as those from detection tests or the determination of attributes for identification. Nevertheless, individual sources of variability, e.g. consistency of reagent performance and analyst interpretation, should be identified and demonstrated to be under control. Additionally, for tests where the limit of detection is an important indication of suitability, the uncertainty associated with the inocula used to determine the limit should be estimated and its significance evaluated. Laboratories should also be aware of the incidence of false positive and false negative results associated with the qualitative tests they use.

6.   EQUIPMENT MAINTENANCE, CALIBRATION, AND PERFORMANCE VERIFICATION

ISO 17025, paragraph 5.5

As part of its quality system, a laboratory is required to operate a documented program for the maintenance, calibration, and performance verification of its equipment.

6.1.   Maintenance

(Guidance on maintenance of equipment can be found in ISO 7218.)

6.1.1.   Maintenance of essential equipment shall be carried out at specified intervals as determined by factors such as the rate of use. Detailed records shall be kept. Examples of maintenance of equipment and intervals are given in Appendix F.

6.1.2. Attention should be paid to the avoidance of cross-contamination arising from equipment, e.g.:

- disposable equipment should be clean and sterile when appropriate;
- reused glassware should be properly cleaned and sterilized when appropriate;
- ideally, laboratories should have a separate autoclave for decontamination. However, one autoclave is acceptable provided that adequate precautions are taken to separate decontamination and sterilization loads, and a documented cleaning program is in place to address both the internal and external environment of the autoclave.

6.1.3. Typically, the following items of equipment will be maintained by cleaning and servicing, inspecting for damage, general verification and, where relevant, sterilizing:

- general service equipment—filtration apparatus, glass or plastic containers (bottles and test tubes), glass or plastic Petri dishes, sampling instruments, wires or loops of platinum, nickel/chromium, or disposable plastic;
- water baths, incubators, microbiological cabinets, autoclaves, homogenizers, fridges, and freezers;
- volumetric equipment—pipettes, automatic dispensers, and spiral platers;
- measuring instruments—thermometers, timers, balances, pH meters, and colony counters.

6.2. Calibration and Performance Verification

6.2.1. The laboratory must establish a program for the calibration and performance verification of equipment which has a direct influence on the test results. The frequency of such calibration and performance verification will be determined by documented experience and will be based on need, type, and previous performance of the equipment. Intervals between calibration and verification shall be shorter than the time, and the equipment has been found to take to drift outside acceptable limits. Examples of calibration intervals and typical performance checks for various laboratory instruments are given in Appendices D and E.

6.2.2. Temperature Measurement Devices

- a. Where temperature has a direct effect on the result of an analysis or is critical for the correct performance of equipment, temperature 'measuring' devices, e.g. liquid-in-glass thermometers, thermocouples, and platinum resistance thermometers (PRTs) used in incubators and autoclaves, shall be of an appropriate quality to achieve the accuracy required.

b. Calibration of devices shall be traceable to national or international standards for temperature. Where the accuracy permits devices that can be demonstrated to conform to an appropriate and nationally or internationally accepted manufacturing specification may be used (e.g. ISO 1770 for liquid-in-glass thermometers). Such devices may, for example, be used for monitoring storage fridges and freezers and also incubators and water baths where acceptable tolerance around the target temperature permits. Verification of the performance of such devices is necessary.

6.2.3. Incubators, Water Baths, and Ovens

The stability of temperature, uniformity of temperature distribution, and time required to achieve equilibrium conditions in incubators, water baths, and ovens, and temperature-controlled rooms shall be established initially and documented, in particular with respect to typical uses (for example position, space between, and height of stacks of Petri dishes). The constancy of the characteristics recorded during initial validation of the equipment shall be checked and recorded after each significant repair or modification.

Laboratories shall monitor the operating temperature of this type of equipment and retain records.

6.2.4. Autoclaves, Including Media Preparators

The following outlines the generally expected approach to calibration and the establishment and monitoring of performance. However, it is recognized that quantitative testing of materials and items processed by autoclaving, able to comment suitably on variation within and between batches may also provide equivalent assurance of quality.

a. Autoclaves should be capable of meeting specified time and temperature tolerances. Pressure cookers fitted only with a pressure gauge are not acceptable. Sensors used for controlling or monitoring operating cycles require calibration and the performance of timers verified.

b. Initial validation should include performance studies (spatial temperature distribution surveys) for each operating cycle and each load configuration used in practice. This process must be repeated after significant repair or modification (e.g. replacement of thermoregulator probe or programmer, loading arrangements, and operating cycle) or where indicated by the results of quality control checks on media. Sufficient temperature sensors should be positioned within the load (e.g. in containers filled with liquid/medium) to enable location differences to be demonstrated. In the case of media preparators, where uniform heating cannot be demonstrated by other means, the use of two sensors, one adjacent to the control probe and the other remote from it, would generally be considered appropriate. Validation and revalidation should consider the suitability of come-up and -down times as well as time at sterilization temperature.

c. Clear operating instructions should be provided based on the heating profiles determined for typical uses during validation/revalidation. Acceptance/rejection criteria should be established and records of autoclave operations, including temperature and time, maintained for every cycle.

d. Monitoring may be achieved by one of the following:

   i. using a thermocouple and recorder to produce a chart or printout;

   ii. direct observation and recording of maximum temperature achieved and time at that temperature.

In addition to directly monitoring the temperature of an autoclave, the effectiveness of its operation during each cycle may be checked by the use of chemical or biological indicators for sterilization/decontamination purposes. Autoclave tape or indicator strips should be used only to show that a load has been processed, not to demonstrate completion of an acceptable cycle.

6.2.5. Weights and Balances

Weights and balances shall be calibrated traceably at regular intervals (according to their intended use).

6.2.6. Volumetric Equipment

a. Volumetric equipment such as automatic dispensers, dispenser/diluters, mechanical hand pipettes, and disposable pipettes may all be used in the microbiology laboratory. Laboratories should carry out initial verification of volumetric equipment and then make regular checks to ensure that the equipment is performing within the required specification. Verification should not be necessary for glassware which has been certified to a specific tolerance. Equipment should be checked for the accuracy of the delivered volume against the set volume (for several different settings in the case of variable volume instruments) and the precision of the repeat deliveries should be measured.

b. For 'single-use' disposable volumetric equipment, laboratories should obtain supplies from companies with a recognized and relevant quality system. After initial validation of the suitability of the equipment, it is recommended that random checks on accuracy are carried out. If the supplier has not a recognized quality system, laboratories should check each batch of equipment for suitability.

6.2.7. Other Equipment

Conductivity meters, oxygen meters, pH meters, and other similar instruments should be verified regularly or before each use. The buffers used for verifications purposes should be stored in appropriate conditions and should be marked with an expiry date.

Where humidity is important to the outcome of the test, hygrometers should be calibrated, the calibration being traceable to national or international standards.

Timers, including the autoclave timer, should be verified using a calibrated timer or national time signal.

Where centrifuges are used in test procedures, an assessment should be made of the criticality of the centrifugal force. Where it is critical, the centrifuge will require calibration.

7. REAGENTS AND CULTURE MEDIA

ISO 17025, paragraph 4.6 and 5.5

7.1. Reagents

Laboratories should ensure that the quality of reagents used is appropriate for the test concerned. They should verify the suitability of each batch of reagents critical for the test, initially and during its shelf life, using positive and negative control organisms, which are traceable to recognized national or international culture collections.

7.2. In-house Prepared Media

7.2.1. The suitable performance of culture media, diluents, and other suspension fluids prepared in-house should be checked, where relevant, with regard to:

- recovery or survival maintenance of target organisms,
- inhibition or suppression of nontarget organisms,
- biochemical (differential and diagnostic) properties,
- physical properties (e.g. pH, volume, and sterility).

Quantitative procedures for evaluation of recovery or survival are to be preferred (see also ISO 11133 Part 1 and 2).

7.2.2. Raw materials (both commercial dehydrated formulations and individual constituents) should be stored under appropriate conditions, e.g. cool, dry, and dark. All containers, especially those for dehydrated media, should be sealed tightly. Dehydrated media that are caked or cracked or show a color change should not be used. Distilled, deionized, or reverse osmosis produced water, free from bactericidal, inhibitory or interfering substances, should be used for preparation unless the test method specifies otherwise.

7.2.3. Shelf life of prepared media under defined storage conditions shall be determined and verified.

7.3. Ready-to-Use Media

7.3.1. All media (diluents and other suspension fluids) procured ready to use or partially complete require validating before use. Evaluation of performance in recovery or survival of target organisms and the inhibition or suppression of nontarget organisms needs to be fully quantitative; attributes (e.g. physical and biochemical properties) should be evaluated using objective criteria.

7.3.2. As part of the validation, the user laboratory needs to have adequate knowledge of the manufacturer's quality specifications, which include at least the following:

- Name of the media and list of components, including any supplements
- Shelf life and the acceptability criteria applied
- Storage conditions
- Sample regime/rate
- Sterility check
- Check of growth of target and nontarget control organisms used (with their culture collection references) and acceptability criteria
- Physical checks and the acceptability criteria applied
- Date of issue of specification

7.3.3. Batches of media should be identifiable. Each one received should be accompanied by evidence that it meets the quality specification. The user laboratory should ensure that it will be notified by the manufacturer of any changes to the quality specification.

7.3.4. Where the manufacturer of media procured ready to use or partially complete is covered by a recognized quality system (e.g. ISO 9000-series registered), checks by the user laboratory of conformance of supplies with the specification defined through initial validation may be applied in accordance with the expectation of consistency. In other circumstances, adequate checks would be necessary on every batch received.

7.4. Labeling

Laboratories shall ensure that all reagents (including stock solutions), media, diluents, and other suspending fluids are adequately labeled to indicate, as appropriate, identity, concentration, storage conditions, preparation date, validated expiry date, and/or recommended storage periods. The person responsible for preparation should be identifiable from records.

8. REFERENCE MATERIALS AND REFERENCE CULTURES

ISO 17025, paragraph 5.6.3

8.1. Reference Materials

Reference materials and certified reference materials (see definition in Appendix A) provide essential traceability in measurements and are used, for example;

- to demonstrate the accuracy of results,
- to calibrate equipment,
- to monitor laboratory performance,
- to validate methods, and
- to enable comparison of methods.

If possible, reference materials should be used in appropriate matrices.

8.2. Reference Cultures

8.2.1. Reference cultures are required for establishing acceptable performance of media (including test kits), for validating methods, and for assessing/ evaluating on-going performance. Traceability is necessary, for example, when establishing media performance for test kit and method validations. To demonstrate traceability, laboratories must use reference strains of micro-organisms obtained directly from a recognized national or international collection, where these exist. Alternatively, commercial derivatives for which all relevant properties have been shown by the laboratory to be equivalent at the point of use may be used.

8.2.2. Following the guidance in ISO 11133–1, reference strains may be subcultured once to provide reference stocks. Purity and biochemical checks should be made in parallel as appropriate. It is recommended to store reference stocks in aliquots either deep-frozen or lyophilized. Working cultures for routine use should be primary subcultures from the reference stock (see Appendix C on preparation of working stocks). If reference stocks have been thawed, they must not be refrozen and reused.

8.2.3. Working stocks should not be subcultured unless it is required and defined by a standard method or laboratories can provide documentary evidence that there has been no change in any relevant property.

Working stocks shall not be subcultured to replace reference stocks. Commercial derivatives of reference strains may only be used as working cultures.

9. SAMPLING

ISO 17025, paragraph 5.7

9.1. In many cases, testing laboratories are not responsible for primary sampling to obtain test items. Where they are responsible, it is strongly recommended that this sampling be covered by quality assurance and ideally by accreditation.

9.2.   Transport and storage should be under conditions that maintain the integrity of the sample (e.g. chilled or frozen where appropriate). The conditions should be monitored and records kept. Where appropriate, responsibility for transport, storage between sampling, and arrival at the testing laboratory shall be clearly documented. Testing of the samples should be performed as soon as possible after sampling and should conform to relevant standards and/or national/international regulations.

9.3.   Sampling should only be performed by trained personnel. It should be carried out aseptically using sterile equipment. Environmental conditions for instance air contamination and temperature should be monitored and recorded at the sampling site. Time of sampling should be recorded.

10.   SAMPLE HANDLING AND IDENTIFICATION

ISO 17025, paragraphs 5.7 and 5.8

10.1.   Microbial flora may be sensitive to factors such as temperature or duration of storage and transport, so it is important to check and record the condition of the sample on receipt by the laboratory.

10.2.   The laboratory should have procedures that cover the delivery of samples and sample identification. If there is insufficient sample or the sample is in poor condition due to physical deterioration, incorrect temperature, torn packaging, or deficient labeling, the laboratory should consult with the client before deciding whether to test or refuse the sample. In any case, the condition of the sample should be indicated on the test report.

10.3.   The laboratory should record all relevant information and particularly the following information:

    a.   date and, where relevant, the time of receipt;
    b.   condition of the sample on receipt and when necessary, temperature;
    c.   characteristics of the sampling operation (sampling date, sampling conditions, etc.).

10.4.   Samples awaiting test shall be stored under suitable conditions to minimize changes to any microbial population present. Storages conditions should be defined and recorded.

10.5.   The packaging and labels from samples may be highly contaminated and should be handled and stored with care so as to avoid any spread of contamination.

10.6.   Subsampling by the laboratory immediately prior to testing is considered as part of the test method. It should be performed according to national or international standards, where they exist, or by validated in-house methods. Subsampling procedures should be designed to take account uneven distribution of micro-organisms (general guidance given in ISO 6887 and ISO 7218).

10.7. A procedure for the retention and disposal of samples shall be written. Samples should be stored until the test results are obtained, or longer if required. Laboratory sample portions that are known to be highly contaminated should be decontaminated prior to being discarded (see Sec. 11.1).

11. DISPOSAL OF CONTAMINATED WASTE

11.1. The correct disposal of contaminated materials may not directly affect the quality of sample analysis, although procedures should be designed to minimize the possibility of contaminating the test environment or materials. However, it is a matter of good laboratory management and should conform to national/international environmental or health and safety regulations (see also ISO 7218).

12. QUALITY ASSURANCE OF RESULTS/QUALITY CONTROL OF PERFORMANCE

ISO 17025, paragraph 5.9

12.1. Internal Quality Control

12.1.1. Internal quality control consists of all the procedures undertaken by a laboratory for the continuous evaluation of its work. The main objective is to ensure the consistency of results day-to-day and their conformity with defined criteria.

12.1.2. A program of periodic checks is necessary to demonstrate that variability (i.e. between analysts and between equipment or materials, etc.) is under control. All tests included in the laboratory's scope of accreditation need to be covered. The program may involve:

- the use of spiked samples
- the use of reference materials (including proficiency testing scheme materials)
- replicate testing
- replicate evaluation of test results.

The interval between these checks will be influenced by the construction of the program and by the number of actual tests. It is recommended that, where possible, tests should incorporate controls to monitor performance.

12.1.3. In special instances, a laboratory may be accredited for a test that it is rarely called on to do. It is recognized that in such cases an ongoing internal quality control program may be inappropriate and that a scheme for demonstrating satisfactory performance which is carried out in parallel with the testing, may be more suitable.

12.2.   External Quality Assessment (Proficiency Testing)

12.2.1.   Laboratories should regularly participate in proficiency testing which are relevant to their scope of accreditation, preference should be given to proficiency testing schemes which use appropriate matrices. In specific instances, participation may be mandatory.

12.2.2.   Laboratories should use external quality assessment not only to assess laboratory bias but also to check the validity of the whole quality system.

13.   TEST REPORTS

ISO 17025, paragraph 5.10

13.1.   If the result of the enumeration is negative, it should be reported as "not detected for a defined unit" or "less than the detection limit for a defined unit." The result should not be given as "zero for a defined unit" unless it is a regulatory requirement. Qualitative test results should be reported as "detected/not detected in a defined quantity or volume." They may also be expressed as "less than a specified number of organisms for a defined unit" where the specified number of organisms exceeds the detection limit of the method and this has been agreed with the client.

13.2.   Where an estimate of the uncertainty of the test result is expressed on the test report, any limitations (particularly if the estimate does not include the component contributed by the distribution of micro-organisms within the sample) have to be made clear to the client.

APPENDIX A

Glossary of Terms

Calibration	Set of operations that establish under specified conditions, the relationship between values of quantities indicated by a measuring instrument or measuring system, or values represented by a material measure or a reference material, and the corresponding values realized by standards.

NOTES:
(1)   The result of a calibration permits either the assignment of values of measure ands to the indications or the determination of corrections with respect to indications.

(2) A calibration may also determine other metrological properties such as the effect of influence quantities.

(3) The result of a calibration may be recorded in a document, sometimes called a calibration certificate or a calibration report.
[VIM: 1993 ISO International vocabulary of basic and general terms in metrology]

Certified reference material	Reference material, accompanied by a certificate, one or more of whose property values are certified by a procedure, which establishes traceability to an accurate realization of the unit in which the property values are expressed, and for which each certified value is accompanied by an uncertainty at a stated level of confidence. [ISO Guide 30:1992]
Limit of determination	Applied to quantitative microbiological tests—the lowest number of micro-organisms within a defined variability that may be determined under the experimental conditions of the method under evaluation.
Limit of detection	Applied to qualitative microbiological tests—the lowest number of micro-organisms that can be detected, but in numbers that cannot be estimated accurately.
Negative deviation	Occurs when the alternative method gives a negative result without confirmation when the reference method gives a positive result. This deviation becomes a false negative result when the true result can be proved as being positive.
Positive deviation	Occurs when the alternative method gives a positive result without confirmation when the reference method gives a negative result. This deviation becomes a false positive result when the true result can be proved as being negative.
Reference cultures	Collective term for reference strain, reference stocks, and working cultures.
Reference strains	Micro-organisms defined at least to the genus and species level, catalogued and described according to its characteristics and preferably stating its origin. [ISO 11133-1:2000] Normally obtained from a recognized national or international collection.
Reference material	Material or substance one or more of whose property values are sufficiently homogeneous and well established to be used for the calibration of an apparatus, the assessment of a

Continued

	measurement method, or for assigning values to materials. [ISO Guide 30:1992]
Reference method	Thoroughly investigated method, clearly and exactly describing the necessary conditions and procedures, for the measurement of one or more property values that has been shown to have accuracy and precision commensurate with its intended use and that can therefore be used to assess the accuracy of other methods for the same measurement, particularly in permitting the characterization of a reference material. Normally a national or international standard method
Reference stocks	A set of separate identical cultures obtained by a single subculture from the reference strain. [ISO 11133-1:2000]
Relative trueness	The degree of correspondence of the results of the method under evaluation to those obtained using a recognized reference method.
Repeatability	Closeness of the agreement between the results of successive measurements of the same measure and under the same conditions of measurement. [VIM: 1993 ISO International vocabulary of basic and general terms in metrology]
Reproducibility	Closeness of the agreement between the results of measurements of the same measurand carried out under changed conditions of measurement. [VIM: 1993 ISO International vocabulary of basic and general terms in metrology]
Sensitivity	The fraction of the total number of positive cultures or colonies correctly assigned in the presumptive inspection. [ISO 13843:2000]
Specificity	The fraction of the total number of negative cultures or colonies correctly assigned in the presumptive inspection. [ISO 13843:2000]
Working culture	A primary subculture from a reference stock. [ISO 11133-1:2000].
Validation	Confirmation, through the provision of objective evidence, that the requirements for a specific intended use or application have been fulfilled. [ISO 9000:2000]
Verification	Confirmation, through the provision of objective evidence, that specified requirements have been fulfilled. [ISO 9000:2000]

APPENDIX B

References

1. ISO/IEC 17025, General requirements for the competence of testing and calibration laboratories.
2. ISO 7218, Microbiology of food and animal feeding stuffs—general rules for microbiological examinations.
3. ISO 6887–1, Preparation of dilutions.
4. ISO Guide 30, Terms and definitions used in connection with reference materials.
5. ISO 9000, Quality management systems–fundamentals and vocabulary.
6. VIM: 1993, ISO international vocabulary of basic and general terms in metrology.
7. ISO (CIPM):1995, Guide to the expression of uncertainty in measurements.
8. Draft ISO/DIS 16140, Food microbiology. Protocol for the validation of alternative methods.
9. ISO 13843, Water quality–guidance on validation of microbiological methods.
10. ISO 11133–1, Microbiology of food and animal feeding stuffs. Guidelines on preparation and production of culture media. Part 1–General guidelines on quality assurance for the preparation of media in the laboratory.
11. Draft ISO/FDIS 11133–2, Microbiology of food and animal feeding stuffs. Guidelines: on preparation and production of culture media. Part 2–Practical guidelines on performance testing on culture media.
12. EN 12741, Biotechnology—laboratories for research, development and analysis—guidance for biotechnology laboratory operations.

APPENDIX C

General Use of Reference Cultures

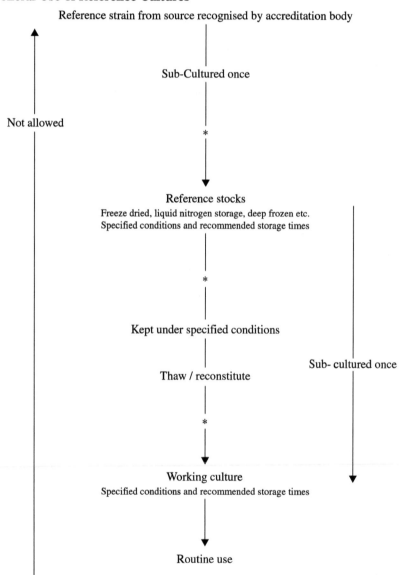

Reference strain from source recognised by accreditation body

Sub-Cultured once

Not allowed

*

Reference stocks

Freeze dried, liquid nitrogen storage, deep frozen etc.
Specified conditions and recommended storage times

*

Kept under specified conditions

Sub- cultured once

Thaw / reconstitute

*

Working culture

Specified conditions and recommended storage times

Routine use

* Parellel purity checks and biochemical tests as appropriate

All parts of the process shall be fully documented and detailed records of all stages must be maintained

APPENDIX D

Guidance of Calibration and Calibration Checks

This information is provided for guidance purposes and the frequency will be based on the need, type, and previous performance of the equipment

Type of equipment	Requirement	Suggested frequency
Reference thermometers (liquid-in-glass)	Full traceable recalibration Single point (e.g. ice-point check)	Every 5 years Annually
Reference thermocouples	Full traceable recalibration Check against reference thermometer	Every 3 years Annually
Working thermometers & working thermocouples	Check against reference thermometer at ice-point and/or working temperature range	Annually
Balances	Full traceable calibration	Annually
Calibration weights	Full traceable calibration	Every 5 years
Check weight(s)	Check against calibrated weight or check on balance immediately following traceable calibration	Annually
Volumetric glassware	Gravimetric calibration to required tolerance	Annually
Microscopes	Traceable calibration of stage micrometer (where appropriate)	Initially
Hygrometers	Traceable calibration	Annually
Centrifuges	Traceable calibration or check against an independent tachometer, as appropriate	Annually

APPENDIX E

Guidance on Equipment Validation and Verification of Performance

This information is provided for guidance purposes and the frequency will be based on the need, type, and previous performance of the equipment

Type of equipment	Requirement	Suggested frequency
Temperature controlled equipment (incubators, baths, fridges, and freezers)	(a) Establish stability and uniformity of temperature (b) Monitor temperature	(a) Initially, every 2 years and after repair/modification (b) Daily/each use
Sterilizing ovens	(a) Establish stability and uniformity of temperature (b) Monitor temperature	(a) Initially, every 2 years and after repair/modification (b) Each use
Autoclaves	(a) Establish characteristics for loads/cycles (b) Monitor temperature/time	(a) Initially, every 2 years and after repair/modification (b) Each use
Safety cabinets	(a) Establish performance (b) Microbiological monitoring (c) Air flow monitoring	(a) Initially, every year and after repair/modification (b) Weekly (c) Each use
Laminar air flow cabinets	(a) Establish performance (b) Check with sterility plates	(a) Initially, and after repair/modification (b) Weekly
Timers	Check against national time signal	Annually
Microscopes	Check alignment	Daily/each use
pH meters	Adjust using at least two buffers of suitable quality	Daily/each use
Balances	Check zero, and reading against check weight	Daily/each use
Deionizers and reverse osmosis units	(a) Check conductivity (b) Check for microbial contamination	(a) Weekly (b) Monthly
Gravimetric diluters	(a) Check weight of volume dispensed (b) Check dilution ratio	(a) Daily (b) Daily
Media dispensers	Check volume dispensed	Each adjustment or replacement
Pipettors/pipettes	Check accuracy and precision of volume dispensed	Regularly (to be defined by taking account of the frequency and nature of use)

Continued

Type of equipment	Requirement	Suggested frequency
Spiral platers	(a) Establish performance against conventional method	(a) Initially and annually
	(b) Check styles condition and the start and end points	(b) Daily/each use
	(c) Check volume dispensed	(c) Monthly
Colony counters	Check against number counted manually	Annually
Centrifuges	Check speed against a calibrated and independent tachometer	Annually
Anaerobic jars/incubators	Check with anaerobic indicator	Each use
Laboratory environment	Monitor for airborne and surface microbial contamination using, e.g. air samplers, settle plates, contact plates, or swabs	Weekly

APPENDIX F

Guidance on Maintenance of Equipment

This information is provided for guidance purposes and the frequency will be based on the need, type, and previous performance of the equipment

Type of equipment	Requirement	Suggested frequency
(a) Incubators	Clean and disinfect internal surfaces	(a) Monthly
(b) Fridges		(b) When required (e.g. every 3 months)
(c) Freezers, ovens		(c) When required (e.g. annually)
Water baths	Empty, clean, disinfect, and refill	Monthly, or every 6 months if biocide used
Centrifuges	(a) Service	(a) Annually
	(b) Clean and disinfect	(b) Each use
Autoclaves	(a) Make visual checks of gasket, clean/drain chamber	(a) Regularly, as recommended by manufacturer

Continued

APPENDIX F (*Continued*)

Type of equipment	Requirement	Suggested frequency
	(b) Full service	(b) Annually or as recommended by manufacturer
	(c) Safety check of pressure vessel	(c) Annually
Safety cabinets	Full service and	Annually or as
Laminar flow cabinets	mechanical check	recommended by manufacturer
Microscopes	Full maintenance service	Annually
pH meters	Clean electrode	Each use
Balances, gravimetric	(a) Clean	(a) Each use
diluters	(b) Service	(b) Annually
Stills	Clean and descale	As required (e.g. every 3 months)
Deionizers, reverse osmosis units	Replace cartridge/membrane	As recommended by manufacturer
Anaerobic jars	Clean/disinfect	After each use
Media dispensers, volumetric equipment, pipettes, and general service equipment	Decontaminate, clean, and sterilize as appropriate	Each use
Spiral platers	(a) Service	(a) Annually
	(b) Decontaminate, clean, and sterilize	(b) Each use
Laboratory	(a) Clean and disinfect working surfaces	(a) Daily, and during use
	(b) Clean floors, disinfect sinks, and basins	(b) Weekly
	(c) Clean and disinfect other surfaces	(c) Every 3 months

Appendix F

EPA MANUAL FOR CERTIFICATION OF LABORATORIES ANALYZING DRINKING WATER

Criteria and Procedures Quality Assurance

CHAPTER I

INTRODUCTION

Public water systems serving at least 25 persons or having at least 15 service connections must comply with the Safe Drinking Water Act (SDWA) and the requirements of the National Primary Drinking Water Regulations (NPDWR) (40 CFR 141). Section 1401(1) (D) of the act defines a National Primary Drinking Water Regulation to include "criteria and procedures ... [for] quality control and testing procedures to insure compliance ..." 40 CFR 142 sets implementation requirements.

The regulations governing primacy at 40 CFR 142.10(b) (4) require a State that has primary enforcement responsibility (primacy) to have laboratory facilities available (the Principal State Laboratory) which have been certified by EPA. In addition, the regulations governing certification (40 CFR 141.28) require that all testing for compliance purposes be performed by laboratories certified by the State or by EPA except that turbidity, free chlorine residual, temperature, and pH may be performed by anyone acceptable to the State. This manual is intended to assist EPA in implementing 40 CFR 142.10(b) (4) by specifying criteria and procedures for certifying principal State laboratories. States with primacy may also choose to use equivalent criteria and procedures similar to those in this manual for their own certification programs.

To obtain and maintain primacy, a State must comply with 40 CFR 142.10, which includes the following provisions:

The establishment and maintenance of a State program for the certification of laboratories conducting analytical measurements of drinking water contaminants pursuant to the requirements of the State primary drinking water regulations including the designation by the State of a laboratory officer, or officers, certified by the administrator, as the official(s) responsible for the State's certification program. The requirements of this paragraph may be waived by the Administrator for any State where all analytical measurements required by the State's primary drinking water regulations are conducted at laboratories operated by the State and certified by the agency. [40 CFR 142.10(b) (3) (i)]

Assurance of the availability to the State of laboratory facilities certified by the administrator and capable of performing analytical measurements of all contaminants specified in the State primary drinking water regulations ... [40 CFR 142.10(b) (4)]

Reference to the Administrator of EPA also refers to his or her designee.

The EPA laboratory certification program extends to its regional laboratories, laboratories on Federal Indian Lands, principal State laboratories in primacy States, and laboratories that perform analyses under the Safe Drinking Water Act in States without primacy. If all analyses are not performed in principal State laboratories, primacy States must have a certification program for certifying other drinking water laboratories.

EPA's National Exposure Research Laboratory in Cincinnati, Ohio (NERL-Ci) is responsible for determining the certification status for EPA regional laboratories in microbiology and chemistry. The National Exposure Research Laboratory in Las Vegas (NERL-LV) has this responsibility for radiochemistry. Regional certification officers are responsible for the certification of the principal State laboratory in each primacy State and are also responsible for certifying all laboratories on Federal Indian Lands and in nonprimacy States. Primacy States with certification programs are responsible for certifying the other drinking water laboratories in their State (i.e., laboratories other than the principal State laboratory).

Regional laboratories must successfully analyze a set of performance evaluation (PE) samples at least annually for all regulated contaminants for which they wish to be certified and pass an on-site evaluation at least every 3 years. At least annually, principal State laboratories must successfully analyze a complete set of unknown PE samples from a source acceptable to the region for the contaminants included in the regulations which the State has adopted, and pass an on-site evaluation every 3 years. For radiochemistry certification, satisfactory performance on additional PE samples may also be required. EPA policy requires the use of the criteria in this manual for the on-site audits of the regional and principal State laboratories.

Chapter II describes the responsibilities of each of the parties involved in the certification program. Chapter III describes how the program operates. Chapters IV, V, and VI cover the technical criteria to be used during the on-site evaluation of a laboratory for chemistry, microbiology, and radiochemistry, respectively. Optional audit forms are also included in Chapters IV, V, and VI. The appendices include the following: a recommended protocol and format, for conducting on-site laboratory evaluations, which may be used by the laboratory auditors; frequently used abbreviations and definitions; EPA's policy on third-party auditors; a list of contaminants a principal State laboratory must have the capability to analyze; a list of contaminants in proposed rules; a list of unregulated chemicals for which systems must monitor under §1445 of the Safe Drinking Water Act; optional record keeping and data audit procedures; and recommended chain-of-custody procedures to be used if necessary.

CHAPTER II

RESPONSIBILITIES

The success of the laboratory certification program depends upon co-operation among the organizations responsible for its implementation. Within the agency, responsibilities for laboratory certification are shared by the Office of Ground Water and Drinking Water (OGWDW), the Office of Research and Development (ORD), and the Regional Offices.

1. OFFICE OF GROUND WATER AND DRINKING WATER AND OFFICE OF RESEARCH AND DEVELOPMENT

The OGWDW and ORD share the responsibility for developing and implementing the national certification program for laboratories that analyze drinking water samples and for implementing the Safe Drinking Water Act. These responsibilities include the following:

- Propose and promulgate regulations;
- Access national laboratory capacity and capability;
- Review the EPA Regional certification programs annually and evaluate the resources and personnel available in each EPA Region to carry out the certification program;
- Develop guidance and respond to questions and comments;
- Develop technical and administrative certification criteria to support future regulations;
- Assess the capacity and capability of the EPA Regional laboratories;
- Revise this manual when necessary;
- Conduct triennial on-site audits of each regional laboratory for chemistry and microbiology and principal State laboratories for radiochemistry (if requested by the region);

- Prepare and distribute performance evaluation (PE) samples (Water Supply Studies) for regulated chemical, microbiological, and radiological contaminants semiannually and evaluate and distribute the results of these studies;
- Develop and participate in training courses to support the certification program;
- Provide technical assistance to EPA and the States;
- Develop and evaluate methods for the analysis of drinking water contaminants.

The National Exposure Research Laboratory in Las Vegas is the lead organization for managing the certification program for laboratories performing radiochemical analyses. At the request of a region, NERL-LV is responsible for conducting triennial on-site audits of principal State laboratory systems for radiochemistry. In these cases, NERL-LV must report the results of its on-site audits to the responsible regional certification authority, who has the final authority to determine certification status.

2. EPA REGIONS

The regions oversee the certification programs in the States. The regions' responsibilities are:

- Determine the certification status for the principal State laboratory system in each primacy State;
- Co-ordinate NERL water supply performance evaluation studies with laboratories in the region (some regions have delegated this responsibility to the States);
- Perform an annual review of State certification programs and performance evaluation results and monitor the adequacy of State programs for certifying laboratories, as described in Chapter III;
- Provide technical assistance to the States' EPA-certified drinking water laboratories, as needed;
- Manage the certification program for drinking water laboratories in nonprimacy States and on Federal Indian Lands using the criteria in this manual.

This last duty may be performed by the State, but the region must retain responsibility for the on-site evaluation of the designated principal State laboratory. Drinking water laboratories may be evaluated by the region or under a region-approved program carried out by a designated State program. In either case, this manual must be the basis for the on-site audits, conducted by EPA, of principal State laboratories, laboratories on Federal Indian Lands, and drinking water laboratories in nonprimacy States.

The regional laboratory should maintain certification for as many regulated contaminants as its resources permit. It enhances both EPA's technical assistance capabilities and credibility with those it certifies. It also ensures the laboratory capability to analyze samples for possible enforcement actions and for States which do not have primacy. Reciprocal agreements with other regions to share scarce resources may be needed.

3. PRIMACY STATES

Primacy States are required to establish and maintain a State program for the certification of laboratories conducting analyses of drinking water compliance samples, unless all compliance samples are analyzed in the State laboratory.

The States must designate a certification officer, or officers, certified by the EPA administrator or his or her designee as the person responsible for the certification program.

States are responsible for the certification of the public and private laboratories in their State. This includes auditing the laboratories and reviewing the PE data. States should also provide technical assistance to laboratories. They may also choose to certify laboratories outside their State either by an on-site evaluation or reciprocity.

CHAPTER III

IMPLEMENTATION

1. EVALUATION OF CERTIFICATION PROGRAMS

The regions and the OGWDW should monitor the certification programs under their preview annually. The adequacy of programs for certifying laboratories is evaluated by assessing each program's scope, staffing, resources, policy, procedures, and effectiveness. This should be done in person during an on-site audit when possible and at least by means of a questionnaire in the other years. The adequacy of these essential program elements is evaluated by:

- Reviewing the program's plan, responsibilities, organizational structure, staff (including educational background and experience), scope and description of the certification process, downgrading criteria and processes, and use of PE samples;
- Requesting an annual program report that includes program highlights, training, continuing education efforts, number of on-site evaluations performed, listing of laboratories certified by discipline or contaminant, and any certification downgrading or upgrading actions along with reasons for those actions;
- Observing on-site audits of drinking water laboratories to allow EPA Regional certification officers to evaluate specific elements of the State certification program;
- Encouraging State and Regional laboratory auditors to observe on-site audits of their own and other laboratories as on-the-job training;
- Sponsoring annual meetings of certification officers to discuss program issues, policies, and problems. Key regional, NERL, and OGWDW and State personnel should be invited to participate.

2. REQUIREMENTS FOR CERTIFICATION OF LABORATORIES

In order to be eligible to analyze compliance samples under the Safe Drinking Water Act, regional and principal State laboratories should meet the minimum criteria specified in this manual, should pass an on-site audit at least once every 3 years, and for chemistry must satisfactorily analyze a set of PE samples or other unknown test samples annually. EPA policy also requires that microbiology and radiochemistry laboratories also successfully analyze PE samples. For those laboratories certified for radiochemistry, satisfactory performance on additional PE samples may also be necessary.

3. INDIVIDUAL(S) RESPONSIBLE FOR THE CERTIFICATION PROGRAM

The NERL (Ci) and NERL (LV) are responsible for certifying the regional laboratories, the regions are responsible for certifying their state laboratories, and the States are responsible for certifying private laboratories.

The certification program personnel in each region should consist of the certification authority(s) (CA), the certification program manager(s), and a certification team comprised of certification officers (COs) and technical experts. Additional third-party auditors and experts may be used. However, third parties must have no authority for certification decisions. Third-party auditing is discussed in Appendix D.

The *Certification Authority* (CA) is the person who has signature authority for all certification decisions. This is the Director, NERL-Ci and NERL-LV and in the regions, it is the Regional Administrator. The CA may delegate this authority to a lower level.

Each EPA Regional Administrator or designee should appoint, by memo, an individual to manage the drinking water laboratory certification activities in that region. This person is the *Certification Program Manager*. This individual(s) should be experienced in quality assurance, hold at least a bachelor's degree or have equivalent experience in either microbiology, chemistry, radiochemistry or a related field and have sufficient administrative and technical stature to be considered a peer of the directors of the laboratories being audited. It is the responsibility of the certification program manager to assure that the regional certification program meets the requirements of this manual.

40 CFR 142.10(b) (3) (i) requires primacy States to designate a person certified by the administrator as the official responsible for the State's certification program. This person would be the State certification authority.

4. ON-SITE LABORATORY AUDIT TEAM

The certification program manager should establish one or more teams of certification officers and auditors to audit laboratories. It is the responsibility of these teams to perform the on-site laboratory audits, review the laboratory PE data, and make recommendations to the CA concerning the certification status of the laboratories.

Team members should be experienced professionals, hold at least a bachelor's degree or equivalent education/experience in the discipline (chemistry, radiochemistry, microbiology, or a related field) for which they certify, and have recent laboratory experience.

Team members should also have experience in laboratory evaluation and quality assurance, be familiar with the drinking water regulations and data reduction and reporting techniques, be technically conversant with the analytical techniques being evaluated and be able to communicate effectively, both orally and in writing.

The on-site team should include at least one CO knowledgeable in each area being audited (e.g. inorganic and organic chemistry, radiochemistry, and microbiology). EPA policy requires COs to successfully complete the appropriate EPA laboratory certification course.

State certification programs may employ third-party auditors who have passed the EPA certification officers' training course (see Appendix D) and meet all the qualifications listed above. Although these third parties may be used to assist the State certification officers, they may not make final certification decisions. These decisions rest with the State.

In areas where experience does not exist within the certification team (e.g. asbestos), outside expert assistance may be obtained in the needed areas to assist the on-site team. Outside experts who have not attended the EPA certification officer training must be accompanied by a certification officer. When using third-party experts, it is critical to avoid conflicts of interest. A third-party auditor who in any way stands to benefit by the certification status of the laboratory audited may not be used.

5. PLANS FOR CERTIFICATION OF LABORATORIES

The certification authority should develop plans for certifying drinking water laboratories under her/his authority. Written plans should include the following:

- Documentation of certification authority and certification officers and their education/experience;
- Schedules of laboratories to be audited;
- Specific types of analyses to be examined;
- Protocol to be followed;
- Strategy for assessing laboratory performance (e.g. PE samples, data audits, etc.);
- Plans for providing technical assistance to laboratories which need upgrading.

6. PRINCIPAL STATE LABORATORIES

To receive and retain primacy, the State must have the availability of laboratory facilities certified by the EPA and capable of performing analytical measurements for all the contaminants specified in the State Primary Drinking Water Regulations. This laboratory or laboratories are considered the Principal State Laboratory and are certified by the region either directly or through a reciprocal agreement.

7. CERTIFICATION PROCESS

The certification process begins when the laboratory director makes a formal request to the certification authority to be certified. This application may be one of the following:

- A request for first-time certification for microbiology, chemistry, or radiochemistry;
- A request for certification to analyze additional or newly regulated contaminants;
- A request to reapply for certification after correction of deficiencies which resulted in the downgrading/revocation of certification status.

The response to a formal application for any of the above requests should be given within 30 days. At this time, a mutually agreeable date and time should be set for the on-site laboratory audit. The recommended protocol for conducting these audits is given in Appendix B. If this is not a first-time certification, an on-site visit may not be necessary.

Drinking water laboratories should verify that they plan to analyze drinking water samples when they request certification. If a laboratory has not been analyzing drinking water samples and does not plan to, the region or State may choose not to renew their certification.

8. TYPES OF CERTIFICATION

After review of PE sample results and an on-site visit, the certification authority should provide a written report within 45 days and classify the laboratory for each contaminant or group of contaminants according to the following rating scheme:

- *Certified*—laboratory that meets the minimum requirements of this manual and all applicable regulatory requirements. "Certified" status may not be granted to any laboratory that has not met performance criteria specified in the National Primary Drinking Water Regulations (NPDWR) and within the policy required by their certification authority.
- *Provisionally certified*—a laboratory that has deficiencies but demonstrates its ability to consistently produce valid data within the acceptance limits specified in the NPDWR, and within the policy required

by their certification authority. A provisionally certified laboratory may analyze drinking water samples for compliance purposes. Provisional certification may not be given if the evaluation team believes that the laboratory cannot perform an analysis within the acceptance limits specified in the regulations.

- *Not certified*—a laboratory that possesses major deficiencies and, in the opinion of the Certification Authority, cannot consistently produce valid data within the acceptance limits specified in the NPDWR and within the policy described by their certification authority.

Interim certification—interim certification may be granted in certain circumstances when it is impossible or unnecessary to perform an on-site audit. Interim certification status may be granted only when the CA judges that the laboratory has the appropriate instrumentation, is using the approved methods, has adequately trained personnel to perform the analyses, and has satisfactorily analyzed PE samples, if available, for the contaminants in question. Performance evaluation samples from a commercial vendor may be used. The CO should perform an on-site audit as soon as possible but in no case later than 3 years. An example of a situation where this type of certification is warranted might be a laboratory that has requested certification for the analysis of additional analytes that involves use of a method for which it already has certification or a very similar method. The CO should review the laboratory's quality control (QC) data before granting this type of certification.

9. PRIMACY

Primacy States, in which all drinking water compliance analyses are not conducted at State-operated laboratories, are required to establish a certification program for drinking water laboratories [see 40 CFR 142.10(b) (3) (i)]. All States, however, are encouraged to develop such programs. EPA encourages the States to base certification of drinking water laboratories either upon criteria contained in this manual or upon state-developed equivalents that are at least as stringent as this manual. All state certification programs must require compliance with all provisions of the NPDWR. Those states required by regulation to develop a certification program must designate "a laboratory officer, or officers, certified by the [Regional] Administrator [or his or her designee], as the official(s) responsible for the State's certification program." [40 CFR §142.10(3) (i).]

10. DRINKING WATER LABORATORIES

Any laboratory which analyzes drinking water compliance samples is considered a drinking water laboratory for the purpose of certification. This includes Federal laboratories that analyze compliance samples and other laboratories that analyze compliance samples for Federal facilities. All such laboratories must be certified by the State or EPA. If requested by the State, a region may certify Federal laboratories in its region.

The region should certify individual laboratories on Federal Indian Lands, if requested by the tribal chairperson. These laboratories must meet the same criteria for certification as specified in the NPDWR and this manual.

The criteria, procedures, and mechanism EPA uses to certify municipal or private drinking water laboratories are the same as those for principal state laboratories.

11. OTHER CONSIDERATIONS FOR LABORATORY CERTIFICATION

11.1. Laboratory Personnel

The laboratory should have sufficient supervisory and other personnel, with the necessary education, training, technical knowledge, and experience for their assigned functions.

11.2. Laboratory Director/Manager or Technical Director

The laboratory director/manager should be a qualified professional with the technical education and experience and managerial capability commensurate with the size/type of the laboratory. The laboratory director/manager is ultimately responsible for ensuring that all laboratory personnel have demonstrated proficiency for their assigned functions and that all data reported by the laboratory meet the required quality assurance (QA) criteria and regulatory requirements.

11.3. Quality Assurance Officer/Manager

The QA officer/manager should be independent from the laboratory management if possible and have direct access to the highest level of management. The QA officer/manager should have a bachelor's degree in science, training in quality assurance principles commensurate with the size and sophistication of the laboratory, and at least 1 year of experience in quality assurance/control. The QA officer/manager should have at least a working knowledge of the statistics involved in Qc of laboratory analysis and a basic understanding of the methods which the laboratory employs.

12. LABORATORY QUALITY ASSURANCE PLAN

All laboratories analyzing drinking water compliance samples must adhere to the QC procedures specified in the methods. This is to ensure that routinely generated analytical data are scientifically valid and defensible and are of known and acceptable precision and accuracy. To accomplish these goals, each laboratory should prepare a written description of its QA activities (a QA plan). It is the responsibility of the QA manager to keep the QA plan up to date. All laboratory personnel must be familiar with the contents of the QA plan. This plan should be submitted to the auditors for review prior to the on-site visit or should be reviewed as part of the on-site visit.

The laboratory QA plan should be a separately prepared text. However, documentation for many of the listed QA plan items may be made by reference to appropriate sections of this manual, the laboratory's standard operating procedures (SOPs), or other literature (e.g. promulgated methods, *Standard Methods for the Examination of Water and Wastewater*, etc.). The QA plan should be updated as necessary.

At a minimum, the following items should be addressed in each QA plan:

1. Laboratory organization and responsibility

   - include a chart or table showing the laboratory organization and lines of responsibility, including QA managers;
   - list the key individuals who are responsible for ensuring the production of valid measurements and the routine assessment of measurement systems for precision and accuracy (e.g. who is responsible for internal audits and reviews of the implementation of the plan and its requirements);
   - reference the job descriptions of the personnel and describe training to keep personnel updated on regulations and methodology, and document that laboratory personnel have demonstrated proficiency for the methods they perform.

2. Process used to identify clients' data quality objectives
3. SOPs with dates of last revision

   - keep a list of SOPs;
   - ensure that current copies of SOPs are in the laboratory and in the QA managers files;
   - ensure that SOPs are reviewed annually and revised as changes are made;
   - ensure that SOPs have signature pages and revisions dated.

4. Field sampling procedures

   - describe the process used to identify sample collectors, sampling procedures and locations, required preservation, proper containers, correct sample container cleaning procedures, sample holding times from collection to analysis, and sample shipping and storage conditions;
   - ensure that appropriate forms are legibly filled out in indelible ink or hard copies of electronic data are available. See Chapters IV, V, and VI for specific items to be included;
   - describe how samples are checked when they arrive for proper containers and temperature and how samples are checked for proper preservation (e.g. pH, chlorine residual) before analysis;
   - ensure that sampling protocol is written and available to samplers.

5. Laboratory sample handling procedures

   • use bound laboratory note books, filled out in ink; entries dated and signed (A secure, password protected, electronic data base is acceptable);
   • store unprocessed and processed samples at the proper temperature, isolated from laboratory contaminants, standards and highly contaminated samples and, sometimes, each other; holding times may not be exceeded;
   • maintain integrity of all samples (e.g. by tracking samples from receipt by laboratory through analysis to disposal);
   • require chain-of-custody procedures for samples likely to be the basis for an enforcement action (see Appendix A);
   • specify criteria for rejection of samples which do not meet shipping, holding time and/or preservation requirements and procedures for notification of sample originators.

6. Calibration procedures for chemistry and radiochemistry (may reference SOP)

   • specify type of calibration used for each method and frequency of use;
   • describe standards' source, age, storage, labeling;
   • perform data comparability checks;
   • use control charts and for radiochemistry, report counting errors with their confidence levels.

7. Analytical procedures (may reference SOP)

   • cite complete method manual;
   • describe QC procedures required by the methods that must be followed.

8. Data reduction, validation, reporting, and verification (may reference SOP)

   • describe data reduction process: method of conversion of raw data to mg/L, pC/L, coliforms/100 mL, etc.;
   • describe data validation process;
   • describe reporting procedures, include procedures and format;
   • describe data verification process;
   • for radiochemistry, describe reporting of counting uncertainties and confidence levels;
   • describe procedure for data corrections.

9. Type of QC checks and the frequency of their use (see Chapters IV, V, and VI) (may reference SOP)
   Parameters for chemistry and radiochemistry should include or reference:

   • instrument performance check standards;
   • frequency and acceptability of method detection limit (MDL) calculations;

- calibration, internal, and surrogate standards;
- laboratory reagent blank, field reagent blank, and trip blank;
- field and laboratory matrix replicates;
- quality control and PE samples;
- laboratory fortified blank and laboratory fortified sample matrix replicates;
- initial demonstration of method capability and use of control charts;
- qualitative identification/confirmation of contaminants.

Parameters for microbiology should include or reference:

- positive and negative culture controls;
- confirmation/verification of presumptive total coliform-positive samples;
- sterility controls;
- performance evaluation and QC samples.

10. List schedules of internal and external system and data quality audits and inter laboratory comparisons (may reference SOP)
11. Perventive maintenance procedures and schedules

   - describe location of instrument manuals and schedules and documentation of routine equipment maintenance;
   - describe availability of instrument spare parts in the laboratory;
   - list any maintenance contracts in place.

12. Corrective action contingencies

   - describe response to obtaining unacceptable results from analysis of PE samples and from internal QC checks;
   - name persons responsible for the various corrective actions;
   - describe how corrective actions taken are documented;

13. Record keeping procedures

   - describe procedures and documentation of those procedures;
   - list length of storage, media type (electronic or hard copy);
   - describe security policy of electronic databases.

If a particular item is not relevant, the QA plan should state this and provide a brief explanation. A laboratory QA plan should be responsive to the above items while remaining brief and easy to follow. Minimizing paperwork, while improving dependability and quality of data, are the intended goals.

13. PERFORMANCE ON ROUTINE WATER SAMPLES

Each certification authority should develop a strategy to assess laboratory performance on routine water samples as part of its certification program.

This strategy may include one or more of the following approaches or some other approach: blind audit samples; analysis of an unknown sample during the on-site evaluation; a split sample program with the laboratory; a data audit; or observation of daily routine during an on-site visit. Each certification authority should develop a written plan that addresses this issue. The regional plan is approved by OGWDW with NERL concurrence. The State plan is approved by the Region.

14. CHAIN-OF-CUSTODY PROCEDURES

Certified laboratories, when requested to process a sample for possible legal action against a supplier, should use an adequate chain-of-custody procedure. An example of such a procedure is found in Appendix A. The State or Region should seek input from its attorney general's office to ensure that the laboratory's procedures are adequate. The procedure used should be documented.

15. REQUIREMENTS FOR MAINTAINING CERTIFICATION STATUS

15.1. Performance Evaluation Samples

All certified drinking water laboratories must satisfactorily analyze PE samples or other unknown test samples to maintain certification for inorganic and organic contaminants. Currently, the regulations stipulate that if a laboratory satisfactorily analyzes 80% of the regulated volatile organic chemicals (VOCs), it may be certified for all of them (except vinyl chloride which is certified separately). If the VOCs are provided in more than one vial, they must be considered collectively. However, it is recommended that a laboratory not be certified for a particular VOC, which is unsatisfactorily analyzed two times in succession.

If the laboratory does not analyze the PE sample, or other unknown test sample, within the acceptance limits specified in the regulations, or within policy described by their CA, the CA must follow the procedure discussed in the section entitled "Criteria and Procedures for Downgrading/Revoking Certification Status."

If a laboratory wishes to be certified for a contaminant by more than one method, it should analyze the PE samples by each method for which it wishes to be certified. If the PE provider cannot accommodate PE data by more than one method, the additional data should be submitted directly to the certification officer. The methods listed on the laboratory's certification certificate must be the methods by which the PE samples were analyzed.

The laboratory should be able to provide documentation to the certification authority that the person(s) analyzing any PE sample is a laboratory employee who routinely analyzes drinking water compliance samples.

To maintain certification in radiochemistry, the NERL-LV requires laboratories to participate satisfactorily in the NERL-LV PE program.

15.2. Methodology

Laboratories must use the methods specified in the drinking water regulations at 40 CFR part 141. These methods are listed in Chapters IV, V, VI, and Appendix G.

15.3. Notification of Certifying Authority of Major Changes

Certified laboratories should notify the appropriate CA (Regional Administrator or designee or the appropriate NERL) in writing, within 30 days of major changes in personnel, equipment, or laboratory location. A major change in personnel is defined as the loss or replacement of the laboratory supervisor or a situation in which a trained and experienced analyst is no longer available to analyze a particular parameter for which certification has been granted. The CA should discuss the situation with the laboratory supervisor and establish a schedule for the laboratory to address major changes. If the CA determines that the laboratory can no longer produce valid data, the CA should follow the procedure for revocation of certification.

15.4. On-Site Evaluation

The CA should be satisfied that a laboratory is maintaining the required standard of quality for certification. Normally, this should be based on a recommendation from an on-site evaluation. If the laboratory undergoes a major change, however, or if it fails a PE sample or other unknown test sample, the CA should consider conducting an evaluation before the usual 3-year period has expired.

16. CRITERIA AND PROCEDURES FOR DOWNGRADING/REVOKING CERTIFICATION STATUS

16.1. Criteria for Downgrading Certification Status

A laboratory should be downgraded to "provisionally certified" status for a contaminant or group of contaminants for any of the following reasons:

- Failure to analyze a PE sample at least annually within the acceptance limits specified in the regulations, or, if there are no requirements specified in the regulations, within policy described by their CA;
- Failure of a certified laboratory to notify the CA within 30 days of major changes (e.g. in personnel, equipment, or laboratory location);
- Failure to satisfy the CA that the laboratory is maintaining the required standard of quality, based upon an EPA on-site evaluation;
- Failure to report compliance data to the public water system or the State drinking water program in a timely manner, thereby preventing compliance with Federal or State regulations and endangering public

health. Data which may cause the system to exceed an MCL should be reported as soon as possible.

16.2. Procedures for Downgrading to "Provisionally Certified" Status

If a laboratory is subject to downgrading on the basis of the above-mentioned criteria, the CA should notify the laboratory director or owner of its intent to downgrade (by registered or certified mail) within 14 days from becoming aware of the situation warranting downgrading. The laboratory director should review the problems cited and, within 30 days of receipt of the letter, send a letter to the CA specifying what immediate corrective actions are being taken and any proposed actions that need the concurrence of the CA. The CA should consider the adequacy of the response and notify the laboratory in writing (by registered or certified mail) of its certification status within 14 days of receipt of its response. The CA should follow up to ensure that corrective actions have been taken.

If a laboratory fails to analyze a PE or other unknown sample within the acceptance limits, the CA should not downgrade certification if the laboratory identifies and corrects the problem to the CA's satisfaction within 30 days of being notified of the failure. If, after a review of the submitted information, the CA determines that the laboratory need not be downgraded, then within 30 days of this decision, the CA should send the laboratory another unknown sample containing the failed contaminant. If the laboratory analyzes this second unknown sample within the acceptance limits established by the EPA or State (using the most recent PE summary statistical compilations), the laboratory should not be downgraded. If the laboratory fails to analyze this second unknown sample within the established limits, the CA should downgrade the laboratory to "provisionally certified" status and notify the laboratory within 14 days (by registered or certified mail). Laboratories should be downgraded only for the analyte failed, except where EPA/State certifies a group of related analytes based on a limited number of analytes in the group. (See "Periodic Performance Evaluation Samples" for additional information.)

During any phase of this procedure, a laboratory may request that the EPA or State provide technical assistance to help identify and resolve any problem.

Once the CA notifies a laboratory, in writing, that it has been downgraded to "provisionally certified" status for procedural, administrative, equipment, or personnel deficiency, the laboratory should correct its problem within 3 months (6 months for major equipment replacement). If the laboratory was downgraded to "provisionally certified" status because of a failure to analyze a PE sample (or other unknown test sample) within the acceptance limits specified in the regulations, or within policy required by their certifying authority, the laboratory should correct its problems and satisfactorily analyze another PE sample (or other unknown sample) within 1 month of receipt of the second PE sample. A provisionally certified laboratory may continue to analyze samples for compliance purposes, but should notify its clients of its downgraded status and provide that information, in writing, on any report.

16.3. Criteria for Revoking Certification Status

A laboratory shall be downgraded from certified, provisionally certified, or interim certified status to "not certified" for a particular contaminant analysis for the following reasons:

- Submission of a PE sample to another laboratory for analysis and reporting the data as its own;
- Falsification of data or other deceptive practices;
- Failure to use the analytical methodology specified in the regulations;
- For provisionally certified laboratories, failure to successfully analyze a PE sample or any other unknown test sample for a particular contaminant within the acceptance limits specified;
- For provisionally certified laboratories, failure to satisfy the CA that the laboratory has corrected deviations identified during the on-site evaluations;
- For provisionally certified laboratories, persistent failure to report compliance data to the public water system or the State drinking water program in a timely manner thereby preventing compliance with Federal and/or State regulations and endangering public health. Data which may cause the system to exceed an MCL should be reported as soon as possible.

16.4. Procedures for Revocation

The CA should notify the laboratory, in writing (by registered or certified mail) of the intent, to revoke certification. If the laboratory wishes to challenge this decision, a notice of appeal should be submitted in writing to the CA within 30 days of receipt of the notice of intent to revoke certification. If no notice of appeal is filed, certification shall be revoked. The notice of appeal should be supported with an explanation of the reasons for the challenge and must be signed by a responsible official from the laboratory such as the president/owner for a commercial laboratory, or the laboratory supervisor in the case of a municipal laboratory or the laboratory director for a State or Regional laboratory.

Within 30 days of receipt of the appeal, the CA should make a decision and notify the laboratory in writing (by registered or certified mail). Denial of the appeal shall result in immediate revocation of the laboratory's certification. Once certification is revoked, a laboratory may not analyze drinking water samples for compliance until its certification has been reinstated.

If the appeal is determined to be valid, the CA should take appropriate measures to re-evaluate the facility and notify the laboratory, in writing (by registered or certified mail), of its decision within 30 days of the re-evaluation.

16.5. Reinstatement of Certification

Through a written request, a laboratory may seek reinstatement of certification, when and if the laboratory can demonstrate to the CA's satisfaction that the deficiencies which produced provisionally certified status or revocation have been corrected. This may include an on-site evaluation, successful analysis of unknown samples, or any other measure the CA deems appropriate.

17. RECIPROCITY

Reciprocity, mutually acceptable certification among regions and/or primacy States, is strongly endorsed by EPA as a highly desirable element in the certification program for drinking water laboratories. EPA also believes that use of a third-party auditing agent by more than one State could promote reciprocity. (See Appendix D.)

States are encouraged to adopt provisions in their laws and regulations to permit reciprocity. Even though ultimate responsibility for reciprocal certification resides with the regions and primacy States, the States may ask for the assistance of EPA in cases involving clarification of what should be considered in a reciprocal agreement. Such requests should be submitted to the region or the OGWDW through the region.

18. TRAINING

Training is an integral part of the laboratory certification process for personnel conducting on-site evaluations of laboratories on behalf of either the regional office or a primacy State.

EPA policy requires that all regional certification officers must initially pass the appropriate EPA laboratory certification training courses for the discipline for which they certify (chemistry, radiochemistry, or microbiology). All laboratory auditors should be experienced professionals and have at least a bachelor's degree or equivalent education/experience in the discipline for which they certify and recent laboratory experience in the field for which they audit laboratories. Third-party auditors (see Appendix D) must also pass the EPA certification training course. Outside experts, retained for their knowledge in a limited area (e.g. asbestos) are not required to pass the laboratory certification course if they are used as part of an on-site audit team which includes a certification officer. Periodic training for both laboratory auditors and analysts should be provided by the regions. Certification officers should attend refresher training programs at least every 5 years to keep their knowledge of the methods and the drinking water program current. It is highly recommended that certification officers have recent bench experience in the methods for which they certify. The OGWDW will notify certification officers of major updates/changes to EPA's certification program. It is recommended that the States use these same criteria in their certification programs.

19. ALTERNATE ANALYTICAL TECHNIQUES

EPA approves analytical methods for all regulated drinking water contaminants. A regulation for a particular contaminant will include approval of one or more methods that must be used to determine that contaminant. Subsequent to publishing a rule (promulgation), the agency may approve new methods or modifications of EPA approved methods through another rule.

> With the written permission of the State, concurred in by the administrator of the U.S. EPA, an alternate analytical technique may be employed. An alternate technique shall be accepted only if it is substantially equivalent to the prescribed test in both precision and accuracy as it relates to the determination of compliance with any MCL. [CFR 141.27(a)]

Anyone can request that EPA approve a new method or modification of a method already approved by EPA, by submitting EPA-specified data and other information to the Director, Analytical Methods Staff, Office of Science and Technology (MS4303), Office of Water, USEPA, 401 M Street, SW, Washington, DC 20460. EPA will evaluate the material to determine whether the method meets EPA criteria. If a method meets the specified criteria, it will be approved. This involves a formal proposal in the *Federal Register*, public comment, and finally (unless otherwise indicated by the public comments or other new information), promulgation in the *Federal Register*. Once approved (i.e. promulgated), any laboratory may use the method.

CHAPTER IV

CRITICAL ELEMENTS FOR CHEMISTRY

1. PERSONNEL

1.1. Laboratory Supervisor

The laboratory supervisor should have at least a bachelor's degree with a major in chemistry or equivalent and at least 1 year of experience in the analysis of drinking water. The laboratory supervisor should have at least a working knowledge of quality assurance principles. The laboratory supervisor has immediate responsibility to insure that all laboratory personnel have demonstrated their ability to satisfactorily perform the analyses to which they are assigned and that all data reported by the laboratory meet the required quality assurance and regulatory criteria.

1.2. Laboratory Analyst

The laboratory analyst should have at least a bachelor's degree with a major in chemistry or equivalent and at least 1 year of experience in the analysis of drinking water. If the analyst is responsible for the operation of analytical instrumentation, he/she should have completed specialized training offered by the manufacturer or another qualified training facility or served a period of apprenticeship under an experienced analyst. The duration of this apprenticeship is proportional to the sophistication of the instrument. Data produced by analysts and instrument operators while in the process of obtaining the required training or experience are acceptable only when reviewed and validated by a fully qualified analyst or the laboratory supervisor.

Before beginning the analysis of compliance samples, the analyst must demonstrate acceptable results for blanks, precision, accuracy, method detection, and specificity and satisfactory analysis on unknown samples.

1.3. Technician

The laboratory technician should have at least a high school diploma or equivalent, complete a method training program under an experienced analyst and have 6 months bench experience in the analysis of drinking water samples.

Before beginning the analysis of compliance samples, the technician must demonstrate acceptable results for blanks, precision, accuracy, method detection, and specificity and satisfactory analysis on unknown samples.

1.4. Sampling Personnel

Personnel who collect samples should have been trained in the proper collection technique for all types of samples which they collect. Their technique should be reviewed by experienced sampling or laboratory personnel.

1.5. Waiver of Academic Training Requirement

The certification officer may waive the need for specified academic training, on a case-by-case basis, for highly experienced analysts.

Training records should be maintained for all personnel. These should include all job-related formal training taken by analyst which pertains to any aspect of his/her responsibilities, including but not limited to analytical methodology, laboratory safety, sampling, quality assurance, data analysis, etc.

2. LABORATORY FACILITIES

The analysis of compliance samples should be conducted in a laboratory where the security and integrity of the samples and the data can be maintained. The laboratory facilities should be clean, have adequate temperature and humidity control, have adequate lighting at the bench top and must meet applicable OSHA standards. The laboratory must have provisions for the proper storage and disposal of chemical wastes. The appropriate type of exhaust hood is required where applicable.

There should be sufficient bench space for processing samples. Workbench space should be convenient to sink, water, gas, vacuum, and electrical sources free from surges. Instruments should be properly grounded. For safety reasons, inorganic and organic facilities should be in separate rooms; organic analysis and sample extraction should also be separated to prevent cross-contamination. The analytical and sample storage areas should be isolated from all potential sources of contamination. There should be sufficient storage space for chemicals, glassware, and portable equipment, sufficient floor and bench space for stationary equipment and areas for cleaning materials.

3. LABORATORY EQUIPMENT AND INSTRUMENTATION

The laboratory must have the instruments and equipment needed to perform the approved methods for which certification has been requested. The checklist on page 43 of this chapter provides more information on the necessary equipment. All instruments should be properly maintained and calibrated.

4. GENERAL LABORATORY PRACTICES

4.1. General

4.1.1. Chemicals/Reagents: Chemicals and reagents used must meet the specifications in the method. If not specified, then "analytical reagent grade" (AR) or American Chemical Society (ACS) grade chemicals or better should be used for analyses in certified laboratories. Consult *Standard Methods for the Examination of Water and Wastewater*, 18th Edition, part 1070 for more detailed information on reagent grades.

4.2. Inorganic Contaminants

4.2.1. Reagent Water: The laboratory should have a source of reagent water having a resistance value of at least 0.5 MΩ (conductivity less than 2μ/cm) at 25°C. High quality water meeting such specifications may be purchased from commercial suppliers. Quality of reagent water is best maintained by sealing it from the atmosphere. Quality checks to meet specifications above should be made and documented at planned intervals based on use. This planned interval should not exceed 1 month. Individual analytical methods may specify additional requirements for the reagent water to be used.

4.2.2. Glassware Preparation: Specific requirements in the methods for the cleaning of glassware must be followed. If no specifications are listed, then glassware should be washed in a warm detergent solution and thoroughly rinsed first with tap water and then with reagent water. This cleaning procedure is sufficient for general analytical needs. It is advantageous to maintain separate sets of suitably prepared glassware for the nitrate, mercury, and lead analyses due to the potential for contamination from the laboratory environment. Refer to Table IV-1 for a summary of glassware cleaning procedures.

4.3. Organic Contaminants

4.3.1. Reagent Water: Reagent water for organic analysis must be free from interferences for the analytes being measured. It may be necessary to treat water with activated carbon to eliminate all interferences. If individual methods specify additional requirements for the reagent water to be used, these must be followed.

Table IV-1 Glassware Cleaning Procedures—Consult the Method for Complete Details; Do Not Overheat Volumetric Glassware

Method	Washing	Drying
502.2/504/ 504.1/524.2	Detergent wash, rinse with tap and distilled water	105°C for 1 hr
505	Detergent wash, rinse with tap and reagent water	400°C for 1 hr or rinse with acetone
506	Rinse immediately with last solvent used, wash with hot water and detergent, rinse with tap and reagent water	400°C for 1 hr or rinse with acetone
507/508	Rinse immediately with last solvent used, wash with hot water and detergent, rinse with tap and reagent water	400°C for 1 hr or rinse with acetone
508.1	Detergent wash, rinse with tap and reagent water or solvent rinse	400°C for 2 hr
508A	No specifications, suggest the same as 515.1/515.2	No specification, suggest the same as 515.1/515.2
509	Detergent wash, rinse with tap and reagent water	105°C for 1 hr for septa, 400°C for 1 hr for vials
515.1/515.2	Rinse immediately with last solvent used, wash with hot water and detergent, rinse with dilute acid, tap and reagent water	400°C for 1 hr or rinse with acetone
525.2	Detergent wash, rinse with tap and distilled water or solvent rinse	Air dry or muffle (no specs) (suggest 400°C for 1 hr)
531.1/6610	Rinse immediately with last solvent used, wash with hot water and detergent, rinse with tap and reagent water	450°C for 1 hr or rinse with acetone
547/548.1	Rinse immediately with last solvent used, wash with hot water and detergent, rinse with tap and reagent water	400°C for several hours or rinse with methanol
549.1	Rinse immediately with last solvent used, wash with hot water and detergent, rinse with tap and reagent water	130°C for several hours or rinse with methanol

(*Continued*)

Table IV-1 (*Continued*)

Method	Washing	Drying
550/550.1	Rinse immediately with last solvent used, wash with hot water and detergent, rinse with tap and reagent water	400°C for 15–30 minutes or rinse with acetone or pesticide quality hexane
1613	Rinse with solvent, sonicate with detergent for 30 min, rinse sequentially with methanol, hot tap water, methanol, acetone, and methylene chloride	Air dry
Metals	Wash with detergent, rinse with tap water, soak 4 hr in 20% (V/V) nitric acid or dilute nitric (~8%)/ hydrochloric (~17%), rinse with reagent water	Air dry
Inorganics	Wash with detergent, rinse with tap and reagent water (use phosphate-free detergent for o-phosphate analysis)	Air dry

4.3.2. Glassware Preparation: Glassware cleaning requirements specified in the methods must be followed. Table IV-11 summarizes the cleaning procedures specified in the EPA methods.

4.4. Laboratory Safety

While safety criteria are not an aspect of laboratory certification, laboratory personnel should apply general and customary safety practices as a part of good laboratory practices. Each laboratory is encouraged to have a safety plan as part of their standard operating procedure (SOP) which includes personnel safety, training, and protection. Where safety practices are included in an approved method (i.e. 515.1), they must be followed. See *Standard Methods for the Examination of Water and Wastewater*, part 1090 for a discussion of laboratory safety.

4.5. Quality Assurance

Laboratories should maintain current quality assurance plans as described in Chapter 3. All laboratory activities including sampling, test methods, instrument operation, data generation, and corrective action should be described in the plan. Plans should be available to all personnel.

5. ANALYTICAL METHODS

5.1. General

A list of approved methods for inorganic and organic contaminants can be found in Tables IV-2 and IV-3, respectively. Methods manuals should be available in-house. Allowed modification to the methods should be documented. All procedural steps in these methods are considered requirements. Other methods cannot be used for compliance samples unless approval has been granted by the agency. Contact the appropriate certifying authority for an alternate test procedure application. Application for the use of an alternate method may require acceptable comparability data. Table IV-4 lists the methods which must be used for contaminants which are regulated for monitoring only (have no MCLs). Table IV-5 lists the methods which must be used for the analysis of disinfectants. Recommended methods for secondary contaminants are listed in Table IV-6. For more information, see Appendix H, Section 2.1.

5.2. Analyses Approved by the State

Measurements for turbidity, pH, temperature, residual disinfectant, calcium, orthophosphate, silica, alkalinity, and conductivity need not be made in certified laboratories, but may be performed by any persons acceptable to the State. However, approved methodology must be used (Tables IV-2–5). The State should institute a quality assurance program to assure validity of data from these measurements.

5.2.1. Turbidity Standards: Sealed liquid turbidity standards purchased from the instrument manufacturer should be calibrated against properly prepared and diluted formazin or styrene divinylbenzene polymer standards at least every 4 months in order to monitor for any eventual deterioration. This calibration should be documented. These standards should be replaced when they do not fall within 15% of the concentration of the standard. Solid turbidity standards composed of plastic, glass, or other materials are not reliable and should not be used.

5.2.2. Residual Chlorine Standards: If visual comparison devices such as color wheels or sealed ampules are used for determining free chlorine residual, the standards incorporated into such devices should be calibrated at least every 6 months. These calibrations should be documented. Directions for preparing temporary and permanent type visual standards can be found in Method 4500-Cl-G of *Standard Methods for the Examination of Water and Wastewater*, 18th Edition, 1992. By comparing standards and plotting such a comparison on graph paper, a corrective factor can be derived and applied to future results obtained on the now calibrated apparatus.

Table IV-2 Approved Methods for Primary Inorganic Chemicals, Parameters in the Lead and Copper Rule, Sodium, and Turbidity [§141.23(k)(l)]

Contaminant	Methodology	EPA	ASTM[c]	SM[d]	Other
Antimony	ICP-MS	200.8[b]			
	Hydride-AA		D3697–92		
	AA-platform	200.9[b]			
	AA-furnace			3113B	
Arsenic	ICP	200.7[b]		3120B	
	ICP-MS	200.8[b]			
	AA-platform	200.9[b]			
	AA-furnace		D2972–93C	3113B	
	Hydride-AA		D2972–93B	3114B	
Asbestos	TEM	100.1[i]			
	TEM	100.2[j]			
Barium	ICP	200.7[b]		3120B	
	ICP-MS	200.8[b]			
	AA-direct			3111D	
	AA-furnace			3113B	
Beryllium	ICP	200.7[b]		3120B	
	ICP-MS	200.8[b]			
	AA-platform	200.9[b]			
	AA-furnace		D3645–93B	3113B	
Cadmium	ICP	200.7[b]			
	ICP-MS	200.8[b]			
	AA-platform	200.9[b]			
	AA-furnace			3113B	

(Continued)

Table IV-2 (*Continued*)

Contaminant	Methodology	EPA	ASTM[c]	SM[d]	Other
Chromium	ICP	200.7[b]		3120B	
	ICP-MS	200.8[b]			
	AA-Platform	200.9[b]			
	AA-Furnace				
Cyanide	Man. Distillation followed by:			3113B	
				4500-CN-C	
	Spec., Amenable		D2036-91B	4500-CN-G	
	Spec. Manual		D2036-91A	4500-CN-E	I-3300–85[e]
	Semi-auto	335.4[f]			
	Ion Sel. Elec. (ISE)			4500CN-F	
Fluoride	Ion Chromatography	300.0[f]	D4327–91	4110B	
	Manual Distill. SPADNS			4500F-B,D	
	Manual ISE		D1179–93B	4500F-C	
	Automated ISE				380–75WE[k]
	Auto. Alizarin			4500F-E	129–71W[k]
Mercury	Manual Cold Vapor	245.1[b]	D3223–91	3112B	
	Auto. Cold Vapor	245.2[a]			
	ICP-MS	200.8[b]			
Nitrate	Ion Chromatography	300.0[f]	D4327–91	4110B	B-1011[h]
	Auto Cd Reduction	353.2[f]	D3867–90A	4500-NO$_3$-F	601[g]
	Ion Selective Elec.			4500-NO$_3$-D	
	Man Cd Reduction		D3867–90B	4500-NO$_3$E	

Analyte	Method				
Nitrite	Ion Chromatography	300.0[f]	D4327-91	4110B	B-1011[h]
	Auto Cd Reduction	353.2[f]	D3867-90A	4500-NO$_3$-F	
	Man Cd Reduction		D3867-90B	4500-NO$_3$-E	
	Spectrophotometric			4500-NO$_2$-B	
Selenium	Hydride-AA		D3859-93A	3114B	
	ICP-MS	200.8[b]			
	AA-Platform	200.9[b]			
	AA-Furnace		D3859-93B	3113B	
Thallium	ICP-MS	200.8[b]			
	AA-Platform	200.9[b]			
	AA-Furnace			3113B	
Lead	ICP-MS	200.8[b]			
	AA-Platform	200.9[b]			
	AA-Furnace		D3559-90D	3113B	
Copper	AA-Furnace		D1688-90C	3113B	
	AA-Direct		D1688-90A	3111B	
	ICP	200.7[b]		3120B	
	ICP-MS	200.8[b]			
	AA-Platform	200.9[b]			
PH	Electrometric	150.1[a]	D1293-84	4500-H$^+$-B.	
		150.2[a]			
Conductivity	Conductance		Dl 125-91A	2510B	
Calcium	EDTA titration		D511-93A	3500-Ca-D	
	AA-Direct		D511-93B	3111B	
	ICP	200.7[b]		3120B	
Alkalinity	Titration		D1067-92B	2320B	I-1030-85[e]
	Elec. titration				

(Continued)

Table IV-2 (*Continued*)

Contaminant	Methodology	EPA	ASTM[c]	SM[d]	Other
Ortho-phosphate unfiltered, no digestion or hydrolysis	Color, automated ascorbic acid	365.1[f]		4500-P-F	
	Color, ascorbic acid		D515-88A	4500-P-E	
	Color, phosphomolybdate				I-1601-85[e]
	Autosegmented Flow				I-2601-90[e]
	Auto discrete				I-2598-85[e]
Silica	Ion Chromatography	300.0[f]	D4327-91	4110	
	Color, molybdate blue; auto seg. flow				I-1700-85[e]
	Color		D859-88		I-2700-85[e]
	Molybdosilicate			4500-Si-D	
	Heteropoly blue			4500-Si-E	
	Auto, molybdate reactive silica			4500-Si F	
	ICP	200.7[b]		3120B	
Temperature	Thermometric			2550B	
Sodium	ICP	200.7[b]		3111B	
Turbidity	AA-Direct Nephelometric[f]	180.1		2130B	GLI Method 2[l]

[a] Methods 150.1, 150.2, and 245.2 are available from U.S. EPA, EMSL, Cincinnati, OH 45268. The identical methods were formerly in "Methods for Chemical Analysis of Water and Wastes," EPA-600/4-79-020, March 1983.

[b] "Methods for the Determination of Metals in Environmental Samples—Supplement I," EPA-600/R-94-111, May 1994. Available at NTIS, PB 94-184942.

[c] *Annual Book of ASTM Standards*, vols. 11.01 and 11.02, American Society for Testing and Materials, 1916 Race Street, Philadelphia, PA 19103.

[d] *Standard Methods for the Examination of Water and Wastewater*, 18th Edition, 1992, American Public Health Association, 1015 Fifteenth Street NW, Washington, DC 20005.

[e] Available from Books and Open-File Reports Section, U.S. Geological Survey, Federal Center, Box 25425, Denver, CO 80225-0425.

[f] "Methods for the Determination of Inorganic Substances in Environmental Samples," EPA-600/R-93-100, August 1993. Available at NTIS, PB94-121811.

[g] Technical Bulletin 601 "Standard Method of Test for Nitrate in Drinking Water," July 1994, PN 221890-001, ATI Orion, 529 Main Street, Boston, MA 02129. This method is identical to Orion WeWWG/5880, which is approved for nitrate analysis. ATI Orion republished the method in 1994, and renumbered it as 601, because the 1985 manual "Orion Guide to Water and Wastewater Analysis," which contained WeWWG/5880, is no longer available.

[h] Method B-1011, "Waters Test Method for Determination of Nitrite/Nitrate in Water Using Single Column Ion Chromatography," Millipore Corporation, Waters Chromatography Division, 34 Maple Street, Milford, MA 01757.

[i] Method 100.1, "Analytical Method for Determination of Asbestos Fibers in Water," EPA-600/4-83-043, EPA, September 1983. Available at NTIS, PB 83-260471.

[j] Method 100.2, "Determination of Asbestos Structure Over 10-µm In Length in Drinking Water," EPA-600/R-94-134, June 1994. Available at NTIS, PB 94-201902.

[k] Industrial Method No. 129-71W, "Fluoride in Water and Wastewater," December 1972, and Method No. 380-75WE, "Fluoride in Water and Wastewater," February 1976, Technicon Industrial Systems, Tarrytown, NY 10591.

[l] GLI Method 2, "Turbidity," November 2, 1992, Great Lakes Instruments, Inc., 8855 North 55th Street, Milwaukee, Wisconsin 53223.

Table IV-3   Approved Methods for Primary Organic Chemicals [§141.24(e)]

Contaminant	Method[c]
Benzene	502.2, 524.2
Carbon tetrachloride	502.2, 524.2, 551
Chlorobenzene	502.2, 524.2
1,2-Dichlorobenzene	502.2, 524.2
1,4-Dichlorobenzene	502.2, 524.2
1,2-Dichloroethane	502.2, 524.2
cis-1,2-Dichloroethylene	502.2, 524.2
trans-1,2-Dichloroethylene	502.2, 524.2
Dichloromethane	502.2, 524.2
1,2-Dichloropropane	502.2, 524.2
Ethylbenzene	502.2, 524.2
Styrene	502.2, 524.2
Tetrachloroethylene	502.2, 524.2, 551
1,1,1 -Trichloroethane	502.2, 524.2, 551
Trichloroethylene	502.2, 524.2, 551
Toluene	502.2, 524.2
1,2,4-Trichlorobenzene	502.2, 524.2
1,1 -Dichloroethylene	502.2, 524.2
1,1,2-Trichloroethane	502.2, 524.2
Vinyl chloride	502.2, 524.2
Xylenes (total)	502.2, 524.2
2,3,7,8-TCDD (dioxin)	1613
2,4-D	515.2, 515.1, 555
Alachlor	505[a], 507, 508.1, 525.2
Atrazine	505[a], 507, 508.1, 525.2
Benzo(a)pyrene	525.2, 550, 550.1
Carbofuran	531.1,6610
Chlordane	505, 508, 508.1, 525.2
Dalapon	515.1, 552.1
Di(2-ethylhexyl)adipate	506, 525.2
Di(2-ethylhexyl)phthalate	506, 525.2
Dibromochloropropane (DBCP)	504.1, 551
Dinoseb	515.2, 515.1, 555
Diquat	549.1
Endothall	548.1
Endrin	505, 508, 508.1, 525.2
Ethylene dibromide (EDB)	504.1, 551
Glyphosate	547, 6651
Heptachlor	505, 508, 508.1, 525.2
Heptachlor Epoxide	505, 508, 508.1, 525.2
Hexachlorobenzene	505, 508, 508.1, 525.2
Hexachlorocyclopentadiene	505, 508, 508.1, 525.2

(Continued)

Table IV-3 (*Continued*)

Contaminant	Method[c]
Lindane	505, 508, 508.1, 525.2
Methoxychlor	505, 508, 508.1, 525.2
Oxamyl	531.1, 6610
PCBs (as decachlorobiphenyl)[b] (as Aroclors)	508A 505, 508
Pentachlorophenol	515.1, 515.2, 525.2, 555
Picloram	515.1, 515.2, 555
Simazine	505[a], 507, 508.1, 525.2
2,4,5-TP (Silvex)	515.1, 515.2, 555
Toxaphene	505, 508, 525.2
Total Trihalomethanes	502.2, 524.2, 551

[a]A nitrogen–phosphorus detector should be substituted for the electron capture detector in Method 505 (or another approved method should be used) to determine alachlor, atrazine, and simazine, if lower detection limits are required.

[b]PCBs are qualitatively identified as Aroclors and measured for compliance purposes as decachlorobiphenyl using Method 508A.

[c]Methods 502.2, 505, 507, 508, 508A, 515.1, and 531.1 are in *Methods for the Determination of Organic Compounds in Drinking Water*, EPA-600/4-88–039, December 1988, Revised, July 1991. Methods 506, 547, 550, 550.1, and 551 are in *Methods for the Determination of Organic Compounds in Drinking Water—Supplement I*, EPA-600–4-90–020, July 1990. Methods 515.2, 524.2, 548.1, 549.1, 552.1, and 555 are in *Methods for the Determination of Organic Compounds in Drinking Water—Supplement II*, EPA-600/R-92–129, August 1992. Method 1613, *Terra- Through Octa-Chlorinated Dioxins and Furans by Isotopic Dilution HRGC/HRMS*, EPA-81/B-94–003, October 1994. These documents are available from the National Technical Information Service, NTIS PB91–231480, PB91–146027, and PB92–207703 and PB95–104774, U.S. Department of Commerce, 5285 Port Royal Road, Springfield, VA 22161. The toll-free number is 800–553–6847. Method 1613 is available from USEPA Office of Water Resource Center (RC-4100), 401 M. Street, SW, Washington, DC 20460. The phone number is 202–260–7786. EPA Methods 504.1, 508.1, and 525.2 are available from U.S. EPA NERL, Cincinnati, OH 45268. The phone number is (513)-569–7586. Method 6651 is contained in the 18th edition *of Standard Methods for the Examination of Water and Wastewater*, 1992, and Method 6610 is contained in the *Supplement to the 18th edition of Standard Methods for the Examination of Water and Wastewater*, 1994, American Public Health Association, 1015 Fifteenth Street NW, Washington, DC 20005.

6. SAMPLE COLLECTION, HANDLING, AND PRESERVATION

The manner in which samples are collected and handled is critical to obtaining valid data. It is important that a written sampling protocol with specific sampling instructions be available to and used by sample collectors and

Table IV-4 Approved Methods for "Unregulated" Contaminants (§141.40)
Regulations specified in §141.40 require monitoring for certain contaminants
to which maximum contaminant levels do not apply. These chemicals are
called "unregulated" contaminants, and presently include sulfate, 34 VOCs,
and 13 SOCs.

	Methods	ASTM	SM
"Unregulated" VOC contaminants[a]			
Chloroform	502.2, 524.2, 551		
Bromodichloromethane	502.2, 524.2, 551		
Bromoform	502.2, 524.2, 551		
Chlorodibromomethane	502.2, 524.2, 551		
Bromobenzene	502.2, 524.2		
Bromomethane	502.2, 524.2		
Chloroethane	502.2, 524.2		
Chloromethane	502.2, 524.2		
o-Chlorotoluene	502.2, 524.2		
p-Chlorotoluene	502.2, 524.2		
Dibromomethane	502.2, 524.2		
m-Dichlorobenzene	502.2, 524.2		
1,1-Dichloroethane	502.2, 524.2		
1,3-Dichloropropane	502.2, 524.2		
2,2-Dichloropropane	502.2, 524.2		
1,1-Dichloropropene	502.2, 524.2		
1,3-Dichloropropene	502.2, 524.2		
1,1,2,2-Tetrachloroethane	502.2, 524.2		
1,1,1,2-Tetrachloroethane	502.2, 524.2		
1,2,3-Trichloropropane	502.2, 524.2, 504.1		
State discretionary contaminants			
Bromochloromethane	502.2, 524.2		
n-Butylbenzene	502.2, 524.2		
sec-Butylbenzene	502.2,524.2		
tert-Butylbenzene	502.2, 524.2		
Dichlorodifluoromethane	502.2, 524.2		
Fluorotrichloromethane	502.2, 524.2		
Hexachlorobutadiene	502.2, 524.2		
Isopropylbenzene	502.2, 524.2		
p-Isopropyltoluene	502.2, 524.2		
Naphthalene	502.2, 524.2		
n-Propylbenzene	502.2, 524.2		
1,2,3-Trichlorobenzene	502.2, 524.2		
1,2,4-Trimethylbenzene	502.2, 524.2		
1,3,5-Trimethylbenzene	502.2, 524.2		

(*Continued*)

Table IV-4 *(Continued)*

	Methods	ASTM	SM
"Unregulated" SOC contaminants[b]			
Aldicarb	531.1, 6610[c]		
Aldicarb sulfone	531.1, 6610[c]		
Aldicarb sulfoxide	531.1, 6610[c]		
Aldrin	505, 508, 525.2, 508.1		
Butachlor	507, 525.2		
Carbaryl	531.1, 6610[c]		
Dicamba	515.1, 515.2, 555		
Dieldrin	505, 508, 525.2, 508.1		
3-Hydroxycarbofuran	531.1, 6610[c]		
Methomyl	531.1, 6610[c]		
Metolachlor	507, 525.2, 508.1		
Metribuzin	507, 525.2, 508.1		
Propachlor	508, 525.2, 508.1		
"Unregulated" inorganic contaminants[d]			
Nickel	200.7		3120B
	200.8		
	200.9		
			3111B
			3113B
Sulfate	300.0	D4327–91	4110B
	375.2		4500-SO$_4$-F
			4500-SO$_4$-C,D

Sources for the Standard Methods and ASTM sulfate methods are referenced above under methods for inorganic chemicals. The EPA methods are contained in "Methods for the Determination of Inorganic Substances in Environmental Samples," EPA-600/R-93-100, August 1993, which is available at NTIS, PB94-121811.

[a]Analysis for the 34 unregulated VOCs listed under paragraphs (e) and (j) of §141.40 shall be conducted using the following recommended methods, or their equivalent as determined by EPA.

[b]Analysis for the 13 unregulated SOCs listed under paragraph (n)(11) of §141.40 shall be conducted using the following recommended methods.

[c]A Standard Methods method.

[d]Analysis for the unregulated inorganic contaminant listed under paragraph (n)(12) of §141.40 shall be conducted using the following recommended methods.

available for inspection by the certification officer (Appendix A, Chain-of-Custody Evaluations).

6.1. Rejection of Samples

The laboratory must reject any sample taken for compliance purposes which does not meet the criteria in 6.2 through 6.6 and notify the authority requesting the analyses. See Appendix H, Section 1.3.

Table IV-5   Approved Methods for Disinfectant Residuals Public water systems must measure residual disinfectant concentrations with one of the analytical methods in the following table. The methods are contained in the 18th edition of *Standard Methods for the Examination of Water and Wastewater.*

Residual[a]	Methodology	SM[c]
Free chlorine[b]	Amperometric titration	4500-Cl D
	DPD ferrous titrimetric	4500-Cl F
	DPD colorimetric	4500-Cl G
	Syringaldahyde (FACTS)	4500-Cl H
Total chlorine[b]	Amperometric titration	4500-Cl D
	Amperometric titration (low level measurement)	4500-Cl E
	DPD ferrous titrimetric	4500-Cl F
	DPD colorimetric	4500-Cl G
	Iodometric electrode	4500-Cl I
Chlorine dioxide	Amperometric titration	4500-ClO$_2$ C
	DPD method	4500-ClO$_2$ D
	Amperometric titration	4500-ClO$_2$ E
Ozone	Indigo method	4500-O$_3$ B

[a]If approved by the State, residual disinfectant concentrations for free chlorine and combined chlorine also may he measured by using DPD colorimetric test kits.
[b]Free and total chlorine residuals may be measured continuously by adapting a specified chlorine residual method for use with a continuous monitoring instrument provided the chemistry, accuracy, and precision of the measurement remain the same. Instruments used for continuous monitoring must be calibrated with a grab sample measurement at least every 5 days, or with protocol approved by the State.
[c]*Standard Methods for the Examination of Water and Wastewater,* 18th Edition, 1992.

6.2.   Sample Containers and Preservation

The type of sample container and the required preservative for each inorganic and organic chemical contaminant are listed in Table IV-7. The use of "blue ice" is discouraged because it generally does not maintain the temperature of the sample at 4°C or less. If blue ice is used, it should be frozen at the time of sampling, the sample should be chilled before packing, and special notice must be taken at sample receipt to be certain the required temperature (4°C) has been maintained.

6.3.   Maximum Holding Times

Samples must be analyzed within the maximum holding times listed in Table IV-7.

Table IV-6 Recommended Methods for Secondary Drinking Water Contaminants Analyses of aluminum, chloride, color, copper, fluoride, foaming agents, iron, manganese, odor, silver, sulfate, total dissolved solids (TDS), and zinc to determine compliance under §143.3 may be conducted with the methods in the following table. Criteria for analyzing aluminum, copper, iron, manganese, silver, and zinc samples with digestion or directly without digestion, and other mandatory procedures are contained in the Technical Notes in Section IV of this document. Measurement of pH may be conducted with one of the methods listed above in Section I under "Methods for Inorganic Chemicals."

Contaminant	EPA	ASTM[a]	SM[b]	Other
Aluminum	200.7[c]		3120B	
	200.8[c]		3113B	
	200.9[c]		3111D	
Chloride	300.0[d]	D4327–91	4110B	
			4500-Cl-D	
Color			2120B	
Foaming agents			5540C	
Iron	200.7[c]		3120B	
	200.9[c]		3111B	
			3113B	
Manganese	200.7[c]		3120B	
	200.8[c]		3111B	
	200.9[c]		3113B	
Odor			2150B	
Silver	200.7[c]		3120B	I-3720–85[f]
	200.8[c]		3111B	
	200.9[c]		3113B	
Sulfate	300.0[d]	D4327–91	4110B	
	375.2[d]		4500-SO_4-F	
			4500-SO_4-C, D	
TDS			2540C	
Zinc	200.7[c]		3120B	
	200.8[c]		3111B	

[a]*Annual Book of ASTM Standards*, Vols. 11.01 and 11.02, American Society for Testing and Materials, 1916 Race Street, Philadelphia, PA 19103.

[b]*Standard Methods for the Examination of Water and Wastewater*, 18th Edition, 1992, American Public Health Association, 1015 Fifteenth Street NW, Washington, DC 20005.

[c]"Methods for the Determination of Metals in Environmental Samples—Supplement I," EPA-600/R-94–111, May 1994. Available at NTIS, PB94–184942.

[d]"Methods for the Determination of Inorganic Substances in Environmental Samples," EPA-600/R-93–100, August 1993. Available at NTIS, PB94–121811.

[e]Industrial Method No. 129–71W, "Fluoride in Water and Wastewater," December 1972, and Method No. 380–75WE, "Fluoride in Water and Wastewater," February 1976, Technicon Industrial Systems, Tarrytown, NY 10591.

[f]Available from Books and Open-File Reports Section, U.S. Geological Survey, Federal Center, Box 25425, Denver, CO 80225–0425.

Table IV-7 Preservation and Holding Times for Regulated Parameters

Parameter/method	Preservative	Sample holding time	Extract holding time	Suggested sample size	Type of container
Metals (except Hg)	HNO_3, pH < 2	6 months		1 L	Plastic or glass
Mercury	HNO_3, pH < 2	28 days		100 mL	Plastic or glass
Alkalinity	Cool, 4°C	14 days		100 mL	Plastic or glass
Asbestos	Cool, 4°C	48 hr			Plastic or glass
Chloride	None	28 days		50 mL	Plastic or glass
Residual disinfectant	None	Immediately		200 mL	Plastic or glass
Color	Cool, 4°C	48 hr		50 mL	Plastic or glass
Conductivity	Cool, 4°C	28 days		100 mL	Plastic or glass
Cyanide	Cool, 4°C, ascorbic acid (if chlorinated), NaOH, pH > 12	14 days		1 L	Plastic or glass
Fluoride	None	28 days		300 mL	Plastic or glass
Foaming agents	Cool, 4°C	48 hr			
Nitrate (chlorinated)	Cool, 4°C	28 days		100 mL	Plastic or glass
Nitrate (nonchlorinated)	Cool, 4°C, H_2SO_4, pH < 2	14 days		100 mL	Plastic or glass
Nitrite	Cool, 4C	48 hr		50 mL	Plastic or glass
Odor	Cool, 4°C	24 hr		200 mL	Glass
Ph	None	Immediately		25 mL	Plastic or glass
o-Phosphate	Filter immediately, Cool, 4°C	48 hr		50 mL	Plastic or glass

Method	Preservation	Holding time	Extract holding	Volume	Container
Silica	Cool, 4°C	28 days		100 mL	Plastic
Solids (TDS)	Cool, 4°C	7 days		100 mL	Plastic or glass
Sulfate	Cool, 4°C	28 days		50 mL	Plastic or glass
Temperature	None	Immediately		1 L	Plastic or glass
Turbidity	Cool, 4°C	48 hr		100 mL	Plastic or glass
502.2	Sodium thiosulfate or ascorbic acid, 4°C, HCl, pH <2	14 days		40–120 mL	Glass with teflon-lined septum
504.1	Sodium thiosulfate, cool, 4°C	14 days	4C, 24 hours	40 mL	Glass with teflon-lined septum
505	Sodium thiosulfate, cool, 4°C	14 days (7 days for heptachlor)	4C, 24 hours	40 mL	Glass with teflon-lined septum
506	Sodium thiosulfate, cool, 4°C, dark	14 days	4C, dark 14 days	1 L	Amber glass with teflon-lined cap
507	Sodium thiosulfate, cool, 4°C, dark	14 days (see method for exceptions)	4C, dark 14 days	1 L	Amber glass with teflon-lined cap
508	Sodium thiosulfate, cool, 4°C, dark	7 days (see method for exceptions)	4C, dark 14 days	1 L	Glass with teflon-lined cap
508A	Cool, 4°C	14 days	30 days	1 L	Glass with teflon-lined cap
508.1	Sodium sulfite, HCl, pH <2, cool, 4°C	14 days (see method for exceptions)	30 days	1 L	Glass with teflon-lined cap
515.1	Sodium thiosulfate, cool, 4°C, dark	14 days	4C, dark 28 days	1 L	Amber glass with teflon-lined cap
515.2	Sodium thiosulfate, HCl, pH <2, cool, 4°C, dark	14 days	≤4C, dark 14 days	1 L	Amber glass with teflon-lined cap

(*Continued*)

Table IV-7 (*Continued*)

Parameter/ method	Preservative	Sample holding time	Extract holding time	Suggested sample size	Type of container
524.2	Ascorbic acid, HCl, pH < 2, cool 4°C	14 days		40–120 mL	Glass with teflon-lined septum
525.2	Sodium sulfite, dark, cool, 4°C, HCl, pH < 2	14 days (see method for exceptions)	30 days from collection	1 L	Amber glass with teflon-lined cap
531.1, 6610	Sodium thiosulfate, monochloroacetic acid, pH < 3, cool, 4°C	Cool, 4°C, 28 days		60 mL	Glass with teflon-lined septum
547	Sodium thiosulfate, cool, 4°C	14 days (18 months frozen)		60 mL	Glass with teflon-lined septum
548.1	Sodium thiosulfate (HCl, pH 1.5–2 if high biological activity), cool, 4°C, dark	7 days	14 days, ≤4°C	≥250 mL	Amber glass with teflon-lined septum
549.1	Sodium thiosulfate, (H_2SO_4, pH < 2 if biologically active) cool, 4°C, dark	7 days	21 days	≥250 mL	High-density amber plastic or silanized amber glass
550, 550.1	Sodium thiosulfate, cool, 4°C, HCl pH < 2	7 days	550, 30 days, 550.1, 40 days, dark, 4°C	1 L	Amber glass with Teflon-lined cap

551	Sodium thiosulfate, Sodium sulfite, ammonium chloride, or ascorbic acid, HCL, pH 4.5–5.0, cool, 4°C	14 days	≥40 mL	Glass with teflon-lined septum
555	Sodium sulfite, HCl, pH ≤ 2, dark, cool, 4°C	14 days	≥100 mL	Glass with teflon-lined cap
1613B	Sodium thiosulfate, cool, 0–4°C, dark	Recommend 40 days	1 L	Amber glass with teflon-lined cap

6.4. Sample Collection and Transport

There must be strict adherence to correct sampling procedures, complete identification of the sample, and prompt transfer of the sample to the laboratory. When the laboratory is not responsible for sample collection and transport, it must verify that the paperwork, preservatives, containers, and holding times are correct or reject the sample. For more information, see Appendix H, Section 1.0.

6.5. Sample Collector

The sample collector should be trained in sampling procedures and have complete written sampling instructions (SOPs) for each type of sample to be collected. The sampler should be able to demonstrate proper sampling technique.

6.6. Sample Report Form

The sample report form should contain the ID, location, date and time of collection, collector's name, preservative added and shipping requirements, container and volume, sample type, analysis, and any special remarks concerning the sample. Indelible ink should be used. See Appendix H, Section 1.1.

6.7. Sample Compositing

Compositing of samples for inorganic and organic analyses must be done in the laboratory. Samples should only be composited if the laboratory detection limit is adequate for the number of samples being composited (up to a maximum of five) and the holding times will not be exceeded.

7. QUALITY CONTROL

Additional information is contained in Appendix H. Specific items are referenced throughout.

7.1. General Requirements

7.1.1. Availability of QC Information: All quality control information should be available for inspection by the certification officer.

7.1.2. Standard Operating Procedures: A manual of analytical methods and the laboratory's QA plan and SOPs should be readily available to the analysts (see Chapter III's discussion of Quality Assurance).

7.1.3. Balances and Weights: Balance range should be appropriate for the application for which it is to be used. Drinking water chemistry laboratories should use balances that weigh to at least 0.0001 g. The balances should be calibrated at least annually with ASTM Type I, Class 1 or 2 weights (ASTM, 1916 Race Street, Philadelphia, PA 19103). This may be done by laboratory personnel or under contract by a manufacturer's representative. We strongly recommend laboratories have a contract to calibrate balances due to the expense of the weights and to serve as an outside QC check of the weights and balances. Weights meeting ASTM Type I, Class 1 or 2 specifications should be recertified if there is reason to believe damage (corrosion, nicks) has occurred or at least every 5 years.

Laboratory personnel should perform at least weekly checks of the balance. Weights meeting ASTM Type 1 specifications may be used but should be calibrated annually against the reference weights at time of balance calibration. A record of all checks should be available for inspection. The checks and their frequency should be as prescribed in the laboratory's QA plan.

7.1.4. Color Standards: Wavelength settings on spectrophotometers should be verified each day they are used with color standards. A record of these checks should be available for inspection. The specific checks and their frequency should be as prescribed in the SOPs.

7.1.5. Thermometers: Thermometers should be traceable to ASTM calibration and recertified whenever the thermometer has been exposed to temperature extremes.

7.1.6. Traceability of Calibration: Calibrations of all measurement devices should be traceable to national standards whenever applicable.

7.2. Specific Requirements

The following are required for each analyte for which a laboratory is certified.

7.2.1. Performance Evaluation Samples: The laboratory must analyze performance evaluation samples (if available) acceptable to the certifying authority at least once per year in order to receive and maintain full certification for an analyte. Results from analysis of the performance evaluation sample must be within the acceptable limits established by the U.S. EPA. These acceptance limits are listed in Table IV-8. The laboratory should document the corrective actions taken when a PE sample is missed. A copy of this documentation should be forwarded to the certification officer. A make-up PE sample must be successfully analyzed. If problems arise, appropriate action must be taken as specified in Chapter 3, Implementation of Certification Program. See Appendix H, Section 2.5.6.

For VOCs, the laboratory may be certified for all VOCs (except vinyl chloride) or for the VOC method if they successfully analyze at least 80% of the regulated VOCs. The intention of the regulation was to allow some flexibility for random misses because the VOC methods include 20 analytes. The intention is not to allow a laboratory to be certified for an analyte which it misses repeatedly.

7.2.2. Quality Control Samples: At least once each quarter, the laboratory should analyze a quality control standard for the analytes they are analyzing in that quarter. The check standard should be prepared from a source other than that from which their standards are prepared. If errors exceed limits specified in the methods, corrective action must be taken and documented,

Table IV-8 Performance Evaluation Sample Acceptance Criteria in the
CFR. Primary and Secondary Drinking Water Regulations [§141.23(k)(3)(ii)
and 141.24(f)(17) and (19)]

Regulated parameter	MCL/[SMCL] (mg/L)	MCLG (mg/L)	Acceptance limit
Metals			
Aluminum	[0.05–0.2]	–	
Antimony	0.006	0.006	± 30%
Arsenic	0.05	–	
Barium	2	2	± 15%
Beryllium	0.004	0.004	± 15%
Cadmium	0.005	0.005	± 20%
Calcium	–	–	
Chromium	0.1	0.1	± 15%
Copper	1.3/90% [1.0]	1.3	± 10%
Iron	[0.3]	–	
Lead	0.015/90%	Zero	± 30%
Manganese	[0.05]	–	
Mercury	0.002	0.002	± 30%
Selenium	0.05	0.05	± 20%
Silica	–	–	
Silver	[0.1]		
Sodium	20[a]	–	
Thallium	0.002	0.0005	± 30%
Zinc	[5.0]	–	
Inorganics			
Alkalinity		–	
Asbestos	7MF/L > 10 u	7MF/L > 10 u	2 S.D.
Chloride	[250]	–	
Residual disinfectant	Detectable	–	
Color	[15 cu]	–	
Conductivity	–	–	
Corrosivity	[Noncorrosive]	–	
Cyanide	0.2	0.2	± 25%
Fluoride	4.0 [2.0]	–	± 10%
Foaming agents	[0.5]	–	
Nitrate (as N)	10	10	± 10%
Nitrite (as N)	1	1	± 15%
Nitrate/nitrite (as N)	10	10	
Odor	[3 ton]	–	
PH	6.5–8.5 [6.5–8.5]	–	
o-Phosphate	–	–	
Solids (TDS)	[500]	–	
Sulfate	Deferred [250]	Deferred	
Temperature	–	–	

(*Continued*)

Table IV-8 *(Continued)*

Regulated parameter	MCL/[SMCL] (mg/L)	MCLG (mg/L)	Acceptance limit
Volatiles			
Trihalomethanes (total)	0.10		±20%
Benzene	0.005	Zero	*
Carbon tetrachloride	0.005	Zero	*
Chlorobenzene	0.1	0.1	*
p-Dichlorobenzene	0.075 [0.005]	0.075	*
o-Dichlorobenzene	0.6	0.6	*
1,2-Dichloroethane	0.005	Zero	*
1,1-Dichloroethylene	0.007	0.007	*
c-1,2-Dichloroethylene	0.07	0.07	*
t-1,2-Dichloroethylene	0.1	0.1	*
Dichloromethane	0.005	Zero	*
1,2-Dichloropropane	0.005	Zero	*
Ethylbenzene	0.7	0.7	*
Styrene	0.1	0.1	*
Tetrachloroethylene	0.005	Zero	*
Toluene	1	1	*
1,2,4-Trichlorobenzene	0.07	0.07	*
1,1,1-Trichloroethane	0.2	0.2	*
1,1,2-Trichloroethane	0.005	0.003	*
Trichloroethylene	0.005	Zero	*
Vinyl chloride	0.002	Zero	±40%
Xylenes (total)	10	10	*
Synthetic organics			
Alachlor	0.002	Zero	±45%
Aldicarb	Postponed	Postponed	2 S.D.
Aldicarb sulfoxide	Postponed	Postponed	2 S.D.
Aldicarb sulfone	Postponed	Postponed	2 S.D.
Atrazine	0.003	0.003	±45%
Carbofuran	0.04	0.04	±45%
Chlordane	0.002	Zero	±45%
2,4-D	0.07	0.07	±50%
Dalapon	0.2	0.2	2 S.D.
Dibromochloropropane (DBCP)	0.0002	Zero	±40%
Dinoseb	0.007	0.007	2 S.D.
Diquat	0.02	0.02	2 S.D.
Endothall	0.1	0.1	2 S.D.
Endrin	0.002	0.002	±30%
Ethylene dibromide (EDB)	0.00005	Zero	±40%
Glyphosate	0.7	0.7	2 S.D.

(Continued)

Table IV-8   (*Continued*)

Regulated parameter	MCL/[SMCL] (mg/L)	MCLG (mg/L)	Acceptance limit
Heptachlor	0.0004	Zero	±45%
Heptachlor epoxide	0.0002	Zero	±45%
Lindane	0.0002	0.0002	±45%
Methoxychlor	0.04	0.04	±45%
Oxamyl (Vydate)	0.2	0.2	2 S.D.
Pentachlorophenol	0.001	Zero	±50%
Picloram	0.5	0.5	2 S.D.
Simazine	0.004	0.004	2 S.D.
Toxaphene	0.003	Zero	±45%
2,4,5-TP (Silvex)	0.05	0.05	±50%
Hexachlorobenzene	0.001	Zero	2 S.D.
Hexachlorocyclopentadiene	0.05	0.05	2 S.D.
Benzo(a)pyrene	0.0002	Zero	2 S.D.
PCBs (as decachlorobiphenyl)	0.0005	Zero	0–200%
2,3,7,8-TCDD (dioxin)	3×10^{-8}	Zero	2 S.D.
Acrylamide	Treatment	Zero	NA
Epichlorohydrin	Treatment	Zero	NA
Di(2-ethylhexyl)adipate	0.4	0.4	2 S.D.
Di(2-ethylhexyl)phthalate	0.006	Zero	2 S.D.

[a]Recommended level.
*The acceptance limits for VOCs are ±20% at ≥0.010 mg/L and ±40% at <0.010 mg/L.
NA—not applicable.

and a follow-up quality control standard analyzed as soon as possible to demonstrate the problem has been corrected.

7.2.3. Calibration Curve: Calibration requirements in the methods must be followed. If there are none, these are some guidelines to follow. At the beginning of each day that samples are to be analyzed, a calibration curve covering the sample concentration range and all target analytes should be generated according to the approved SOP. The curve should be composed of at least three concentrations, although some methods recommend that five concentrations be included. Some inorganic methods require at least a blank and three concentrations for each analyte. The calibration standards should be from a source different from the quality control standard used for 7.2.2.   Field measurements (e.g. pH and chlorine residual) should also be made on instruments which have been properly calibrated as specified in the method or instrument manual and checked daily. The less precise the measurement, the greater the number of concentrations which should be included in the calibration curve. EPA gas chromatography methods recommend using five standards. See Appendix H, Section 2.3.1.

7.2.4. Calibration Check: The calibration for some methods is so time consuming that 7.2.3 is impractical. For these methods, the calibration curve should be initially developed as specified in 7.2.3. Thereafter, at the beginning of each day on which analyses are performed, this curve should be verified by analysis of at least a reagent blank and one standard for each of the target analytes in the expected concentration range of the samples analyzed that day. All checks must be within the control limits specified in the method or the system recalibrated as specified in 7.2.3.

It is recommended that a calibration standard of one multicomponent analyte (PCBs, toxaphene, and chlordane) also be analyzed each day or work shift. By rotating the analyte chosen, continuing calibration data can be obtained on all the multicomponent analytes over a period of 1–2 weeks. If a positive for a multicomponent analyte is found in a sample, a calibration check for that analyte should be performed as soon as possible. Calibration requirements in the methods must be followed if different. See Appendix H, Section 2.4.

7.2.5. Blanks: A laboratory reagent blank should be carried through the full analytical procedure with every sample batch. In general, results from laboratory reagent blanks should not exceed the laboratory's method detection limit (MDL); see Section 7.2.8 and Appendix H, Section 2.5.4.

7.2.6. Laboratory Fortified Sample Matrix: The laboratory should add a known quantity of analytes to a minimum of 10% of the routine samples (except when the method specifies a different percentage, e.g. furnace methods or hone as in 524.2) to determine sample matrix interference. The fortified concentration should not be less than the background concentration of the sample selected for fortification unless specified by the method. If the sample concentration is unknown or less than detectable, the analyst should choose an appropriate concentration (e.g. a percentage of the MCL or midpoint in the calibration range). Over time, samples from all routine sample sources should be fortified as much as is practical. If any of these checks are not within the control limits specified in 7.2.7, and the laboratory performance is in control, the result for that sample must be labeled to inform the data user that the results are suspect due to matrix effects. See Appendix H, Section 2.5.2.

7.2.7. Control Charts: Control charts, generated from laboratory fortified blanks (LFBs), should be maintained by the laboratory. Until sufficient data are available from the laboratory, usually a minimum of 20–30 test results on a specific analysis, the laboratory should use the control limits (if specified) in the methods.

When sufficient data become available, the laboratory should develop FRB control charts from the mean percentage recovery ($\bar{x}$) and the standard deviation (S) of the percentage recovery for the QC checks specified above (see Chapter VI of the *Handbook for Analytical QC in Water and Wastewater Laboratories*, EPA-600/4-79-019 or *Standard Methods for the Examination of Water and Wastewater*, part 1020B, or similar QC reference texts for further information).

These data are used to establish upper and lower control limits as follows:

upper control limit $= \bar{x} + 3S$ (upper warning limit $+ 2S$)

lower control limit $= \bar{x} - 3S$ (lower warning limit $- 2S$)

After each 5–10 new recovery measurements, new control limits should be calculated using the most recent 20–30 data points. These calculated control limits should not exceed those established in the method. If any of these control limits are tighter than the method specifications, the laboratory should use the tighter criteria.

7.2.8. Initial Demonstration of Capability: Before beginning the analysis of compliance samples, an initial demonstration of capability (IDC) must be performed for each method. It is recommended that an IDC be performed for each analyst and instrument. In addition, it is recommended that the IDC also address the variability introduced if more than one sample preparation technician is used. Precision, accuracy, and MDL should be similar for each technician. The IDC includes a demonstration of the ability to achieve a low background, the precision and accuracy required by the method, the MDL in accordance with the procedure given in 40 CFR 136, Appendix B and satisfactory performance on an unknown sample. This definition is more encompassing than the definition of IDC in the methods. The IDC should be redone if there is a change in analyst, instrument, or a significant change in the method. Minor changes should prompt a check to ascertain that the precision, accuracy, and sensitivity have been maintained.

7.2.9. Quantitation of Multicomponent Analytes (Toxaphene, Chlordane and PCBs): The quantitation of multicomponent analytes requires professional judgment on the part of the analyst. This is required due to the complex nature of the chromatography involved, sample weathering and degradation and interferences that may be present in the samples. The pattern of peaks found in the sample should be examined carefully and compared to a standard. The peaks in the sample that match the peak ratios in the standard can be used in quantitation. Peaks that have obvious interferences (such as pesticides or phthalates or peaks exhibiting poor peak shape) or appear to have been degraded or weathered should not be used for quantitation. A representative number (5–9) of peaks is suggested. Peak area should be used for quantitation and the analyst should ensure that the samples and standards have been integrated in the same manner. Quantitation can be done by using the total peak area (comparing the area of the 5–9 peaks used for quantitation of the sample to the area of the standard) or by calculating each peak separately (using area) and taking the average concentration of the 5–9 peaks. Because of factors such as peak shape and baseline rise, the most accurate quantitation is obtained when the concentration of the sample closely matches that of the standard (e.g. within 20% of the standard). See EPA Method 8081, *Organochlorine Pesticides and PCRs as Aroclors by Gas Chromatography; Capillary Column Technique*, for a more detailed discussion of quantitation of multicomponent analytes.

Note: PCBs are qualitatively identified as Aroclors and measured for com- pliance purposes as decachlorobiphenyl. Chlordane is regulated as tech- nical chlordane, a mixture of at least 11 major components and 30 minor ones.

7.2.10. Periodic MDL Calculation: Method detection limit calculations are required by regulation for VOCs, lead, and composited samples. Table IV-9 lists the MDL requirements in the drinking water regulations. The CFR requires that an MDL of 0.0005 mg/L be attained for VOCs. For inorganics, an MDL of 1/5 of the MCL must be attained for compositing. Otherwise, the MCL is the required detection limit for inorganics. For synthetic organic che- micals (SOCs), the detection limits listed in the CFR must be attained for compositing, but are currently not required for certification. This is some- what contradictory, because the detection limits listed are the required mon- itoring triggers. Most methods require initial MDL calculations for all analytes. It is therefore recommended that the certification officers require the laboratories to calculate their detection limits for all regulated contami- nants until a change is made in the regulation. See Appendix H, Section 2.3 for more guidelines on MDLs.

It is recommended that sample preparation and analyses for the MDL calculation be made over a period of at least 3 days to provide a more realistic MDL. The analyst should recalculate MDLs when a significant change in the method, analyst, or instrument is made which would affect the MDL. In addi- tion, the analyst should regularly check the MDLs by analyzing LFBs at the same level as those that were used to calculate the MDL.

The calculation of MDLs by the CFR procedure may not be adequate for Aroclors, toxaphene, and chlordane, because they require pattern or peak profile recognition for identification. Presently, no standard procedure exists, so it is recommended that the MDL be defined as the lowest concentration for which pattern recognition is possible.

7.2.11. Low Level Quantitation: The laboratory's detection limits should be reported to the client along with the data. Many laboratories are uneasy reporting all detections above their MDLs because they realize that statisti- cally, contaminants present at the MDL concentration may be missed 50% of the time. Also, contaminants may be reported which cannot be confirmed by a second technique. Laboratories may prefer not to report contaminants at levels less than two to three times their MDL or below the level at which they routinely analyze their lowest standard. While this may be a scientifically sound practice, whether it is an acceptable practice will depend on State and Federal reporting requirements. It is important for users of data to under- stand the statistical and qualitative significance of the data. Laboratories may be required by the States to achieve a specific MDL or quantitation limit more stringent than that required by EPA.

7.2.12. Laboratory Fortified Blanks: The analyst should routinely verify the reporting limit (if one is used) for each analyte by analyzing an LFB at the reporting level. In addition, most methods require that an LFB be ana-

Table IV-9 Detection Limit Requirements in the CFR (mg/L)

Inorganics	MCL[a]	MCLG	Detection limit required to composite [§141.23(a)(4)]
Asbestos	7 MFL	7 MFL	1.4 MFL
Cyanide	0.2	0.2	0.04
Fluoride	4.0		0.8
Nitrate	10	10	2
Nitrite	1	1	0.2

Metals	MCL[b]	MCLG	Detection limit required to composite [§141.23(a)(4)]
Antimony	0.006	0.006	0.001
Arsenic	0.05	–	0.01
Barium	2	2	0.4
Beryllium	0.004	0.004	0.0008
Cadmium	0.005	0.005	0.001
Chromium	0.1	0.1	0.02
Copper[c]	1.3	1.3	0.001
			0.02 (for direct aspiration AA)
Lead[c]	0.015	Zero	0.001
Mercury	0.002	0.002	0.0004
Selenium	0.05	0.05	0.01
Thallium	0.002	0.0005	0.0004

Volatile organics[d]	MCL	MCLG	Required MDL
THMs	0.10	NA	NA
Benzene	0.005	Zero	0.0005
Carbon tetrachloride	0.005	Zero	0.0005
Chlorobenzene	0.1	0.1	0.0005
o-Dichlorobenzene	0.6	0.6	0.0005
p-Dichlorobenzene	0.075	0.075	0.0005
1,2-Dichloroethane	0.005	Zero	0.0005
1,1-Dichloroethylene	0.007	0.007	0.0005
c-1,2-Dichloroethylene	0.07	0.07	0.0005
t-1,2-Dichloroethylene	0.1	0.1	0.0005
Dichloromethane	0.005	Zero	0.0005
1,2-Dichloropropane	0.005	Zero	0.0005
Ethylbenzene	0.7	0.7	0.0005
Styrene	0.1	0.1	0.0005
Tetrachloroethylene	0.005	Zero	0.0005
Toluene	1	1	0.0005
1,2,4-Trichlorobenzene	0.07	0.07	0.0005
1,1,1-Trichloroethane	0.2	0.2	0.0005
1,1,2-Trichloroethane	0.005	0.003	0.0005

(Continued)

Table IV-9 *(Continued)*

Inorganics	MCL[a]	MCLG	Detection limit required to composite [§141.23(a)(4)]
Trichloroethylene	0.005	Zero	0.0005
Vinyl chloride	0.002	Zero	0.0004
Xylenes	10	10	0.0005

SOCs	MCL	MCLG	Monitoring trigger[e]
Alachlor	0.002	Zero	0.0002
Atrazine	0.003	0.003	0.0001
Benzo(a)pyrene	0.0002	Zero	0.00002
Carbofuran	0.04	0.04	0.0009
Chlordane	0.002	Zero	0.0002
2,4-D	0.07	0.07	0.0001
Di(2-ethylhexyl)adipate	0.4	0.4	0.0006
Di(2-ethylhexyl)phthalate	0.006	Zero	0.0006
Dibromochloropropane (DBCP)	0.0002	Zero	0.00002
Dalapon	0.2	0.2	0.001
Dinoseb	0.007	0.007	0.0002
Dioxin (2,3,7,8-TCDD)	3×10^{-8}	Zero	5×10^{-9}
Diquat	0.02	0.02	0.0004
Endothall	0.1	0.1	0.009
Endrin	0.002	0.002	0.00001
Ethylene dibromide (EDB)	0.00005	Zero	0.00001
Glyphosate	0.7	0.7	0.006
Heptachlor	0.0004	Zero	0.00004
Heptachlor epoxide	0.0002	Zero	0.00002
Hexachlorobenzene	0.001	Zero	0.0001
Hexachlorocyclopentadiene	0.05	0.05	0.0001
Lindane	0.0002	0.0002	0.00002
Methoxychlor	0.04	0.04	0.0001
Oxamyl	0.2	0.2	0.002
PCBs (as decachlorobiphenyl)	0.0005	Zero	0.0001
Pentachlorophenol	0.001	Zero	0.00004
Picloram	0.5	0.5	0.0001
Simazine	0.004	0.004	0.00007
Toxaphene	0.003	Zero	0.001
2,4,5-TP (Silvex)	0.05	0.05	0.0002

[a]The monitoring trigger for the inorganics is the MCL except for both nitrate and nitrite, which are 1/2 the MCL

[b]The monitoring trigger for metals is the MCL unless compositing, then 1/5 MCL is required.

[c]Action level.

[d]A laboratory must be able to achieve an MDL of 0.5 μg/L to be certified to analyze samples for compliance monitoring [§141.24(f)(17)(i)(E) and (ii)(C)]. This is also the monitoring trigger for VOCs [§141.24(f)(H)].

[e]The monitoring triggers for SOCs listed in the regulation are also required for compositing but are not required by regulation for certification [§141.24(g)(7), (10)(i) and (18)].

lyzed with each batch of samples at 10 times the MDL or a midlevel standard. Precision and accuracy data must be documented for this determination.

7.2.13. Qualitative Identification: The analyst should also verify at what concentration specificity (qualitative identification/confirmation) can be determined (e.g. a spectrum should be identifiable or second column confirmation should be possible).

8. RECORDS AND DATA REPORTING
8.1. Legal Defensibility

Compliance monitoring data should be made legally defensible by keeping thorough and accurate records. The QA plan and/or SOPs should describe the policies and procedures used by the facility for record retention and storage. If samples are expected to become part of a legal action, chain-of-custody procedures should be used (see Appendix A).

CHEMISTRY

Laboratory	
Street	
City, State	
Zip	
Telephone No. Fax No.	
Audit Team Leader	
Audit Team Members	
Audit Team Affiliation	
Date	

8.2. Maintenance of Records

Public water systems are required to maintain records of chemical analyses of compliance samples for 10 years (40 CFR 141.33) and lead and copper for 12 years (40 CFR 141.91). The laboratory should maintain easily accessible records for 5 years or until the next certification data audit is complete, whichever is longer. The client water system should be notified before disposing of records so they may request copies if needed. This includes all raw data, calculations, and quality control data. These data files may be either hard copy, microfiche or electronic. Electronic data should always be backed up by protected tape or disk or hard copy. If the laboratory changes its computer hardware or software, it should make provisions for transferring old data to the new system so that it remains retrievable within the time frames specified above. Data that are expected to become part of a legal action will probably need to be maintained for a longer period of time. Check with your legal counsel. See Appendix H, Section 3.0, and *Good Automated Laboratory Practices*, EPA 2185, Office of Information Management, Research Triangle Park, NC 27711, 8/10/95.

8.3. Sampling Records

Data should be recorded in ink with any changes lined through such that original entry is visible. Changes should be initialed and dated. The following information should be readily available in a summary or other record(s):

1. Date, location (including name of utility and PWSS ID #), site within the system, time of sampling, name, organization and phone number of the sampler, and analyses required;
2. Identification of the sample as to whether it is a routine distribution system sample, check sample, raw or finished water sample, repeat or confirmation sample, or other special purpose sample;
3. Date of receipt of the sample;
4. Sample volume/weight, container type, preservation and holding time, and condition on receipt;
5. pH and disinfectant residual at time of sampling (from plant records);
6. Transportation and delivery of the sample (person/carrier, conditions).

8.4. Analytical Records

Data should be recorded in ink with any changes lined through such that original entry is visible. Changes should be initialed and dated. The following information should be readily available:

1. Laboratory and persons responsible for performing analysis;
2. Analytical techniques/methods used;
3. Date and time of analysis;
4. Results of sample and quality control analyses;
5. Calibration and standards information.

8.5. Personnel Records

Resumes and training records should be maintained for all personnel. Analyst and technician IDC documentation should be kept on file as well as results of proficiency testing.

8.6. Reconstruction of Data

Adequate information should be available to allow the auditor to reconstruct the final results for compliance samples and performance evaluation samples. See Appendix H.

8.7. Computer Programs

Computer programs should be verified initially and periodically by manual calculations and the calculations should be available for inspection. Access to computer programs and electronic data should be limited to appropriate personnel.

9. ACTION RESPONSE TO LABORATORY RESULTS

When a laboratory is responsible, either by contract or State policy, to report sample results which would cause a system to be out of compliance, the proper authority should be promptly notified and a request should be made for resampling from the same sampling point immediately. See Chapter III.

Sample checklists for on-site evaluation of laboratories involved in analysis of public water supplies.

Laboratory _____ Evaluator _____

Location _____ Date _____

PHYSICAL FACILITY

Item	Acceptable Yes No	Comments
Environment		
Heating/Cooling/Humidity		
Lighting		
Ventilation/Exhaust hoods		
Cleanliness		
Electrical and water services		
Work Space		
Separation of incompatible testing areas		
Controlled access where appropriate		
Housekeeping		
Unencumbered access		
Adequate work space		
Storage		
Chemicals properly stored and dated		
Standards properly stored, dated and labeled with concentration, preparer's name and solvent, origin, purity & traceability		
Computers & automated equipment		
Safety procedures		

Laboratory _____ Evaluator _____

Location _____ Date _____

PERSONNEL (Use additional paper if necessary.)

Position/ Title	Name	Education Level Degree/Major*	Specialized Training	Present Specialty	Experience
Laboratory Director					
Manager					
Supervisors					
Instrument Operators					
AA					
TEM					
HPLC					
GC					
ICP					
GC/MS					
IC					
Other analysts					

	Yes	No	Comments
An organization chart available			
QA manager has line authority			
Personnel job descriptions and resumes available			
Personnel training documented			

*If the major is not in chemistry, list hours of college level courses in chemistry.

Laboratory _____ Evaluator _____

Location _____ Date _____

QUALITY ASSURANCE AND DATA REPORTING

Item	Comments	Satisfactory Yes	No
QA plan			
Organization			
Sampling SOPs available and used Preservation Containers Holding times Samplers trained			
Sample Rejection			
Laboratory sample handling Log in procedure Bound log book or secure computer log in Storage Tracking			
Analytical Methods Written methods available Approved methods used SOPs available and used			
Calibration Type and frequency Source of standards Data comparability Instrument tuning			
Blanks Trip Field Method			
Method Detection Limits Initial Frequency Acceptability			
Precision and Accuracy Initial Frequency Acceptability Control charts Laboratory fortified blanks Matrix duplicates			

Item	Comments	Satisfactory Yes	No
Other QC Checks Performance check samples Internal and surrogate standards Matrix spikes and replicates			
Qualitative Identification/ Confirmation			
Performance Evaluation Samples Analyzed			
Data Reduction and Validation Calculations Transcription Significant Figures Validation			
Preventive Maintenance			
Records Retention			
Corrective Action			

Laboratory ——————— Evaluator ———————

Location ——————— Date ———————

Item	No. of Units	Method	Manufacturer	Model	Satisfactory Yes No	
Analytical Balance 0.1 mg readability Stable base ASTM type 1 or 2 weights (formerly Class S) Service contracts						
Magnetic Stirrer Variable speed, TFE coated stir bar						
pH Meter Accuracy ±0.1 units Line or battery Usable with specific ion electrodes						
Conductivity Meter Readable in ohms or mhos Range of 2 ohms to 2 mhos Line or battery						
Hot Plate - temp control						
Centrifuge To 3000 rpm, Option of 4 x 50 mL						
Color Standards To verify wavelengths photometers Should cover 200-800 nm						
Refrigerator/Freezer Standard laboratory, explosion proof for organics Capable of maintaining nominal temperature of 4C						

Item	No. of Units	Method	Manufacturer	Model	Satisfactory Yes	No
Drying Oven Gravity or convection Controlled from room temp to 180°C or higher(±2°C)						
Muffle Furnace To 450°C for cleaning organic glassware						
Thermometer Mercury filled Celsius 1°C or finer subdivision to 180°C NBS Certified or traceable						
Glassware Borosilicate Volumetrics should be Class A						
Spectrophotometer Range 400 - 700 nm Band width - < 20 nm Use several size & shape cells Path length 1 - 5 cm		Cyanide, Fluoride Disinfectants Mercury Nitrate/Nitrite o-Phosphate Sulfate , Silica				
Filter Photometer Range 400 - 700 nm Band width 10 -70 nm Use several size & shape cells Pathlength 1 - 5 cm		same as above				
Amperometric Titrator		Disinfectants				
Specific Ion Meter Accuracy ± 1 mV		Cyanide Fluoride Nitrate				

Item	No. of Units	Method	Manufacturer	Model	Satisfactory Yes	No
Inductively Coupled Plasma (sequential, simultaneous) Computer controlled Background correction Radio frequency generator Argon gas supply Mass Spectrometer Range 5-250 amu Resolution 1 amu peak width at 5% peak height		200.7, 3120B 200.8				
Water Bath Electric or steam heat Controllable within 5°C to 100°C		Mercury Nitrate Pesticides				
Ion Chromatograph Conductivity detector, UV detector Suppressor column, Separator column		Fluoride , Chloride Nitrate/Nitrite o-Phosphate Sulfate				
Atomic Absorption Spectrophotometer Single channel, Single or double beam Grating monochrometer Photo multiplier detector Adjustable slits, Range 190-800 nm Readout system: Response time compatible with AA Able to detect positive interference for furnace Chart recorder, CRT or hard copy printer		Metals				
Air/Acetylene commercial grade		Barium, Calcium Nickel, Sodium Copper				

Item	No. of Units	Method	Manufacturer	Model	Satisfactory Yes No
Nitrous Oxide - comm. grade		Barium			
Graphite Furnace Argon or Nitrogen (commercial grade) Reach required temperatures Background corrector provision for offline analysis Pipets and tips microliter capacity with disposable tips 5-100 microliters metal free tips		Antimony, Lead Arsenic, Barium Beryllium Cadmium, Nickel Chromium Selenium Thallium Copper			
Arsine Generator		Arsenic, Selenium			
Hydride Generator hydrogen, commercial grade		Antimony Arsenic, Selenium			

Mercury Analyzer
Spectrophotometer
Dedicated analyzer having a mercury lamp
  acceptable
Adsorption cell:
10 cm quartz cell with quartz end windows or
  11.5 cm plexiglass cell with 2.5 cm ID
Air pump to deliver flow of at least
  1 L/min
Aeration tube with coarse glass frit
Flowmeter to measure air flow of 1 L/min
Drying unit:
6 in. tube with 20 grams magnesium
  perchlorate or heating device or lamp to
  prevent condensation on cell
		Mercury			

Glassware Separatory funnels Kuderna Danish (K-D) concentrators		SOCs			

Item	No. of Units	Method	Manufacturer	Model	Satisfactory Yes	No
Gas Chromatograph Split/splitless injection Oven temp. control ± 0.2°C Recorder, hard copy Oven temp. programmer Sub-ambient accessory Variable-constant differential flow control		Organics				
Electron Capture detector Linearized		504.1, 505 508, 508.1 508A, 515.1 515.2, 551, 552.1				
Electrolytic Conductivity/Photoionization detector		502.2 506 (PID only)				
Nitrogen Phosphorus detector		507				
Mass spectrometer (quadrupole or ion trap) All glass enrichment device All glass transfer line Electron ionization at ≥ 70 eV Scanning 35-260 amu ≤ 2 sec Interfaced data system		524.2, 525.2 548.1				
Purge & Trap system All glass purger 5/25 mL sample size		502.2, 524.2				

Item	No. of Units	Method	Manufacturer	Model	Satisfactory Yes No
High Performance Liquid Chromatograph Constant flow Capable of injecting 20-500 μL					
Gradient system post-column reactor fluorescence detector excitation at 330 nm (230?) detection at >418 nm		531.1, 6610			
Gradient system UV detector at 254 nm fluorescence detector excitation at 280 nm detection at > 389 nm		550, 550.1			
Isocratic system photodiode array detector excitation at 257 nm detection at > 308 nm		549.1			
Isocratic system post-column reactor fluorescence detector excitation at 340 nm detection at >455 nm*		547			
Gradient system photodiode array/UV detector 210-310 nm		555			

Item	No. of Units	Method	Manufacturer	Model	Satisfactory Yes No	
Auto Analysis System multi-channel pump manifold, colorimeter		Cyanide, Silica Fluoride Nitrate/nitrite o-Phosphate Sulfate				
Transmission Electron Microscope 80 kV 300-100,000X magnification 1 mm resolution calibrate screen SAED and ED		Asbestos				

Laboratory _____ Evaluator _____

Location _____ Date _____

METHODOLOGY

Contaminant	Method(s) Name/Number and revision	Reference Cite source, year, page	Samples/Mo	Satisfactory Yes	No
Antimony					
Arsenic					
Barium					
Beryllium					
Cadmium					
Chromium					
Copper					
Lead					
Mercury					
Selenium					
Thallium					
Alkalinity					
Calcium					
o-Phosphate					
Silica					
Temperature					

Contaminant	Method(s) Name/Number and revision	Reference Cite source, year, page	Samples/Mo	Satisfactory Yes No
Asbestos				
Cyanide				
Fluoride				
Nitrate				
Nitrite				
Total THMs				
VOCs				
Herbicides				
Pesticides				
EDB/DBCP				
Dioxin				
Other SOCs				
PCBs				
Carbamates				
Diquat				
Endothall				
Glyphosate				
Chlorine				
Chlorine dioxide				
Ozone				

Contaminant	Method(s) Name/Number and revision	Reference Cite source, year, page	Samples/Mo	Satisfactory Yes No
Unregulated VOCs				
Unregulated Pesticides				
Unregulated Herbicides				
Unregulated Carbamates				
Aluminum				
Chloride				
Color				
Foaming agents				
Iron				
Manganese				
Nickel				
Odor				
Silver				
Sodium				
Sulfate				
TDS				
Turbidity				
Zinc				

Laboratory _____ Evaluator _____

Location _____ Date _____

SAMPLE COLLECTION

Item	Comments	Satisfactory Yes	No
Trained Sample Collector			
Representative sampling			
Complete sample form			
Appropriate sampling and preservation			
Samples exceeding holding times discarded			
VOCs & THMs Hermetic seal			

Laboratory _____ Evaluator _____

Location _____ Date _____

SAMPLE HANDLING AND PRESERVATION

Contaminant	Container material & size	Preservatives	Holding Time Sample	Extract	Satisfactory Yes	No
Mercury						
Metals						
Silica						
Asbestos						
Cyanide						
Fluoride						
Nitrate						
Nitrite						
Alkalinity						
o-Phosphate						
Total THMs						
VOCs						
Herbicides						
Pesticides						
EDB/DBCP						

Contaminant	Container material & size	Preservatives	Holding Time Sample	Holding Time Extract	Satisfactory Yes	Satisfactory No
Dioxin						
Other SOCs						
Carbamates						
Diquat						
Endothall						
Glyphosate						
Residual Disinfectants						
Conductivity						
pH						
Temperature						
Turbidity						
Chloride						
Color						
Foaming agents						
Odor						
Sulfate						
TDS						

CHAPTER V

CRITICAL ELEMENTS FOR MICROBIOLOGY

Note: This chapter uses the term "must" to refer to certification criteria that are required by the National Primary Drinking Water Regulations which include the approved drinking water methods. The term "should" is used for procedures that, while not specifically required by the regulations, are considered good laboratory practice for quality assurance. To assure the validity of the data, it is critical that laboratories observe both the regulatory and nonregulatory criteria. Certification Officers have the prerogative to refuse certification if the quality control data are judged unsatisfactory or insufficient.

Note: References to *Standard Methods for the Examination of Water and Wastewater* are to the 18th Edition (1992).

1. PERSONNEL

1.1. Supervisor/Consultant

The supervisor of the microbiology laboratory should have a bachelor's degree in microbiology, biology, or equivalent. Supervisors who have a degree in a subject other than microbiology should have had at least one college-level microbiology laboratory course in which environmental microbiology was covered. In addition, the supervisor should have a minimum of 2 weeks training at a Federal agency, State agency, or academic institution in microbiological analysis of drinking water or, 80 hr of on-the-job training in water microbiology at a certified laboratory, or other training acceptable to the State or EPA. If a supervisor is not available, a consultant having the same qualifications may be substituted, as long as the laboratory can document that the consultant is acceptable to the State and is present on-site frequently enough to satisfactorily perform a supervisor's duties.

The laboratory supervisor has the responsibility to insure that all laboratory personnel have demonstrated their ability to satisfactorily perform the analyses to which they are assigned and that all data reported by the laboratory meet the required quality assurance and regulatory criteria.

1.2. Analyst (Or Equivalent Job Title)

The analyst should perform microbiological tests with minimal supervision, and have at least a high school education. In addition, the analyst should have a minimum of at least 3 months of bench experience in water, milk, or food microbiology. The analyst should also have training acceptable to the State (or EPA for nonprimacy States), in microbiological analysis of drinking water and a minimum of 30 days of on-the-job training under an experienced analyst. Analysts should take advantage of workshops and training programs that may be available from State regulatory agencies and professional societies. Before analyzing compliance samples, the analyst must demonstrate acceptable results for precision, specificity, and satisfactory analysis on unknown samples.

1.3. Waiver of Academic Training Requirement

The certification officer may waive the need for the above-specified academic training, on a case-by-case basis, for highly experienced analysts.

1.4. Personnel Records

Personnel records which include academic background, specialized training courses completed and types of microbiological analyses conducted, should be maintained on laboratory analysts.

2. LABORATORY FACILITIES

Laboratory facilities should be clean, temperature and humidity controlled, and have adequate lighting at bench tops. They should have provisions for disposal of microbiological waste. Laboratory facilities should have sufficient bench-top area for processing samples; storage space for media, glassware, and portable equipment; floor space for stationary equipment (incubators, water baths, refrigerators, etc.); and associated area(s) for cleaning glassware and sterilizing materials.

3. LABORATORY EQUIPMENT AND SUPPLIES

The laboratory must have the equipment and supplies needed to perform the approved methods for which certification has been requested.

3.1. pH Meter

3.1.1. Accuracy and scale graduations must be within ± 0.1 units.

3.1.2. pH buffer aliquots should be used only once.

3.1.3. Electrodes should be maintained according to the manufacturer's recommendations.

QC 3.1.4. pH meters should be standardized before each use period with pH 7.0 and either pH 4.0 or 10.0 standard buffers, whichever range covers the desired pH of the media or reagent. The date and buffers used should be recorded in a logbook.

QC 3.1.5. Commercial buffer solution containers should be dated upon receipt, and when opened. Buffers should be discarded before the expiration date.

3.2. Balance (Top Loader or Pan)

3.2.1. Balances should have readability of 0.1 g

QC 3.2.2. Balances should be calibrated monthly using ASTM type 1, 2, or 3 weights (minimum of three traceable weights which bracket laboratory weighing needs). (ASTM, 1916 Race Street, Philadelphia, PA 19103) Nonreference weights should be calibrated every 6 months with reference weights.

QC 3.2.3. Service contracts or internal maintenance protocols and maintenance records should be available. Maintenance should be conducted annually at a minimum. A record of the most recent calibration should be available for inspection. Correction values should be on file and used. A reference weight should be recertified if it is damaged or corroded.

3.3. Temperature Monitoring Device

3.3.1. Glass, dial, or electronic thermometers must be graduated in 0.5°C increments (0.2°C increments for tests which are incubated at 44.5°C) or less. The fluid column in glass thermometers should not be separated. Dial thermometers that cannot be calibrated should not be used.

QC 3.3.2. Calibrations of glass and electronic thermometers should be checked annually and dial thermometers quarterly, at the temperature used, against a reference National Institute of Standards and Technology [formerly National Bureau of Standards (NBS)] thermometer or one that meets the requirements of NBS Monograph SP 250–23. The calibration factor should be indicated on the thermometer. Also, the laboratory should record the date the thermometer was calibrated and the calibration factor in a QC record book.

QC 3.3.3. If a thermometer differs by more than 1°C from the reference thermometer, it should be discarded. Reference thermometers should be recalibrated every 3 years.

QC 3.3.4. Continuous recording devices that are used to monitor incubator temperature should be recalibrated at least annually. A reference thermometer that meets the specifications described in Paragraph 3.3.2. should be used for calibration.

3.4. Incubator Unit

3.4.1. Incubator units must have an internal temperature monitoring device and maintain a temperature of 35 ± 0.5°C, and if used, 44.5 ± 0.2°C. For nonportable incubators, thermometers should be placed on the top and bottom shelves of the use area with the thermometer bulb immersed in liquid (except for electronic thermometers). If an aluminum block incubator is used, culture dishes and tubes should fit snugly. Laboratories which use the chromogenic/fluorogenic substrate tests with air-type incubators should note the caution indicated in 5.6.8.

QC   3.4.2.   Calibration-corrected temperature should be recorded for days in use at least twice per day with readings separated by at least 4 hr.

3.4.3.   An incubation temperature of 44.5 ± 0.2°C can best be maintained with a water bath equipped with a gable cover and a pump or paddles to circulate water.

3.5. Autoclave

3.5.1.   The autoclave should have an internal heat source, a temperature gauge with a sensor on the exhaust, a pressure gauge, and an operational safety valve. The autoclave should maintain a sterilization temperature during the sterilizing cycle and complete an entire cycle within 45 min when a 12–15 min sterilization period is used. The autoclave should depressurize slowly enough to ensure that media will not boil over and bubbles will not form in inverted tubes.

3.5.2.   Because of safety concerns and difficulties with operational control, pressure cookers should not be used.

QC   3.5.3.   The date, contents, sterilization time and temperature, total time for each cycle, and analyst's initials should be recorded each time the autoclave is used. A copy of the service contract or internal maintenance protocol and maintenance records should be kept. Maintenance should be conducted annually at a minimum. A record of the most recent service performed should be available for inspection.

QC   3.5.4.   A maximum-temperature-registering thermometer or continuous recording device should be used during each autoclave cycle to ensure that the proper temperature was reached, and the temperature recorded. Overcrowding should be avoided. Spore strips or ampules should be used monthly to confirm sterilization.

QC   3.5.5.   Automatic timing mechanisms should be checked quarterly with a stopwatch or other accurate timepiece or time signal.

3.5.6.   Autoclave door seals should be clean and free of caramelized media. Also, autoclave drain screens should be cleaned frequently and debris removed.

3.6. Hot Air Oven

3.6.1.   The oven should maintain a stable sterilization temperature of 170–180°C for at least 2 hr. Only dry items should be sterilized with a hot air oven.

Overcrowding should be avoided. The oven thermometer should be graduated in 10°C increments or less, with the bulb placed in sand during use.

QC   3.6.2.   The date, contents, sterilization time and temperature of each cycle, and analyst's initials should be recorded.

QC   3.6.3.   Spore strip or ampule should be used on a monthly basis to ensure sterility of items.

3.7. Colony Counter

A dark field colony counter should be used to count heterotrophic plate count colonies.

3.8. Conductivity Meter

3.8.1.   Meters should be suitable for checking laboratory reagent-grade water and readable in appropriate M units (micromhos or microsiemens per centimeter). Use an instrument capable of measuring conductivity with an error not exceeding 1% or $1\,\mu$/cm, whichever is more lenient.

QC   3.8.2.   Cell constant should be determined monthly using a method indicated in Section 2510, "Conductivity," in *Standard Methods*. Monthly calibration checks using an appropriate certified and traceable low-level standard may be substituted for determining the cell constant.

3.8.3.   If an in-line unit cannot be calibrated, it should not be used to check reagent-grade water.

3.9. Refrigerator

3.9.1.   Refrigerators should maintain a temperature of 1–5°C. Thermometers should be graduated in at least 1°C increments and the thermometer bulb immersed in liquid.

QC   3.9.2.   The temperature should be recorded for days in use at least once per day.

3.10. Inoculating Equipment

Sterile metal or disposable plastic loops, wood applicator sticks, sterile swabs, or sterile plastic disposable pipet tips should be used. If wood applicator sticks are used, they should be sterilized by dry heat. The metal inoculating loops and/or needles should be made of nickel alloy or platinum. (For the coliform test and any other oxidase test used for the verification of membrane filter colonies, nickel alloy loops must not be used because they may interfere with the oxidase test.)

3.11. Membrane Filtration Equipment (If MF Procedure Is Used)

3.11.1.   MF units must be stainless steel, glass, or autoclavable plastic, not scratched or corroded, and must not leak.

QC   3.11.2.   If graduation marks on clear glass or plastic funnels are used to measure sample volume, their accuracy should be checked with a standard graduated cylinder, and a record of this calibration check retained. Tolerance should be $\leq 2.5\%$.

3.11.3.   A $10-15 \times$ stereomicroscope with a fluorescent light source must be used to count sheen colonies.

3.11.4.   Membrane filters must be approved by the manufacturer for total coliform water analysis. Approval is based on data from tests for toxicity, recovery, retention, and absence of growth-promoting substances. Filters must be cellulose ester, white, grid marked, 47 mm diameter, and $0.45 \mu m$ pore size, or alternate pore sizes if the manufacturer provides performance data equal to or better than the $0.45 \mu m$ pore size. Membrane filters must be purchased presterilized or autoclaved for 10 min at 121°C before use.

QC   3.11.5.   The lot number for membrane filters and the date received should be recorded.

3.12. Culture Dishes (Loose or Tight Lids)

3.12.1.   Presterilized plastic or sterilizable glass culture dishes must be used. To maintain sterility of glass culture dishes, stainless steel or aluminum canisters, or a wrap of heavy aluminum foil or char-resistant paper, must be used.

3.12.2.   Loose-lid petri dishes should be incubated in a tight-fitting container, e.g. plastic vegetable crisper containing a moistened paper towel to prevent dehydration of membrane filter and medium.

3.12.3.   Opened packs of disposable culture dishes should be resealed between use periods.

3.13. Pipets

3.13.1.   To sterilize and maintain sterility of glass pipets, stainless steel or aluminum canisters should be used, or individual pipets should be wrapped in char-resistant paper or aluminum foil.

3.13.2.   Pipets must have legible markings and should not be chipped or etched.

3.13.3. Opened packs of disposable sterile pipets should be resealed between use periods.

3.13.4. Pipets delivering volumes of 10 mL or less must be accurate within a 2.5% tolerance.

3.13.5. Calibrated micropipetters may be used if tips are sterile. Micropipetters should be calibrated annually and replaced if the tolerance is greater than 2.5%.

3.14. Culture Tubes and Closures

3.14.1. Tubes should be made of borosilicate glass or other corrosion-resistant glass or plastic.

3.14.2. Culture tubes and containers should be of sufficient size to contain medium plus sample without being more than three quarters full.

3.14.3. Tube closures should be stainless steel, plastic, aluminum, or screw caps with nontoxic liners. Cotton plugs should not be used.

3.15. Sample Containers

3.15.1. Sample containers must be wide-mouth plastic or noncorrosive glass bottles with nonleaking ground glass stoppers or caps with nontoxic liners that should withstand repeated sterilization, or sterile plastic bags containing sodium thiosulfate. Other appropriate sample containers may be used. The capacity of sample containers should be at least 120 mL (4 oz.).

3.15.2. Glass stoppers must be covered with aluminum foil or char-resistant paper for sterilization.

3.15.3. Glass and plastic bottles that have not been presterilized should be sterilized by autoclaving or, for glass bottles, by dry heat. Empty containers should be moistened with several drops of water before autoclaving to prevent an "air lock" sterilization failure.

3.15.4. If chlorinated water is to be analyzed, sufficient sodium thiosulfate ($Na_2S_2O_3$) must be added to the sample before sterilization to neutralize any residual chlorine in the water sample. Dechlorination is addressed in Section 9060A of Standard Methods.

3.16. Glassware and Plasticware

3.16.1. Glassware must be borosilicate glass or other corrosion-resistant glass and free of chips and cracks. Markings on graduated cylinders and pipets must be legible. Plastic items must be clear and nontoxic to microorganisms.

QC   3.16.2.   Graduated cylinders for measurement of sample volumes must have a tolerance of 2.5% or less. In lieu of graduated cylinders, precalibrated containers that have clearly marked volumes of 2.5% tolerance may be used. The calibration of each new lot of precalibrated containers should be validated by selecting at least one container at random and checking the calibration using a previously verified graduated cylinder.

3.17.   Ultraviolet Lamp (If Used)

3.17.1.   The unit should be disconnected monthly and the lamps cleaned by wiping with a soft cloth moistened with ethanol.

QC   3.17.2.   If a UV lamp (254 nm) is used for sanitization, the lamp should be tested quarterly with a UV light meter or agar spread plate. The lamp should be replaced if it emits less than 70% of its initial output or if an agar spread plate containing 200–250 micro-organisms, exposed to the UV light for 2 min, does not show a count reduction of 99%. Other methods may be used to test a lamp if data demonstrate that they are as effective as the two suggested methods.

4.   GENERAL LABORATORY PRACTICES

Although safety criteria are not covered in the laboratory certification program, laboratory personnel should be aware of general and customary safety practices for laboratories. Each laboratory is encouraged to have a safety plan available.

4.1.   Sterilization Procedures

4.1.1.   Required times for autoclaving at 121°C are listed below. The items must be at temperature for this required amount of time. Except for membrane filters and pads and carbohydrate-containing media, indicated times are minimum times which may necessitate adjustment depending upon volumes, containers, and loads.

Item	Time (min)
Membrane filters and pads	10
Carbohydrate containing media	12–15
Contaminated test materials	30
Membrane filter assemblies	15
Sample collection bottles	15
Individual glassware	15
Dilution water blank	15
Rinse water (0.5–1 L)	15–30[a]

[a]Time depends upon water volume per container and autoclave load.

4.1.2. Autoclaved membrane filters and pads and all media should be removed immediately after completion of the sterilization cycle.

4.1.3. Membrane filter equipment must be autoclaved before the beginning of the first filtration series. A filtration series ends when 30 min or longer elapses after a sample is filtered.

4.1.4. Ultraviolet light (254 nm) may be used as an alternative to sanitize equipment, if all supplies are presterilized and QC checks are conducted as indicated in Paragraph 3.17.2. Ultraviolet light may also be used to control bacterial carry-over between samples during a filtration series.

4.2. Sample Containers

4.2.1. See Section 6.2 for sample preservation.

QC 4.2.2. At least one sample container should be selected at random from each batch of sterile sample bottles or other containers, and sterility confirmed by adding approximately 25 mL of a sterile nonselective broth (e.g. tryptic soy, trypticase soy, or tryptone broth). The broth should be incubated at 35 $\pm$ 0.5°C for 24 hr and checked for growth. Resterilize if growth is detected.

4.3. Reagent-Grade Water

4.3.1. Only satisfactorily tested reagent water from stills or deionization units may be used to prepare media, reagents, and dilution/rinse water for performing bacteriological analyses.

QC 4.3.2. The quality of the reagent water should be tested and should meet the following criteria:

Parameter	Limits	Frequency
Conductivity	$< 2\mu$ J/ cm (μS/cm) at 25°C	Monthly
Pb, Cd, Cr, Cu, Ni, Zn	Not greater than 0.05 mg/L per contaminant. Collectively, no greater than 0.1 mg/L	Annually
Total chlorine residual[a]	< 0.1 mg/L	Monthly
Heterotrophic plate count[b]	< 500/mL	Monthly
Bacteriological quality of reagent water[c]	Ratio of growth rate 0.8:3.0	Annually

[a]DPD Method should be used. Not required if source water is not chlorinated.
[b]Pour Plate Method. See Standard Methods 9215B.
[c]See Standard Methods, Section 9020B. This bacteriological quality test is not needed for ASTM (ASTM, 1916 Race Street, Philadelphia, PA 19103) Types 1 reagent water, as defined in Standard Methods, Section 1080.

4.4. Dilution/Rinse Water

4.4.1. Stock buffer solution or peptone water should be prepared, as specified in Standard Methods, Section 9050C.

4.4.2. Stock buffers should be autoclaved or filter sterilized, and containers should be labeled and dated. Stock buffers should be refrigerated. Stored stock buffers should be free from turbidity.

QC   4.4.3.   Each batch of dilution/rinse water should be checked for sterility by adding 50 mL of water to 50 mL of a double strength nonselective broth (e.g. tryptic soy, trypticase soy, or tryptose broth). Incubate at 35 $\pm$ 0.5°C for 24 hr and check for growth. Discard if growth is detected.

4.5. Glassware Washing

4.5.1. Distilled or deionized water should be used for final rinse.

QC   4.5.2.   A glassware inhibitory residue test (Standard Methods, Section 9020B) should be performed before the initial use of a washing compound and whenever a different formulation of washing compound, or washing procedure, is used. In addition, batches of dry glassware should be spot checked occasionally for pH reaction, especially if glassware is soaked in alkali or acid (Standard Methods, Section 9020B). These tests will ensure that glassware is at a neutral pH and is free of toxic residue.

4.5.3. Laboratory glassware should be washed with a detergent designed for laboratory use.

5. ANALYTICAL METHODOLOGY
5.1. General

5.1.1. For compliance samples, laboratories must only use the analytical methodology specified in the Total Coliform Rule [40 CFR 141.21(f)] and the Surface Water Treatment Rule [40 CFR 141.74(a)].

5.1.2. A laboratory must be certified for all analytical methods, indicated below, that it uses for compliance purposes. At a minimum, the laboratory must be certified for one total coliform method and one fecal coliform or *E. coli* method. A laboratory should also be certified for a second total coliform method if one method cannot be used for some drinking waters (e.g. where the water usually produces confluent growth on a plate). In addition, for principal State laboratories and other laboratories that may enumerate heterotrophic bacteria (HPC) for compliance with the Surface Water Treatment Rule, the laboratory must be certified for the Pour Plate Method, the only method approved for heterotrophic bacteria.

5.1.3. Absorbent pads must be saturated with a liquid medium (at least 2 mL of broth) and excess medium removed by "decanting" the plate.

5.1.4. Water samples should be shaken vigorously about 25 times before analyzing.

QC   5.1.5. If no total coliform-positive result occurs during a quarter, the laboratory should perform the coliform procedure using a known coliform-positive, fecal coliform and/or *E. coli*-positive control to spike the sample.

5.1.6. Sample volume analyzed for total coliforms in drinking water must be 100 ± 2.5 mL.

5.1.7. Media

5.1.7.1. The use of dehydrated or prepared media manufactured commercially is strongly recommended due to concern about quality control. Dehydrated media should be stored in a cool, dry location. Caked or discolored dehydrated media should be discarded.

QC   5.1.7.2. For media prepared in the laboratory, the date of preparation, type of medium, lot number, sterilization time and temperature, final pH, and the technician's initials should be recorded.

QC   5.1.7.3. For liquid media prepared commercially, the date received, type of medium, lot number, and pH verification should be recorded. Medium should be discarded by manufacturer's expiration date.

QC   5.1.7.4. Each new lot of dehydrated or prepared commercial medium should be checked before use with positive and negative culture controls. In addition, each batch of laboratory-prepared medium should include positive and negative culture controls. These control organisms can be stock cultures (periodically checked for purity) or commercially available disks impregnated with the organism. Results should be recorded.

5.1.7.5. Prepared plates may be refrigerated in sealed plastic bags or containers. Because of potential evaporation, they may not be kept for more than 2 weeks. Each bag or container should include the date prepared or an expiration date. Broth in loose-cap tubes should be stored at < 30°C no longer than 2 weeks. Broth in tightly capped tubes should be stored at < 30°C no longer than 3 months.
    When ready to use, the refrigerated sterilized medium should be incubated overnight at room temperature; media with growth should be discarded.

QC   5.1.8. Laboratories are encouraged to perform parallel testing between a newly approved test and another EPA-approved procedure for enumerating total coliforms for at least several months and/or over several seasons to assess the effectiveness of the new test for the wide variety of water types submitted for analysis. During this testing, spiking the samples occasionally with sewage or a pure culture may be necessary to ensure that some of the tests are positive.

5.2. Membrane Filter (MF) Technique (For Total Coliforms in Drinking Water)

5.2.1. Media

5.2.1.1. M-Endo medium broth or agar (also known as M-Endo broth MF and M-Coliform broth) or LES Endo agar (also known as M-Endo Agar LES) must be used in the single step or enrichment techniques. Ensure that ethanol used in the rehydration procedure is not denatured. Medium must be prepared in a sterile flask and a boiling water bath must be used or, if constantly attended, a hot plate with a stir bar may be used, to bring the medium just to the boiling point. The medium must not be boiled. pH must be 7.2 ± 0.2 for LES Endo agar and 7.2 ± 0.1 for M-Endo medium.

5.2.1.2. MF broth must be refrigerated no longer than 96 hr, poured MF agar plates no longer than 2 weeks, and ampuled M-Endo broth in accordance with the manufacturer's expiration date. Broth, plates, or ampules should be discarded earlier if growth or surface sheen is observed.

QC 5.2.1.3. MF sterility check should be conducted on each funnel in use at the beginning and the end of each filtration series by filtering 20–30 mL of dilution water through the membrane filter and testing for growth. If the control indicates contamination, all data from affected samples must be rejected and an immediate resampling should be requested. A filtration series ends when 30 min or more elapse between sample filtrations.

5.2.2. To prevent carry-over, the filtration funnels must be rinsed with two or three 20–30 mL portions of water after each sample filtration.

5.2.3. Inoculated medium must be incubated at 35 ± 0.5°C for 22–24 hr.

5.2.4. All samples resulting in confluent or too numerous to count (TNTC) growth must be invalidated unless total coliforms are detected. If no total coliforms are detected, record as "confluent growth" or "TNTC" and request an additional sample from the same sampling site. Confluent growth is defined as a continuous bacterial growth covering the entire membrane filter without evidence of sheen colonies (total coliforms). TNTC is defined as greater than 200 colonies on the membrane filter in the absence of detectable coliforms. Laboratories must not invalidate samples when the membrane filter contains at least one sheen colony. (Before invalidation, the laboratory may perform a verification test on the total coliform-negative culture, i.e. on confluent or TNTC growth, and a fecal coliform/*E. coli* test. If the verification test is total coliform-positive, the sample must be reported as total coliform-positive. If the test is total coliform-negative, the sample must be invalidated. A fecal coliform/*E. coli*-positive result is considered a total coliform-positive, fecal coliform/*E. coli*-positive sample, even if the initial and/or verification total coliform test is negative.)

5.2.5.   All sheen colonies (pick all sheen colonies up to a maximum of five) must be verified using either single strength lactose broth (LB) or lauryl tryptose broth (LTB) and single strength brilliant green lactose bile broth (BGLBB), or EPA-approved cytochrome oxidase and β-galactosidase rapid test procedure. Individual colonies can be transferred with a sterile needle or loop, or applicator stick. When picking individual colonies, different morphological types of up to five red questionable sheen colonies and/or red nonsheen colonies per sample must be verified to include different types. Alternatively, wipe the entire surface of the membrane filter with a sterile cotton swab.

5.2.6.   When EC medium or EC medium + MUG is used, the colonies must be transferred by employing one of the options specified by paragraph 141.21(f)(5). For the swab technique, a single swab can be used to inoculate a presumptive total coliform-positive culture into up to three different media (e.g. EC or EC–MUG medium, LTB, and BGLBB, in that order).

5.3.   Multiple Tube Fermentation Technique (MTF or MPN)(For Total Coliforms in Drinking Water)

5.3.1.   Various testing configurations can be used [CFR 141.21(f)(3), see Appendix G], as long as a total sample volume of 100 mL is examined for each test.

5.3.2.   Media

5.3.2.1.   LTB (also known as lauryl sulfate broth) must be used in the presumptive test and BGLBB in the confirmed test. LB may be used in lieu of LTB [40 CFR 141.21(f)(3)] if the laboratory conducts at least 25 parallel tests between this medium and LTB using the waters normally tested and this comparison demonstrates that the false-positive rate and false-negative rate for total coliforms, using LB, is less than 10%. This comparison should be documented and the records retained. The pH must be 6.8 $\pm$ 0.2 for LTB, and 7.2 $\pm$ 0.2 for BGLBB.

5.3.2.2.   The test medium concentration must be adjusted to compensate for the sample volume so that the resulting medium after sample addition is single strength. If a single 100-mL sample volume is used, the inverted vial should be replaced with an acid indicator (bromcresol purple) to prevent problems associated with gas bubbles in large inverted tubes. The media must be autoclaved at 121°C for 12–15 min.

5.3.2.3.   Sterile medium in tubes must be examined to ensure that the inverted vials, if used, are free of air bubbles and are at least one-half to two-thirds covered after the water sample is added.

5.3.2.4. If MTF media are refrigerated after sterilization, they should be incubated overnight at room temperature before use. Tubes/bottles showing growth and/or bubbles should be discarded. If prepared broth media are stored, they should be maintained in the dark at < 30°C no longer than 3 months for screw-cap tubes/bottles and 2 weeks for tubes/bottles with loose-fitting closures. Media should be discarded if evaporation exceeds 10% of the original volume.

5.3.3. After the medium is inoculated, it must be incubated at 35 ± 0.5°C for 24 ± 2 hr. If no gas or acid is detected, it must be incubated for another 24 hr.

5.3.4. All samples that produce a turbid culture (i.e. heavy growth, opaque) in the absence of gas/acid production, in LTB or LB, must be invalidated. The laboratory must collect, or request that the system collect, another sample from the same location as the original invalidated sample. (Before invalidation, the laboratory may perform a confirmed test on the total coliform-negative culture. If the confirmed test is total coliform-positive, the sample must be reported as such. If the test is total coliform-negative, the sample must be invalidated.)

5.3.5. 24- and 48-hr gas-positive or acid-positive tubes must be confirmed using BGLBB.

5.3.6. A completed test is not required

5.3.7. If the MTF test is used on water supplies that have a history of confluent growth or TNTC by the MF procedure, all presumptive tubes with heavy growth without gas/acid production must be submitted to the confirmed test and a fecal coliform/*E. coli* test to check for coliform suppression. (The Total Coliform Rule requires that laboratories invalidate presumptive tubes with heavy growth without gas production and request that the system provide another water sample within 24 hr. However, if the confirmed test is coliform-positive, the laboratory may consider the first sample as coliform-positive. In contrast, if the confirmed test is coliform-negative, the laboratory must not consider this sample as coliform-negative, because high levels of non-coliform bacteria in the presumptive tubes may have injured, killed, or suppressed the growth of any coliforms in the sample. A fecal coliform/*E. coli*-positive result is considered a total coliform-positive, fecal coliform/*E. coli*-positive sample, even if the presumptive and/or confirmed total coliform test is negative.)

5.4. Presence-Absence (P-A) Coliform Test (For Drinking Water)

5.4.1. Medium

5.4.1.1. Six-times formulation strength may be used if the medium is filter sterilized rather than autoclaved.

5.4.1.2. The medium must be autoclaved for 12 min at 121°C. Total time in the autoclave should be less than 30 min. Space should be allowed between bottles. The pH must be 6.8 ± 0.2.

5.4.1.3. If prepared medium is stored, it should be maintained in a culture bottle at < 30°C in the dark for no longer than 3 months. If evaporation exceeds 10% of original volume earlier, the medium should be discarded.

5.4.2. A 100-mL sample must be inoculated into a P–A culture bottle.

5.4.3. Medium must be incubated at 35 ± 0.5°C and observed for a yellow color (acid) after 24 and 48 hr.

5.4.4. Yellow cultures must be confirmed in BGLBB and a fecal coliform/*E. coli* test must be conducted.

5.4.5. All samples which produce a nonyellow turbid culture in P–A medium must be invalidated. The laboratory must collect, or request that the system collect, another sample from the same location as the original invalidated sample. (Before invalidation, the laboratory may perform a confirmed test on the total coliform negative culture and/or a fecal coliform/*E. coli* test. If the confirmed test is total coliform-positive, the sample must be reported as such. If the confirmed test is negative, the sample must be invalidated. A fecal coliform/*E. coli*-positive result is considered a total coliform-positive, fecal coliform/*E. coli*-positive sample, even if the presumptive and/or confirmed total coliform test is negative.)

5.5. Fecal Coliform Test (Using EC Medium for Fecal Coliforms in Drinking Water or Source Water, or A-1 Medium for Fecal Coliforms in Source Water Only)

5.5.1. EC medium must be used to determine whether a total coliform-positive culture taken from the distribution system contains fecal coliforms, in accordance with the Total Coliform Rule. The laboratory must transfer each total coliform-positive culture from a presumptive tube/bottle, or each presumptive total coliform-positive colony (five such colonies minimum) unless a cotton swab is used, to at least one tube containing EC medium with an inverted vial, as specified by §141.21(f)(5).

5.5.2. EC medium may be used to enumerate fecal coliforms in source water, in accordance with the Surface Water Treatment Rule. Initially, conduct an MTF test (presumptive phase). Three sample volumes of source water (10, 1, and 0.1 mL), 5 or 10 tubes/sample volume, should be used. A culture from each total coliform-positive tube must be transferred to a tube containing EC medium with an inverted vial.

5.5.2.1. Medium must be autoclaved for 12–15 min at 121°C. The pH must be 6.9 ± 0.2.

5.5.2.2. Inverted vials should be examined to ensure that they are free of air bubbles. The inverted vial must be at least one-half to two-thirds covered after the sample is added.

5.5.2.3. If prepared medium is stored, it should be maintained in the dark at < 30°C. Prepared medium stored in tubes with loose-fitting closures should be used within 2 weeks. Prepared medium stored in tightly closed screw type tubes may be kept up to 3 months. If the medium is stored in a refrigerator, it should be incubated overnight at room temperature before use; tubes that show growth and/or bubbles should be discarded.

5.5.3. A-1 medium may be used as an alternative to EC medium to enumerate fecal coliforms in source water, in accordance with the Surface Water Treatment Rule. A-1 medium must not be used for drinking water samples. Three sample volumes of source water (10, 1, and 0.1 mL), 5 or 10 tubes/sample volume, should be used. Unlike EC medium, A-1 medium can be directly inoculated with a water sample.

5.5.3.1. Medium must be sterilized by autoclaving at 121°C for 10 min. The pH must be 6.9 ± 0.1.

5.5.3.2. Inverted vials should be examined to ensure that they are free of air bubbles.

5.5.3.3. Loose-cap tubes should be stored in dark at room temperature not more than 2 weeks. A-1 medium may be held up to 3 months in a tightly closed screw-cap tube in the dark at < 30°C.

5.5.4. The water level of the water bath must be above the upper level of the medium in the culture tubes.

5.5.5. EC medium must be incubated at 44.5 ± 0.2°C for 24 ± 2 hr. A-1 medium must be incubated at 35 ± 0.5°C for 3 hr, then at 44.5 ± 0.2°C for 21 ± 2 hr.

5.5.6. Any amount of gas detected in the inverted vial of a tube that has turbid growth must be considered a fecal coliform-positive test.

5.6. Chromogenic/Fluorogenic Substrate Tests [MMO-MUG Test (Colilert Test) for Total Coliforms in Source Water and Total Coliforms and *E. coli* in Drinking Water; Colisure Test for Total Coliforms and *E. coli* in Drinking Water]

5.6.1. Media

5.6.1.1. These media must not be prepared from basic ingredients, but rather purchased from a commercially available source.

5.6.1.2.   The media must be protected from light. Colisure medium must be refrigerated until use and brought to room temperature before adding the sample.

5.6.1.3.   Some lots of fluorogenic media have been known to autofluoresce. Therefore, each lot of medium should be checked before use with a 366-nm ultraviolet light with a 6-W bulb. If the media exhibit faint fluorescence, the laboratory should use another lot that does not fluoresce. If the samples plus a medium exhibit a color change before incubation, it should be discarded and another batch of medium used.

QC   5.6.1.4.   For each lot of medium, a quality control check must be performed by inoculating sterile water containing the medium with an MUG-positive *E. coli* strain, an MUG-negative coliform, and a noncoliform and analyzing them.

QC   5.6.1.5.   Laboratories may also use Quanti-Tray test or Quanti-Tray 2000 test for drinking water and source waters. Both tests use the Colilert medium. If the Quanti-Tray or Quanti-Tray 2000 test is used, the sealer should be checked monthly by adding a dye (e.g. bromcresol purple) to the water. If dye is observed outside the wells, another sealer should be obtained.

5.6.2.   A glass bottle that contains inoculated medium should be checked with a 366-nm ultraviolet light source with a 6-W bulb. If fluorescence is observed before incubation, do not use.

5.6.3.   For enumerating total coliforms in source water with the Colilert test, 5- or 10-tube MTF, Quanti-Tray or Quanti-Tray 2000 must be used for each sample dilution tested. Dilution water (for the chromogenic/fluorogenic substrate test only), if used, must be sterile dechlorinated tap water, deionized water, or distilled water.

5.6.4.   For determining the presence of total coliforms in drinking water by a chromogenic/fluorogenic substrate test, laboratories must use 10 tubes, each containing 10 mL of water sample, or a single vessel containing 100 mL of water sample.

5.6.5.   For the Colilert test, samples must be incubated at 35 $\pm$ 0.5°C for 24 hr. A yellow color in the medium equal to or greater than the reference comparator indicates the presence of total coliforms and must be reported as a total coliform-positive. If the sample is yellow, but lighter than the comparator, it must be incubated for another 4 hr (do not incubate more than 28 hr total). If the color is still lighter than the reference comparator at 28 hr, the sample should be reported as negative. Laboratories that use the Colilert-18 test must incubate for 18 hr.

5.6.6.   For the Colisure test, samples must be incubated at 35 $\pm$ 0.5°C for 28 hr. If an examination of the results at 28 hr is not convenient, then results may be examined at any time between 28 and 48 hr. If the medium changes from a yellow color to a magenta color, the sample must be reported as *E. coli* positive.

5.6.7. For *E. coli* testing, the laboratory must place all total coliform-positive bottles/tubes under an ultraviolet lamp (366 nm, 6-W) in a darkened room. If *E. coli* is present, the medium will emit a blue fluorescence.

QC 5.6.8. Air-type incubators, especially small ones, may not bring a cold 100-mL water sample(s) to the specified incubation temperature of 35°C for several hours. This problem may be further aggravated if several cold water samples are placed in the incubator at the same time. The problem may cause false-negative results with the chromogenic/fluorogenic substrate tests. Therefore, laboratories with air-type incubators should check the time it takes for a 100-mL water sample (or a set of 100-mL water samples, depending on normal use) to reach 35°C, and ensure that the specified incubation period at that temperature is followed. This check should be repeated whenever there is a significant change in the sample load.

5.6.9. The Colilert/Colisure tests must not be used to confirm total coliforms on membrane filters. The filtration step not only concentrates coliforms, but also noncoliforms and turbidity, which at high levels, can suppress coliforms or cause false-positive results in the chromogenic/fluorogenic substrate test.

5.6.10. The Colilert/Colisure tests must not be used to confirm total coliforms in the MTF or Presence–Absence (P–A) Coliform Test. High densities of noncoliforms in the inoculum may overload the chromogenic/fluorogenic substrate test suppressant reagent system and cause false positive results.

5.7. EC Medium + MUG Test (For *E. coli*)

5.7.1. If EC medium + MUG is used, a total coliform-positive culture must be transferred from a presumptive tube/bottle or colony to EC medium + MUG, as specified by §141.21(f)(5).

5.7.2. Medium

5.7.2.1. MUG may be added to EC medium before autoclaving. EC medium + MUG is also available commercially. The final MUG concentration must be 50 μg/mL. The pH must be 6.9 $\pm$ 0.2.

5.7.2.2. The inverted vial may be omitted, because gas production is not relevant to the test, and the use of an inverted vial may cause confusion on test interpretation.

5.7.2.3. Test tubes and autoclaved medium should be tested before use with a 366-nm ultraviolet light to ensure they do not fluoresce. If fluorescence is exhibited, nonfluorescing tubes or another lot of medium that does not fluoresce should be used; alternatively, an MUG-positive *E. coli* and MUG-negative (e.g. uninoculated) control should be performed for each analysis.

5.7.2.4. If prepared medium is stored, it should be maintained in the dark at < 30°C. Prepared medium stored in tubes with loose-fitting closures should be used within 2 weeks. Prepared medium stored in tightly closed screw type tubes may be kept up to 3 months. Tubes with growth should be discarded.

QC   5.7.2.5. In accordance with Paragraph 5.1.7.5, control cultures should be incubated at 35 ± 0.5°C for 24 hr in LTB. A loopful should be transferred to EC medium ± MUG and then incubated at 44.5 ± 0.2°C for 24 hr. The results should be read and recorded.

5.7.3. The water level of the water bath must be above the upper level of the medium.

5.7.4. The medium must be incubated at 44.5 ± 0.2°C for 24 ± 2 hr.

5.7.5. Fluorescence must be checked using an ultraviolet lamp (366 nm) with a 6-W bulb in a darkened room. Laboratories should ensure that a weak autofluorescence of medium, if present, is not misinterpreted as positive for *E. coli*. [If uncertain, an MUG-positive *E. coli* and MUG negative (e.g. uninoculated) control for each analysis should be used whenever the medium autofluoresces.]

5.8. Nutrient Agar + MUG Test (For *E. coli*)

5.8.1. Medium

5.8.1.1. Medium must be autoclaved in 100-mL volumes at 121°C for 15 min. MUG may be added to Nutrient Agar before autoclaving. Nutrient Agar + MUG is also available commercially. The final MUG concentration must be 100 μg/mL. The pH must be 6.8 ± 0.2.

5.8.1.2. If sterile medium is stored, the medium should be refrigerated in petri dishes, in a plastic bag or tightly closed container, and used within 2 weeks. Before use, refrigerated sterilized medium should be incubated overnight at room temperature; plates with growth should be discarded.

QC   5.8.1.3. Positive and negative controls should be tested as stated in Paragraph 5.1.7.5. Filter or spot-inoculate control cultures onto a membrane filter on M-Endo LES agar or M-Endo broth or agar, and incubate at 35°C for 24 hr. Then transfer the filter to Nutrient Agar + MUG and incubate at 35°C for another 4 hr. The results should be read and recorded.

5.8.2. The membrane filter containing coliform colony(ies) must be transferred from the total coliform medium to the surface of Nutrient Agar + MUG medium. Each sheen colony should be marked with a permanent marker on the lid. Also, the lid and the base should be marked with a line to realign the lid should it be removed. A portion of the colony may be trans-ferred with a needle to the total coliform verification test before transfer to Nutrient Agar + MUG or after the 4-hr incubation time. Another method is to swab the entire membrane filter surface after the 4-hr incubation time onto the Nutrient Agar + MUG medium, with a sterile cotton swab, and transfer to a total coliform verification test.

5.8.3.   Inoculated medium must be incubated at 35 ± 0.5°C for 4 hr.

5.8.4.   Fluorescence must be checked using an ultraviolet lamp (366 nm) with a 6-W bulb in a darkened room. Any amount of fluorescence in a halo around a sheen colony should be considered positive for *E. coli.*

5.9.   Heterotrophic Plate Count (For Enumerating Heterotrophs in Drinking Water)

5.9.1.   The Pour Plate Method must be used for enumerating heterotrophic bacteria in drinking water under §141.74(a)(3), (also listed in Appendix G) and should be used for testing reagent grade water. For systems that have been granted a variance from the Total Coliform Rule's maximum contami-nant level (see variance criteria in the preamble of FR 56:1556–1557, January 15, 1991), any method in Standard Methods, Section 9215, "Hetero-trophic Plate Count," may be used with R2A medium, for enumerating het-erotrophic bacteria in drinking water.

5.9.2.   Media [includes plate count agar (tryptone glucose extract agar) and R2A agar]

5.9.2.1.   Final pH values must be 7.0 ± 0.2 for plate count agar and 7.2 ± 0.2 for R2A agar.

5.9.2.2.   (For Pour Plate Method) Melted agar must be tempered at 44–46°C in water bath before pouring. Melted agar should be held no longer than 3 hr. Sterile agar medium should not be melted more than once.

5.9.2.3.   (For Spread Plate Method) 15 mL of R2A agar medium (or other medium) should be poured into a petri dish (100 × 15 mm or 90 × 15 mm) and allowed to solidify.

5.9.2.4.   Refrigerated medium may be stored in bottles or in screw-capped tubes for up to 6 months, or in petri dishes for up to 2 weeks. Prepared petri dishes with R2A medium may be stored for up to 1 week.

5.9.3.   For most potable water samples, countable plates can be obtained by plating 1.0 and/or 0.1 mL volumes of the undiluted sample. At least duplicate plates per dilution should be used.

5.9.4.   (For Pour Plate Method) The sample must be aseptically pipetted onto the bottom of a 100 mm × 15 mm petri dish (100 × 15 mm or 90 × 15 mm).

Then, 12–15 mL of tempered melted (44–46°C) agar must be added to each petri dish. The sample and melted agar must be mixed carefully to avoid spillage. After agar plates have solidified on a level surface, the plates must be inverted and incubated at 35° ± 0.5°C for 48 ± 3 hr. Plates should be stacked no more than four high and arranged in the incubator to allow proper air circulation and to maintain uniform incubation temperature.

5.9.5. (For Spread Plate Method) 0.1 or 0.5 mL of the sample (or dilution) must be pipetted onto the surface of the predried agar in the plate, and then spread over the entire surface of the agar using a sterile bent glass rod. The inoculum must be absorbed completely by the agar before the plate is inverted and incubated. The plate must be incubated at 20–28°C for 5–7 days.

5.9.6. (For Membrane Filter Technique) The volume to be filtered must yield between 20 and 200 colonies. The filter is transferred to a petri dish containing 5 mL of solidified R2A medium, and incubated at 20–28°C for 5–7 days. If plates with loose-fitting lids are used, plates must be placed in a plastic box with a close-fitting lid containing moistened paper towels. Paper towels should be rewetted as necessary to maintain moisture. Colonies must be counted using a stereoscopic microscope at 10–15× amplification.

5.9.7. (For Pour Plate and Spread Plate Techniques) Colonies must be counted manually using a dark field colony counter. In determining sample count, laboratories must only count plates having 30–300 colonies, except for plates inoculated with 1.0 mL of undiluted sample. Counts less than 30 for such plates are acceptable. (Fully automatic colony counters are not suitable because of the size and small number of colonies observed when potable water is analyzed for heterotrophic bacteria).

QC   5.9.8. Each batch or flask of agar should be checked for sterility by pouring a final control plate. Data should be rejected if control is contaminated.

5.10. Membrane Filter Technique (For Enumerating Total Coliforms in Source Water)

5.10.1. The same as Paragraphs 5.2.1–5.2.4 except that in Paragraph 5.2.4, laboratories must invalidate any sample which results in confluent growth or TNTC, even when total coliform colonies are present and because coliform density must be determined.

5.10.2. To optimize counting, appropriate sample dilutions must be used to yield 20–80 total coliform colonies per membrane.

5.10.3. Initial counts must be adjusted based upon verified data, as in Standard Methods, Section 9222B.

QC   5.10.4.   If two or more analysts are available, each analyst should count the total coliform colonies on the same membrane, monthly. Colony counts should agree within 10%.

5.11.   Multiple Tube Fermentation Technique (For Enumerating Total Coliforms in Source Water)

5.11.1.   Laboratories must use at least three series of 5 tubes each with appropriate sample dilutions of source water (e.g. 0.1, 0.01, and 0.001 mL).

5.11.2.   Follow the instructions in Paragraphs 5.3.1–5.3.6, except for 5.3.4 on sample invalidation.

5.11.3.   All samples that produce a turbid culture (i.e. heavy growth, opaque) in the absence of gas/acid production, in LTB or LB, must be invalidated. The laboratory must collect, or request that the system collect, another sample from the same location as the original invalidated sample. The laboratory may use another method to test the second sample. Alternatively, if a sample produces a turbid culture in the absence of gas production, a confirmed test may be performed. If the confirmed test is total coliform-positive, the MPN should be reported. If a confirmed test is total coliform-negative, the sample must be invalidated and another one requested.

5.12.   Fecal Coliform Membrane Filter Procedure (For Enumerating Fecal Coliforms in Source Water)

5.12.1.   Medium (m-FC broth/agar)

5.12.1.1.   m-FC broth (with or without agar) must be prepared by bringing it just to the boiling point. However, medium must not be autoclaved. Final pH must be 7.4 ± 0.2.

5.12.1.2.   When stored, prepared medium should be refrigerated. Broth medium must be discarded after 96 hr and poured an agar medium in petri dishes should be discarded after 2 weeks. Medium should be discarded earlier if growth is observed.

5.12.2.   Appropriate sample volumes to yield 20–60 fecal coliform colonies per membrane for at least one dilution must be used.

QC   5.12.3.   To prevent carry-overs, the filtration funnels must be rinsed with two or three 20–30 mL portions of sterile rinse water after each sample filtration.

QC   5.12.4.   A sterility check should be conducted at the beginning and end of each filtration series by filtering 20–30 mL of dilution water through the membrane filter. If the control indicates contamination, all data from affected samples should be rejected and immediate resampling requested.

5.12.5.   Inoculated medium must be incubated at 44.5° ± 0.2°C for 24 ± 2 hr.

QC   5.12.6.   If two or more analysts are available, each analyst should count fecal coliform colonies on the same membrane monthly. Colony counts should agree within 10%.

6.   SAMPLE COLLECTION, HANDLING, AND PRESERVATION

Paragraphs 6.1–6.5 are applicable to those laboratories that collect samples. However, all laboratories should make an effort to ensure proper sample collection; all laboratories are responsible for Paragraph 6.6.

6.1.   Sample Collector

The sample collector should be trained in aseptic sampling procedures and, if required, approved by the appropriate regulatory authority or its designated representative.

6.2.   Sampling

(For drinking water) Samples must be representative of the water distribution system. Water taps used for sampling should be free of aerators, strainers, hose attachments, mixing type faucets, and purification devices. Cold water taps should be used. The service line must be cleared before sampling by maintaining a steady water flow for at least 2 min (until the water changes temperature). At least 100 mL of sample must be collected, allowing at least a 1-in. air space to facilitate mixing of the sample by shaking. Immediately after collection, a sample information form should be completed (see Paragraph 6.5). See Section 3.15.4 regarding sample dechlorination.

Source water samples must be representative of the source of supply, collected not too far from the point of intake, but at a reasonable distance from the bank or shore. The sample volume should be sufficient to perform all the tests required.

6.3.   Sample Icing

Samplers are encouraged, but not required, to hold drinking water samples at < 10°C during transit to the laboratory. Source water samples must be held at < 10°C (see Standard Methods, Section 9060B).

6.4.   Sample Holding/Travel Time

The time from sample collection to initiation of analysis for total coliforms, fecal coliforms, or *E. coli* in drinking water must not exceed 30 hr. The time from sample collection to initiation of analysis for total coliforms and fecal coliforms in source water, and heterotrophic bacteria in drinking water must not exceed 8 hr (see Section 9060B in *Standard Methods*). All samples received in the laboratory should be analyzed on the day of receipt. If the laboratory receives the sample late in the day, the sample may be refrigerated overnight as long as analysis begins within 30 hr of sample collection.

6.5. Sample Information Form

After collection, the sampler should enter on a sample information form, in indelible ink, the following information:

- Name of system (public water system site identification number, if available)
- Sample identification (if any)
- Sample site location
- Sample type (e.g. routine distribution system sample, repeat sample, raw or process water, other special purpose sample)
- Date and time of collection
- Analysis required
- Disinfectant residual
- Name of sampler and organization (if not the water system)
- Sampler's initials
- Person(s) transporting the samples from the system to the laboratory (if not the sampler)
- Transportation condition (e.g. < 10°C, protection from sunlight). If a commercial shipper was used, shipping records should be available.
- Any remarks

6.6. Chain-of-Custody

Sample collectors and laboratories must follow applicable State regulations pertaining to chain-of-custody. An example of such a plan is provided in Appendix A.

7. QUALITY ASSURANCE

A written QA plan should be prepared and followed (see Chapter III). The QA plan should be available for inspection by the certification officer.

8. RECORDS AND DATA REPORTING

8.1. Legal Defensibility

Compliance monitoring data should be made legally defensible by keeping thorough and accurate records. The QA plan and/or SOPs should describe the policies and procedures used by the facility for record retention and storage. If samples are expected to become part of a legal action, chain-of-custody procedures should be used (see Appendix A).

8.2. Maintenance of Records

Public Water Systems are required to maintain records of microbiological analyses of compliance samples for 5 years (40 CFR 141.33). The laboratory should maintain easily accessible records for 5 years or until the next certification data audit is complete, whichever is longer. The client water system should be notified before disposing of records so they may request copies if needed. This includes all raw data, calculations, and quality control data. These data files may be either hard copy, microfiche, or electronic. Electronic data should always be backed up by protected tape or disk or hard copy. If the laboratory changes its computer hardware or software, it should make provisions for transferring old data to the new system so that it remains retrievable within the time frames specified above. Data that are expected to become part of a legal action will probably need to be maintained for a longer period of time. Check with your legal counsel. See Appendix H, Section 3.0, and *Good Automated Laboratory Practices*, EPA 2185, Office of Information Management, Research Triangle Park, NC 27711, 8/10/95.

8.3. Sampling Records

Data should be recorded in ink with any changes lined through such that original entry is visible. Changes should be initialed and dated. The following information should be readily available in a summary or other record(s):

1. Sample information form, from Section 6.5 above;
2. Date and time of sample receipt by the laboratory; name of carrier
3. Name of laboratory person receiving the sample;
4. If a deficiency in the condition of the sample is noted, the sample, at a minimum, is flagged;
5. If sample transit time exceeds 30 hr (8 hr for source water samples), sample must be tagged;

8.4. Analytical Records

Data should be recorded in ink with any changes lined through such that original entry is visible. Changes should be initialed and dated. The following information should be readily available in a summary or other record(s):

1. Laboratory sample identification;
2. Date and time analysis begins;
3. Laboratory and person(s) responsible for performing analysis;
4. Analytical technique or method used;
5. All items marked QC;
6. Results of analyses.

8.5. Preventive Maintenance

Laboratories should maintain preventive maintenance and repair activities records for all instruments and equipment (including pH meters, analytical balances, incubators, refrigerators, autoclaves, and water baths). Records should be kept for 5 years.

9. ACTION RESPONSE TO LABORATORY RESULTS

9.1. Testing Total Coliform-Positive Cultures

For the Total Coliform Rule, laboratories must test all total coliform-positive cultures for the presence of either fecal coliforms or *E. coli*.

9.2. Notification of Positive Results

For the Total Coliform Rule, laboratories must promptly notify the proper authority of a positive total coliform, fecal coliform, or *E. coli* result, so that appropriate follow-up actions (e.g. collection of repeat samples) can be conducted [see CFR 141.21(b) and (e), 40 CFR 141.31, etc.]. A total coliform-positive result is based on a confirmed phase for the Multiple Tube Fermentation Technique and Presence–Absence (P–A) Coliform Test, or verified test for Membrane Filter Technique. No requirement exists for confirmation of positive Colilert/Colisure tests, fecal coliform test, or *E. coli* tests. In those rare cases where a presumptive total coliform-positive culture does not confirm/verify as such, but is found to be fecal coliform or *E. coli* positive, the sample is considered total coliform-positive and fecal coliform/*E. coli* positive.

9.3. Invalidation of Total Coliform-Negative Sample

For the Total Coliform Rule, the laboratory must promptly notify the proper authority (usually the water system) when results indicate that noncoliforms may have interfered with the total coliform analysis, as described in 40 CFR 141.21(c)(2).

LABORATORY_____

ADDRESS_____

_____

_____

_____

TELEPHONE NUMBER/FAX NUMBER_____

CONDUCTED BY_____

DATE_____

NAMES/TITLES/RESPONSIBILITIES OF KEY PERSONNEL INTERVIEWED

_____

_____

_____

_____

_____

_____

_____

Microbiology laboratory analysis review checklist.

Element	Yes	No	Comments
1. PERSONNEL			
1.1 Supervisor/Consultant			
Supervisor of analyst has a bachelor's degree in microbiology, biology, or equivalent with at least one college-level laboratory course in environmental microbiology, and has a minimum of two weeks course training or 80 hours of on-the-job training in water microbiology at a certified laboratory, or other training acceptable to the State or EPA			
If supervisor not available, consultant with same training and experience substituted, acceptable to the State, and present on-site frequently enough to satisfactorily perform a supervisor's duties			
1.2 Analyst (or equivalent job title)			
Analyst has a high school education, 3 months bench experience in microbiology, training in microbiological analysis of drinking water acceptable to the State (or EPA) and a minimum of 30 days on-the-job training under an experienced analyst			
Analyst demonstrated acceptable results for precision, specificity, and satisfactory analysis on unknown samples before analyzing compliance samples			
1.3 Waiver of Academic Training Requirement			
Need for specified academic training waived for highly experienced analysts			
1.4 Personnel Records			
Personnel records maintained on laboratory analysts include academic background, specialized training courses completed and types of microbiological analyses conducted			
2. LABORATORY FACILITIES			
Laboratory facilities clean, temperature and humidity controlled, with adequate lighting at bench top			
Sufficient space available for processing samples, bench top equipment, storage, cleaning glassware and sterilizing materials			
Provisions made for disposal of microbiological wastes			
3. LABORATORY EQUIPMENT AND SUPPLIES			
3.1 pH meter			
Accuracy and scale graduations within $\pm$ 0.1 units			
Buffer aliquot used only once			

Element	Yes	No	Comments
Electrodes maintained according to manufacturer's recommendations			
QC Meter standardized each use period with pH 7.0 and either 4.0 or 10.0 buffers, with date and buffers used recorded in log book			
QC Commercial buffer solutions dated when received and opened and discarded before expiration date			
3.2. Balance (top loader or pan)			
Readability of 0.1 g			
QC Calibrated monthly using ASTM type 1, 2, or 3 weights (minimum 3 traceable weights which bracket laboratory weighing needs)			
QC Non-reference weights calibrated every six months with reference weights			
QC Annual service contract or internal maintenance protocol established, records available of most recent recalibration, and correction values on file and used			
QC Reference weight recertified if damaged or corroded			
3.3 Temperature Monitoring Device			
Temperature monitoring devices graduated in 0.5°C increments (0.2°C increments for tests which are incubated at 44.5°C) or less			
No separation in fluid column of glass thermometer			
No dial thermometers used which cannot be adjusted			
QC Glass and electronic thermometers calibrated annually, dial thermometers quarterly, at the temperature used against reference NIST thermometer or one meeting the requirements of NBS Monograph SP 250-23			
QC Calibration factor marked on thermometer and calibration date and calibration factor recorded in QC record book			
QC Thermometer discarded if off more than 1°C from reference thermometer, reference thermometers recalibrated every 3-5 years			
QC Continuous recording devices used to monitor incubator temperature recalibrated annually as above			
3.4 Incubator Unit			
Incubator units have an internal temperature monitoring device and maintain temperature of 35 ± 0.5°C, and if used, 44.5 ± 0.2°C			

Element	Yes	No	Comments
Thermometers placed on top and bottom shelves of use area in non-portable incubators, with thermometer bulb immersed in liquid (except for electronic thermometers)			
For aluminum block incubator, culture dishes and tubes fit snugly			
QC Calibration-corrected temperature recorded twice daily for days in use, readings separated by at least four hours			
Water bath equipped with gable cover and pump or paddles used to circulate water (recommended for maintaining 44 ± 0.2°C)			
3.5 Autoclave			
Autoclave has internal heat source, temperature gauge with sensor on exhaust, pressure gauge, and operational safety valve			
Maintains sterilization temperature during cycle and completes entire cycle within 45 minutes when 12-15 minute sterilization period used			
Depressurizes slowly enough to ensure media will not boil over and bubbles will not form in inverted tubes			
Pressure cookers not used			
QC Date, contents, sterilization time, temperature, total cycle time, and analyst's initials recorded for each cycle			
QC Copy of service contract or internal maintenance protocol and maintenance records kept			
QC Maintenance conducted annually at a minimum, with record of most recent service performed available for inspection			
QC Maximum-temperature-registering thermometer or continuous recording device used each autoclave cycle and temperature recorded			
QC Overcrowding avoided			
QC Spore strips or ampules used monthly			
QC Automatic timing mechanism checked quarterly with stopwatch or other accurate timepiece or time signal			
Autoclave door seals clean and free of caramelized media			
Autoclave drain screen cleaned frequently			
3.6 Hot Air Oven			
Maintains stable sterilization temperature of 170-180°C for at least 2 hours			
Only dry items sterilized in hot air oven			

Element	Yes	No	Comments
Overcrowding avoided			
Oven thermometer graduated in 10°C increments or less, with bulb placed in sand during use			
QC Date, contents, sterilization time, temperature, and analyst's initials recorded for each cycle			
QC Spore strip or ampule used monthly			
3.7 Colony Counter			
Colony counter, dark field model, used to count Heterotrophic Plate Count colonies			
3.8 Conductivity Meter			
Suitable for checking laboratory reagent-grade water, readable in micromhos/cm or microsiemens/cm with measurement error not exceeding 1% or 1 micromhos/cm, whichever is more lenient			
QC Cell constant determined monthly			
In-line unit which cannot be calibrated not used to check reagent-grade water			
3.9 Refrigerator			
Maintains 1-5°C			
Thermometer graduated in 1°C increments or less, with thermometer bulb immersed in liquid			
QC Temperature recorded for days in use at least once per day			
3.10 Inoculating Equipment			
Sterile metal or disposable plastic loops, wood applicator sticks, sterile swabs, or sterile plastic disposable pipet tips used			
Wood applicator sticks sterilized by dry heat			
Metal inoculating loops and needles made of nickel alloy or platinum (nickel alloy loops not used for oxidase test)			
3.11 Membrane Filtration (MF) Equipment			
MF units of stainless steel, glass, or autoclavable plastic, not scratched or corroded and do not leak			
QC Graduations on funnels used to measure sample volume checked for accuracy have tolerance of ≤2.5%, and a record of this calibration check retained			

Element	Yes	No	Comments
10x to 15x stereo microscope with fluorescent light source used to count sheen colonies			
Membrane filters approved by manufacturer for use in total coliform analysis of water			
Membrane filters of cellulose ester, white, gridmarked, 47 mm diameter, and 0.45 μm pore size			
Membrane filters and pads purchased presterilized or autoclaved before use			
Lot number and date received recorded for membrane filters			
3.12 Culture Dishes (loose or tight lids)			
Presterilized plastic or sterilizable glass culture dishes used			
Sterility of glass culture dishes maintained by placement in stainless steel or aluminum canisters or wrapped in heavy aluminum foil or char-resistant paper			
Loose-lid dishes incubated in tight-fitting container with moistened paper towel			
Opened packs of disposable culture dishes resealed between use periods			
3.13 Pipets			
Glass pipets sterilized and maintained in stainless steel or aluminum canisters or wrapped individually in char-resistant paper or aluminum foil			
Pipets with legible markings, not chipped or etched			
Opened packs of disposable sterile pipets resealed between use periods			
Pipets delivering volumes of 10 mL or less accurate within 2.5% tolerance			
Micropipetters used with sterile tips, calibrated annually, and replaced if tolerance greater than 2.5%			
3.14 Culture Tubes and Closures			
Tubes of borosilicate glass or other corrosion-resistant glass or plastic			
Culture tubes and containers of sufficient size to contain medium plus sample without being more than three quarters full			
Tube closures used of stainless steel, plastic, aluminum, or screw caps with non-toxic liner; cotton plugs not used			

Element	Yes	No	Comments
3.15 Sample Containers			
Wide-mouth plastic or non-corrosive glass bottles, with non-leaking ground glass stoppers or caps with non-toxic liners, or sterile plastic bags containing sodium thiosulfate used			
Sample container capacity at least 120 mL (4 oz)			
Glass stoppers covered with aluminum foil or char-resistant paper for sterilization			
Sample containers sterilized by autoclaving or (for glass bottles) dry heat			
Containers moistened with several drops of water before autoclaving to prevent "air lock" sterilization failure			
Sufficient sodium thiosulfate added to sample containers before sterilization, if laboratory analyzes chlorinated water			
3.16 Glassware and Plasticware			
Glassware made of borosilicate glass or other corrosion-resistant glass, free of chips and cracks, with markings legible			
Plastic items clear and non-toxic to microorganisms			
QC Graduated cylinders and pre-calibrated containers used to measure samples volumes accurate with a tolerance of 2.5% or less			
QC New lots of pre-calibrated containers validated to have 2.5% tolerance			
3.17 Ultraviolet Lamp (if used)			
Unit cleaned monthly by wiping with soft cloth moistened with ethanol			
QC If used for sanitization, tested quarterly with UV light meter or by agar spread plate method (other methods acceptable if data demonstrates they are as effective)			
4. GENERAL LABORATORY PRACTICES			
Laboratory facilities clean, temperature and humidity controlled, and adequate lighting			
4.1 Sterilization Procedures			

Element	Yes	No	Comments

Required times for autoclaving material at 121°C (except for membrane filters and pads and carbohydrate-containing media, indicated times represent minimum times, dependent upon volumes, containers, and loads):
    - membrane filters and pads   10 min
    - carbohydrate containing media   12-15 min
    - contaminated test materials   30 min
    - membrane filter assemblies   15 min
    - sample collection containers   15 min
    - individual glassware   15 min
    - dilution water blank   15 min
    - rinse water (0.5 - 1 L)   15-30 min*
 	
* time depends upon water volume per container and autoclave load			

Autoclaved membrane filters and pads and all media removed immediately after completion of sterilization cycle			
Membrane filter equipment autoclaved before beginning of first filtration series (filtration series ends when 30 minutes or longer elapses after a sample filtered)			
When UV light (254 nm) used to sanitize equipment, all supplies presterilized and QC checks conducted on UV lamp			
UV light used to control bacterial carry-over between samples during filtration series (optional)			
4.2 Sample Containers			
QC Sterility of each lot of sample containers or bags confirmed by adding 25 mL of a sterile non-selective broth to at least one container, incubating at 35 ± 0.5°C for 24 hours and checking for growth			
4.3 Reagent-Grade Water			
Only satisfactorily tested reagent water from stills or deionization units used to prepare media, reagents and dilution/rinse water			

Element	Yes	No	Comments
QC 'Quality of reagent water should be tested and meets the following criteria:			
- conductivity <2 micromhos/cm monthly (microsiemens/cm) at 25°C			
- Pb, Cd, Cr not greater than 0.05 mg/L per annually Cu, Ni, Zn contaminant, and no greater than 0.1 mg/L total			
- total chlorine <0.1 mg/L monthly residual*			
- heterotrophic <500/mL monthly plate count*			
- bacteriological ratio of growth rate 0.8:3.0 annually quality of reagent water*			
*See section 4.3.2 of this chapter for additional details			
4.4 Dilution/Rinse Water			
Stock buffer solution or peptone water prepared as specified in Standard Methods			
Stock buffers autoclaved or filter-sterilized and containers labeled, dated, and refrigerated			
Stored stock buffer free of turbidity			
QC Each batch of dilution/rinse water checked for sterility by adding 50 mL of water to 50 mL double strength non-selective broth, incubating at 35 ± 0.5°C for 24 hours, and checking for growth			
4.5 Glassware Washing			
Distilled or deionized water used for final rinse			
QC Glassware inhibitory residue test performed on initial use of washing compound and whenever different formulation or washing procedure used			
QC Batches of dry glassware spot-checked for pH reaction			
Laboratory glassware washed with detergent designed for laboratory use			
5. ANALYTICAL METHODOLOGY			
5.1 General			

Element	Yes	No	Comments
Only analytical methodology specified in Total Coliform Rule and Surface Water Treatment Rule used for compliance samples			
Laboratory certified for all analytical methods it uses for compliance purposes			
Laboratory certified for at least one total coliform method and one fecal coliform or *E. coli* method			
Laboratory certified for a second total coliform method, if one method cannot be used for some drinking waters			
Laboratory that enumerates heterotrophic bacteria (i.e., HPC) for compliance with the Surface Water Treatment Rule certified for the Pour Plate Method			
Absorbent pads, when used, saturated with liquid medium and excess removed			
Water sample shaken vigorously (about 25 times) before analysis			
QC If no total coliform-positive results occur during a quarter, laboratory performs coliform procedure using a known coliform-positive, fecal coliform- and/or *E. coli*-positive control to spike the sample			
Sample volume analyzed for total coliforms in drinking water is 100 ± 2.5 mL			
Media			
Dehydrated or prepared media manufactured commercially used (strongly recommended)			
Dehydrated media stored in cool dry location and caked or discolored dehydrated media discarded			

QC   Laboratory media preparation records include:
  - date of preparation
  - type of medium
  - lot number
  - sterilization time and temperature
  - final pH
  - technician's initials			

QC For liquid media prepared commercially, the following are recorded:
  - date received
  - type of medium
  - lot number
  - pH verification			

Element	Yes	No	Comments
QC Liquid media prepared commercially discarded by manufacturer's expiration date			
QC Each new lot of dehydrated and prepared commercial medium checked before use with positive and negative culture controls and results recorded			
QC Each new batch of laboratory-prepared medium checked before use with positive and negative culture controls and results recorded			
Prepared plates refrigerated in sealed plastic bags or containers not longer than two weeks, with bag or container dated with preparation or expiration date			
Loose-cap tubes of broth stored at $<30\,^{\circ}C$ no longer than two weeks, tightly capped tubes no longer than 3 months at $<30\,^{\circ}C$			
Refrigerated medium incubated at room temperature overnight before use and discarded if growth observed			
QC Parallel testing performed between a newly approved test procedure and another EPA-approved procedure for several months and/or several seasons (recommended)			
5.2 Membrane Filter (MF) Technique (for total coliforms in drinking water)			
Media			
M-Endo broth or agar or LES Endo agar in single step or enrichment technique used			
Ethanol not denatured			
Medium prepared in sterile flask and dissolved using boiling water bath or hot plate with stir bar			
Medium not boiled			
LES Endo agar medium pH 7.2 ± 0.2 M-Endo medium pH 7.2 ± 0.1			
MF broth refrigerated no longer than 96 hours, poured MF agar plates no longer than 2 weeks, ampuled M-Endo broth as per manufacturer's expiration date			
Uninoculated media discarded if growth or surface sheen observed			
QC Sterility check conducted on each funnel in use at beginning and end of each filtration series (filtration series ends when 30 minutes or more elapse between sample filtrations)			

Element	Yes	No	Comments
QC If sterility control indicates contamination, all data rejected and another sample requested			
Funnels rinsed with two or three 20-30 mL portions of sterile rinse water after each sample filtration to prevent carry-over			
Inoculated medium incubated at 35° ± 0.5°C for 22-24 hours			
Samples resulting in confluent or too numerous to count (TNTC) growth invalidated unless total coliforms detected (if laboratory performs verification test before invalidation and test is total coliform-positive, sample is reported as such, but if test is total coliform-negative, sample is invalidated)			
Sample not invalidated if membrane filter contains at least one sheen colony			
All sheen colonies verified (up to a maximum of five) using either single strength (LB) or (LTB) and single strength (BGLBB) or an EPA-approved cytochrome oxidase and beta-galactosidase rapid test procedure			
When picking individual colonies, up to five red questionable sheen colonies and/or red non-sheen colonies verified to include different types or entire MF surface is swabbed			
When EC medium or EC medium + MUG used, colonies transferred by employing one option specified by 141.21 (f)(5)			
Swab used to transfer presumptive total coliform-positive culture can inoculate up to three different media (e.g., EC medium, LTB, and BGLBB in that order)			
5.3 Multiple Tube Fermentation Technique (MTF or MPN) (for total coliforms in drinking water)			
Total sample volume of 100 mL examined by test configuration found in 141.21 (f)(3) or Appendix G			
Media			
LTB used in presumptive test and BGLBB in confirmed test			
LB used if system conducts at least 25 parallel tests between this medium and LTB and demonstrates false-positive rate and false-negative rate for total coliforms of less than 10%, with comparison documented and records retained			
LTB pH 6.8 ± 0.2			
BGLBB pH 7.2 ± 0.2			
Test medium concentration adjusted to compensate for sample volume so resulting medium single strength after sample addition			

Element	Yes	No	Comments
If single 100 mL sample volume used, inverted vial replaced with acid indicator			
Medium autoclaved at 121°C for 12-15 minutes			
Inverted vials in sterile medium free of bubbles and at least one-half to two-thirds covered after water sample added			
Refrigerated sterile MTF media incubated overnight at room temperature before use, with tubes/bottles showing growth and/or bubbles discarded			
Prepared broth media stored in dark at <30°C for no longer than 3 months in screw-cap tubes/bottles, two weeks for those with loose-fitting closures			
Media discarded if evaporation exceeds 10% of original volume			
Inoculated medium incubated at 35°C ± 0.5°C for 24 ± 2 hours			
If no gas or acid detected, inoculated medium incubated for another 24 hours			
All samples showing turbid culture (i.e., heavy growth, opaque) in the absence of gas/acid production invalidated and another sample collected from the same location (if laboratory performs confirmed test on turbid culture and confirmed test is total coliform-positive, sample reported as such, but if total coliform-negative, sample is invalidated)			
All 24- and 48-hour gas-positive or acid-positive tubes confirmed using BGLBB			
Completed Test not required			
When MTF test used on water supplies that have a history of confluent growth or TNTC by the MF procedure, all presumptive tubes with heavy growth without gas/acid production submitted to confirmed test and fecal coliform/*E. coli* test to check for coliform suppression			
5.4 Presence-Absence (P-A) Coliform Test (for drinking water)			
Medium			
When six-times formulation strength medium used, medium filter-sterilized, not autoclaved			
Medium autoclaved for 12 minutes at 121°C with total time in autoclave less than 30 minutes and with space between bottles			
Medium pH 6.8 ± 0.2			

Element	Yes	No	Comments
Prepared medium stored in the dark at <30°C for no longer than 3 months			
Stored medium discarded if evaporation exceeds 10% of original volume			
100 mL sample inoculated into P-A culture bottle			
Medium incubated at 35° ± 0.5°C and observed for yellow color (acid) after 24 and 48 hours			
Yellow cultures confirmed in BGLBB and fecal coliform/*E. coli* test conducted			
Non-yellow turbid culture in P-A medium invalidated and another sample obtained from the same location (if confirmed test performed and sample is total coliform-positive, sample is reported as such, but if confirmed test is negative, sample invalidated)			
5.5 Fecal Coliform Test (using EC Medium for fecal coliforms in drinking or source water, or A-1 Medium for fecal coliforms in source water only)			
EC medium used to determine whether total coliform-positive culture taken from distribution system contains fecal coliforms, in accordance with Total Coliform Rule			
EC medium used to enumerate fecal coliforms in source water, in accordance with Surface Water Treatment Rule, using cultures transferred from each total coliform-positive tube			
Three sample volumes (10, 1, and 0.1 mL) and 5 or 10 tubes/sample volume used			
Autoclaved at 121°C for 12-15 minutes			
Medium pH 6.9 ± 0.2			
Inverted vials free of bubbles and at least one-half to two-thirds covered after sample added			
Tubes with loose-fitting closures used within two weeks, tightly closed screw-cap tubes no longer than 3 months when held in the dark at <30°C			
Refrigerated medium incubated at room temperature overnight before use and tubes with growth or bubbles in vials discarded			
Alternatively, A-1 Medium used to enumerate fecal coliforms in source water, in accordance with Surface Water Treatment Rule			
A-1 medium not used for drinking water samples			

Element	Yes	No	Comments
Three sample volumes of source water (10, 1, and 0.1 mL) and 5 or 10 tubes/sample volume used			
Autoclaved at 121°C for 10 minutes			
Medium pH 6.9 ± 0.1			
Inverted vials free of air bubbles and at least one-half to two-thirds covered after water sample added			
Loose-cap tubes stored in dark at room temperature no longer than 2 weeks, tightly closed screw-cap tubes no longer than 3 months when held in the dark at <30°C			
Water level in water bath above upper level of medium in culture tubes			
EC Medium incubated at 44.5°C ± 0.2°C for 24 ± 2 hours			
A-1 Medium incubated at 35°C ± 0.5°C for 3 hours, then at 44.5°C ± 0.2°C for 21 ± 2 hours			
Any gas detected in inverted vial considered fecal coliform positive			
5.6 Chromogenic/Fluorogenic Substrate Tests (MMO-MUG Test [Colilert] for total coliforms in source water and total coliforms and *E. coli* in drinking water; Colisure Test for total coliforms and *E. coli* in drinking water)			
Media			
Purchased from commercially available source only			
Media protected from light			
Colisure medium refrigerated until use, brought to room temperature before adding sample			
Each lot of medium checked for autofluoresence before use with 366-nm ultraviolet light with 6 watt bulb			
Medium which exhibits faint fluorescence discarded and another lot used			
Medium plus sample which exhibits color change before incubation discarded and another batch of medium used			
QC Each lot of medium checked by inoculating sterile water containing the medium with a MUG-positive *E. coli* strain, a MUG-negative coliform, and a non-coliform and analyzing them			
If Quanti-Tray or Quanti-Tray 2000 test used with Colilert medium, sealer checked monthly to determine leakage			

Element	Yes	No	Comments
Glass bottles that contain inoculated medium checked with 366-nm ultraviolet light source with 6 watt bulb and discarded if fluorescence observed before incubation			
For enumeration of total coliforms in source water with Colilert Test, 5 or 10 tube MTF, Quanti-Tray, or Quanti-Tray 2000 used for each sample dilution tested			
For chromogenic/fluorogenic substrate test only, sterile dechlorinated tap water, deionized water, or distilled water used as dilution water			
For determining presence of total coliforms in drinking water by chromogenic/fluorogenic substrate test, 10 tubes each containing 10 mL water sample or single vessel containing 100 mL sample used			
For Colilert Test:			
Sample incubated at 35° ± 0.5° for 24 hours (for Colilert-18 test, sample incubated 18 hours)			
Yellow color in medium equal to or greater than reference comparator indicates total coliform presence			
Medium with yellow color lighter than comparator and incubated for another 4 hours (28 hours total)			
Yellow color in medium lighter than comparator incubated for 28 hours recorded as negative			
For Colisure Test:			
Sample incubated at 35° ± 0.5°C for 28 to 48 hours			
Total coliform positive sample indicates color change from yellow to magenta			
For *E. coli* determination, UV lamp (366-nm, 6-watt) shone on total coliform-positive bottles/tubes in darkened room with blue fluorescence indicating *E. coli* presence			
QC Air-type incubators tested to determine time necessary for cold 100 mL water sample (or set of 100 mL water samples) to reach incubation temperature of 35°C, ensuring specified incubation time at that temperature is followed			
Colilert/Colisure Test not used to confirm total coliforms on membrane filters			
Colilert/Colisure Test not used to confirm total coliforms in MTF or P-A tests			
5.7 EC Medium + MUG (for *E. coli*)			

Element	Yes	No	Comments
Total coliform-positive culture transferred to EC medium + MUG			
Medium			
MUG added to EC medium before autoclaving or commercially available EC + MUG used			
Final MUG concentration 50 μg/mL			
Medium pH 6.9 $\pm$ 0.2			
Inverted vial omitted (optional)			
Test tubes and autoclaved medium checked for autofluorescence before use with 366-nm UV light			
If fluorescence exhibited, non-fluorescing tubes or another lot of medium that does not fluoresce used or MUG-positive (*E. coli*) and a MUG-negative (e.g. uninoculated) control included for each analysis			
Prepared medium in tubes with loose-fitting closures used within two weeks, or three months for tightly closed screw-cap tubes when held in the dark at <30°C			
Uninoculated medium with growth discarded			
QC Each lot of commercially prepared medium and each batch of laboratory-prepared medium checked by inoculating LTB with positive and negative culture controls, incubating at 35°C $\pm$ 0.5°C for 24 hours and then transferring to EC Medium + MUG for further incubation at 44.5°C $\pm$ 0.2°C for 24 hours, with results read and recorded			
Water level of water bath above upper level of medium			
Incubated at 44.5° $\pm$ 0.2°C for 24 $\pm$ 2 hours			
Fluorescence checked using UV lamp (366-nm) with 6 watt bulb in a darkened room			
5.8 Nutrient Agar + MUG Test (for *E. coli*)			
Medium			
Medium autoclaved in 100 mL volumes at 121°C for 15 minutes			
MUG added to Nutrient Agar before autoclaving or Nutrient Agar + MUG purchased commercially			
Final MUG concentration 100 μg/L			
Medium pH 6.8 $\pm$ 0.2			

Element	Yes	No	Comments
Medium in petri dishes stored refrigerated in plastic bag or tightly closed container and used within two weeks			
Refrigerated sterilized medium incubated at room temperature overnight and plates with growth discarded			
QC Quality of medium lot/batch evaluated by filtering or spot-inoculating positive and negative control cultures onto membrane filter on M-Endo medium, incubating at 35°C for 24 hours, then transferring filter to NA + MUG and further incubating at 35°C for 4 hours, with results read and recorded			
Filter containing total coliform colony(ies) transferred to surface of Nutrient Agar + MUG medium			
Before incubation, presence of each sheen colony marked on petri dish lid with permanent marker, and lid and base marked to realign lid when removed			
For total coliform verification test, portion of colony transferred with needle before or after NA + MUG incubation			
Alternatively, membrane filter surface swabbed with sterile cotton swab after 4 hour incubation and transferred to total coliform verification test			
Inoculated medium incubated at 35 ± 0.5°C for 4 hours			
Fluorescence checked using UV lamp (366 nm) with 6 watt bulb in a darkened room, with any fluorescence in halo around sheen colony considered positive for *E. coli*			
5.9 Heterotrophic Plate Count for enumerating heterotrophs in drinking water			
Pour Plate Method used for enumerating heterotrophic bacteria in drinking water and for testing reagent grade water			
For systems granted a variance from Total Coliform Rule's maximum contaminant level, any method in Standard Methods used with R2A medium for enumerating heterotrophic bacteria in drinking water			
Media (plate count agar [tryptone glucose extract agar] and R2A agar)			
Plate count agar pH 7.0 ± 0.2			
R2A agar pH 7.2 ± 0.2			
(For Pour Plate Method) melted agar tempered at 44-46°C in waterbath before pouring, held no longer than 3 hours, and melted only once			

Element	Yes	No	Comments
(For Spread Plate Method) 15 mL of R2A medium or other medium poured into petri dish and solidified			
Refrigerated medium in bottles or screw-capped tubes stored for up to 6 months, petri dishes with medium for up to 2 weeks (one week for R2A prepared petri dishes)			
Countable plates obtained for most potable waters by plating 1.0 mL and/or 0.1 mL volume of undiluted sample			
At least duplicate plates per dilution used			
(For Pour Plate Method)			
Sample pipetted aseptically into bottom of petri dish and then 12-15 mL tempered melted agar added			
Sample mixed with spillage avoided			
After solidification on level surface, plates inverted and incubated at 35°C ± 0.5°C for 48 ± 3 hours			
Plates stacked no more than four high			
(For Spread Plate Method)			
0.1 or 0.5 mL of sample or dilution pipetted onto surface of pre-dried agar plate and inoculum spread over entire agar surface using sterile bent glass rod			
Inoculum absorbed completely before plates inverted and incubated at 20-28°C for 5-7 days			
(For Membrane Filter Technique)			
Volume filtered to yield between 20-200 colonies			
Filter transferred to petri dish containing 5 mL solidified R2A medium and incubated at 20-28°C for 5-7 days			
Petri dishes with loose-fitting lids placed in container with close fitting lid and moistened paper towels			
Colonies counted using stereoscopic microscope at 10-15X magnification			
(For Pour Plate and Spread Plate Techniques)			
Colonies counted manually using dark field colony counter			
Only plates with 30 to 300 colonies counted, except for plates inoculated with 1.0 mL of undiluted sample			
Fully automatic colony counters not used			

Element	Yes	No	Comments
QC Medium sterility verified by pouring final control plate and data rejected if control contaminated			
5.10 Membrane Filter Technique (for enumerating total coliforms in source water)			
Same as Section 5.2, Membrane Filter Technique (for total coliforms in drinking water), except invalidation does not apply			
Appropriate sample dilutions used to yield 20 to 80 total coliform colonies per membrane			
Initial counts adjusted based upon verified data			
QC If two or more analysts available, each counts total coliform colonies on same membrane monthly and agree within 10%			
5.11 Multiple Tube Fermentation Technique (for enumerating total coliforms in source water)			
At least three series of 5 tubes each with appropriate sample dilutions of source water used			
Same as Section 5.3, Multiple Tube Fermentation Technique (for total coliforms in drinking water) except on sample invalidation			
All samples invalidated which produce turbid growth in the absence of gas/acid production in LTB or LB and another sample obtained, which may be tested using another method			
Alternatively, confirmed test performed on turbid culture in the absence of gas/acid production and, if total coliform-positive, most probable number reported, or if total coliform-negative, sample invalidated and another requested			
5.12 Fecal Coliform Membrane Filter Procedure (for enumerating fecal coliforms in source water)			
Medium			
m-FC broth (with or without agar) sterilized by bringing to boiling point, not autoclaved			
Medium final pH 7.4 ± 0.2			
Prepared medium refrigerated and broth discarded after 96 hours, poured agar medium in petri dishes after 2 weeks			
Uninoculated medium discarded if growth observed			
Sample volumes yield 20-60 fecal coliform colonies per membrane for at least one dilution			

Element	Yes	No	Comments
QC Funnels rinsed with two or three 20-30 mL portions of sterile rinse water after each sample filtration to prevent carry-over			
QC Sterility checked at beginning and end of each filtration series and all data rejected from affected samples and resampling requested if controls contaminated			
Inoculated medium incubated at 44.5°C ± 0.2°C for 24 ± 2 hours			
QC If two or more analysts available, each counts fecal coliform colonies on same membrane monthly and counts agree within 10%			
6. SAMPLE COLLECTION, HANDLING, AND PRESERVATION			
6.1 Sample Collector			
Trained in aseptic sampling procedures and, if required, approved by appropriate regulatory authority or designated representative			
6.2 Sampling			
Sample representative of water distribution system			
Water taps used for sampling free of aerators, strainers, hose attachments, mixing type faucets, and purification devices			
Cold water tap used			
Service line cleared before sampling by maintaining steady water flow for at least 2 minutes			
At least 100 mL sample volume collected, allowing one inch air space in container			
Sample information form completed immediately after sample collection			
Source water representative of supply, collected not too far intake at a reasonable distance from shore			
6.3 Sample Icing			
Samples held at < 10°C during transit to laboratory (recommended for drinking water, required for source water)			
6.4 Sample Holding/Travel Time			
Time from sample collection to initiation of analysis for total coliforms, fecal coliforms, or *E. coli* does not exceed 30 hours for drinking water samples			
Time from sample collection to initiation of analysis for total coliforms and fecal coliforms in source water and heterotrophic bacteria in drinking water does not exceed 8 hours			

Element	Yes	No	Comments
All samples analyzed on day of receipt by laboratory, unless laboratory receives sample late in day and then refrigerates sample overnight and begins analysis within holding time			
6.5 Sample Information Form			

Entered on sample information form in indelible ink:
  - name of system (PWSS identification number if available)
  - sample identification (if any)
  - sample site location
  - sample type (e.g. routine, repeat, raw or process)
  - date and time of collection
  - analysis required
  - disinfectant residual
  - name of sampler and organization (if not water system)
  - sampler's initials
  - person(s) transporting sample from system to laboratory (if not sampler)
  - transportation condition (e.g. $<10°C$, protection from sunlight), if shipper used, shipping records available
  - any remarks			

6.6 Chain-of-Custody			
Applicable regulations followed by collectors and laboratory			
7. QUALITY ASSURANCE			
Written QA Plan prepared, followed, and available for inspection			
8. RECORDS AND DATA REPORTING			
8.1 Legal Defensibility			
Compliance monitoring data legally defensible by keeping thorough and accurate records			
QA plan and/or SOPs describe policies and procedures used by facility for record retention and storage			
Chain-of-custody procedures used if samples expected to become part of legal action			
8.2 Maintenance of Records			
Microbiological analyses records kept by or accessible to laboratory for at least 5 years or until next certification data audit completed, whichever is longer			
Client water system notified before disposal of records			
8.3 Sampling Records			

Element	Yes	No	Comments
Data recorded in ink with changes lined through such that original entry visible and changes initialed and dated			
Sampling records include: - sample information form, from Section 6.5 - date and time of sample receipt by laboratory - name of laboratory person receiving sample - if any deficiency in sample condition noted, sample, at a minimum, flagged - if sample transit time exceeds 30 hours (8 hours for source water samples), sample tagged			
8.4 Analytical Records			
Data recorded in ink with changes lined through such that original entry visible and with changes initialed and dated			
Analytical records include: - laboratory sample identification - date and time analysis begins - laboratory and person(s) responsible for performing analysis - analytical technique or method used - all items marked QC - results of analysis			
8.5 Preventive Maintenance			
Preventive maintenance and repair records for all instruments and equipment kept for 5 years			
9. ACTION RESPONSE TO LABORATORY RESULTS			
9.1 Testing Total Coliform-Positive Cultures			
For the Total Coliform Rule, all total coliform positive cultures tested for presence of either fecal coliforms or *E. coli*			
9.2 Notification of Positive Results			
For Total Coliform Rule, proper authority notified promptly by laboratory of positive total coliform, fecal coliform or *E. coli* results			
Total coliform positive result based on confirmed phase for MTF Technique and P-A Coliform Test or verified test for MF Technique (no requirement for confirmation of positive Colilert/Colisure, fecal coliform or *E. coli* tests)			
9.3 Invalidation of Total Coliform-Negative Sample			
For Total Coliform Rule, proper authority notified when results indicate non-coliforms may have interfered with total coliform analysis			

CHAPTER VI

CRITICAL ELEMENTS FOR RADIOCHEMISTRY

1. PERSONNEL

1.1. Laboratory Supervisor

The laboratory supervisor should have at least a bachelor's degree with a major in chemistry or equivalent and at least 1 year of experience in the analysis of drinking water for radiochemicals. The laboratory supervisor should have at least a working knowledge of quality assurance principles. The laboratory supervisor has immediate responsibility to insure that all laboratory personnel have demonstrated their ability to satisfactorily perform the analyses to which they are assigned and that all data reported by the laboratory meet the required quality assurance criteria.

1.2. Laboratory Analyst

The laboratory analyst should have at least a bachelor's degree with a major in chemistry or equivalent and at least 1 year of experience in the analysis of drinking water for radiochemicals. If the analyst is responsible for the operation of analytical instrumentation, he or she should have completed specialized training offered by the manufacturer or another qualified training facility or served a period of apprenticeship under an experienced analyst. The duration of this apprenticeship is proportional to the sophistication of the instrument.

Before beginning the analysis of compliance samples, the analyst should demonstrate acceptable results for blanks, blind spikes, precision, accuracy, method detection and specificity and satisfactory analysis on unknown samples.

1.3. Technician

The laboratory technician should have at least a high school diploma or equivalent, complete a method training program under an experienced analyst and have 6 months bench experience in the analysis of drinking water samples.

Before beginning the analysis of compliance samples, the technician should demonstrate acceptable results for blanks, blind spikes, precision, accuracy, method detection and specificity and satisfactory analysis on unknown samples.

1.4. Sampling Personnel

Personnel who collect samples should have training in the proper collection technique for all types of samples which they collect. Their technique should be reviewed by experienced sampling personnel.

1.5. Analysts and Operators in Training

Data produced by analysts and instrument operators while in the process of obtaining the required training or experience are acceptable only when reviewed and validated by a fully qualified analyst or the laboratory supervisor.

1.6. Waiver of Academic Training Requirement

The certification officer may waive the need for specified academic training, on a case-by-case basis, for highly experienced analysts.

Training records should be maintained for all personnel. These should include all job-related formal training taken by the analyst which pertains to any aspect of his/her responsibilities, including but not limited to analytical methodology, laboratory safety, sampling, quality assurance, data analysis, etc.

2. LABORATORY FACILITIES
2.1. General

The analysis of compliance samples should be conducted in a laboratory where the security and integrity of the samples and the data can be maintained. The laboratory facilities should be clean, have adequate temperature and humidity control and adequate lighting at the bench top. The laboratory must have provisions for the proper disposal of chemical and radiological wastes, including liquid scintillation cocktail mixtures. The appropriate type of exhaust hoods are required where applicable.

A minimum of 150–200 ft^2 of laboratory space and 15 linear feet of usable bench space per analyst are recommended. There should be sufficient bench space for processing samples. Workbench space should be convenient to sink, water, gas, vacuum, and electrical outlets free from surges.

Analytical and sample storage areas should be isolated from all potential sources of contamination. Any sample having an emission rate in excess of 0.5 mrem/hr should be stored in a secured location away from drinking water samples. There should be sufficient storage space for chemicals, glassware and portable equipment, sufficient floor and bench space for stationary equipment and areas for cleaning materials.

2.2. Instrumentation

Instruments must be properly grounded. Counting instruments must be located in a room other than the one in which samples and standards are being prepared and in which other types of wet chemical analyses are being performed.

An uninterrupted power supply should be available for radiation counting equipment.

2.3. Preparation of Standards

In areas where radioactive standards are being prepared, care should be taken to minimize contamination of surfaces, other samples, and personnel.

Either bench surfaces of an impervious material covered with adsorbent paper, or plastic or fiberglass trays lined with adsorbent paper are acceptable.

3. LABORATORY EQUIPMENT AND INSTRUMENTATION

The laboratory is required to have the equipment, supplies, and instrumentation necessary to perform the approved methods for which it is certified.

3.1. Radiation Counting Instruments

The types of radiation counting systems needed to comply with measurements described in the regulations are as follows:

3.1.1. Liquid scintillation system: A liquid scintillation system is essential if the laboratory is to be certified for the measurement of tritium in drinking water samples. It is recommended that the liquid scintillation system have spectral analysis capabilities to establish proper regions of energy discrimination. The system must have a sensitivity adequate to meet or exceed the detection limit requirements of CFR 141.25(c).

3.1.2. Gas-flow proportional counting system: A gas-flow proportional counting system may be used for the measurement of gross alpha and gross beta activities, radium-226, radium-228, strontium-89, strontium-90, radioactive cesium, and iodine-131 as described in the Reference in CFR 141.25(a). The detector may be either a "windowless" (internal proportional counter) or a "thin window" type. A combination of shielding and a cosmic (guard) detector operated in anticoincidence with the main detector should be used to achieve low background beta counting capability. The alpha and beta background count of the system must be low enough so that the sensitivity of the radioanalysis of water samples will meet or exceed the requirement of 40 CFR 141.25(c) with reasonable counting time (not more than 1000 min).

3.1.3. Alpha scintillation counting system: For measurement of gross alpha activities and radium-226, a scintillation system designed for alpha counting may be substituted for the gas-flow proportional counter described. In such a system, a Mylar disc coated with a phosphor (silver-activated zinc sulfide) is either placed directly on the sample or on the face of a photomultiplier tube, enclosed within a light-tight container, along with the appropriate electronics (high voltage supply, preamplifier, amplifier, timer, and sealer).

3.1.4. Scintillation cell system: A scintillation system designed to accept scintillation flasks ("Lucas cells") should be used for the specific measurement of radium-226 by the radon emanation method. The system consists of a light-tight enclosure capable of accepting the scintillation flasks, a detector (phototube), and the appropriate electronics (high voltage supply, amplifier, timers, and sealers). The flasks (cells) needed for this measurement may either be purchased from commercial suppliers or constructed by the laboratory.

3.1.5. Gamma spectrometer systems: Either a solid state lithium drifted germanium Ge(Li) detector or a high purity intrinsic Ge(Hp) germanium detector connected to a multichannel analyzer is needed if the laboratory is to be certified for analyses of manmade photon emitters.

A system with a lithium drifted germanium, or a high purity intrinsic germanium detector may be used for measurement of manmade photon emitters if the efficiency of the detector is adequate to meet the detection limits required at 40 CFR 141.25(c). These detectors should be shielded with a minimum of 10 cm of iron or equivalent. The multichannel analyzer, in addition to appropriate electronics should contain a memory of not less than 4096 channels and at least one readout device.

3.1.6. Alpha Spectrometer System and Beta Scintillation: These counters and others as mentioned in legislation.

4. GENERAL LABORATORY PRACTICES
4.1. Chemicals/Reagents

Chemicals and reagents used must meet the specifications in the methods. If not specified, then "analytical reagent grade" (AR) or American Chemical Society (ACS) grade chemicals or better should be used. Radioactive standards should be certified by the National Institute of Standards and Technology (NIST, formerly NBS) or traceable to a certified source.

4.2. Reagent Water

The laboratory should have a source of reagent water, ASTM type 1, 2, and 3 or equivalent, having a minimum resistance of $10\,M\Omega/cm$ at $25°C$. The background radioactivity should be checked periodically and should not be at a level which interferes with the radionuclide tests.

4.3. Glassware/Plasticware

Specific requirements in the methods for the cleaning of glassware must be followed. If there are no specifications, glassware should be acid washed then washed in detergent solution and thoroughly rinsed first with tap water and then with reagent water.

4.4. Safety

Guidelines in the Laboratory Safety Manual, the Chemical Hygiene Plan or the Standard Operating Procedures should include proper worker safety training and protection. When circumstances warrant the use of protective equipment, this should include the use of gloves, laboratory coats, eye protection, and the proper pipetting techniques.

5. ANALYTICAL METHODS

The approved methods cited in the CFR at §141.25(a) and (b) must be used for the analysis of drinking water compliance samples. These are listed in Table VI-1 .

6. SAMPLE COLLECTION, HANDLING, AND PRESERVATION

Sample containers, preservatives, and holding times in the methods must be followed. Table VI-2 lists critical elements for sample handling including preservation. Sample preservatives should be checked for radioactive content.

7. QUALITY ASSURANCE

Additional information is contained in Appendix H. Specific items are referenced throughout.

7.1. General Requirements

7.1.1. Availability of Records and Documents: The analytical methods, quality assurance manual, and standard operating procedures should be readily available to the analysts. All quality control data and records should be available for inspection by the certification officer.

7.2. Performance Evaluation Studies

Two types of performance evaluation studies are administered by NERL-LV:

7.2.1. Blind performance evaluation studies: Two water samples (A and B) are distributed twice each year. These samples contain a mixture of alpha, beta, and gamma analytes. A laboratory must successfully analyze at least one set of performance samples each year (either those distributed by NERL-LV or other PE samples acceptable to the State) within the acceptance limits for each analyte for which the laboratory wishes to be certified.

7.2.2. Other performance evaluation samples: Water samples containing a varied number of analytes are distributed several times per year. A laboratory should successfully analyze at least two sets each year within the acceptance limits in the regulations for each analyte for which the laboratory wishes to be certified.

7.2.3. Acceptance Limits: The radionuclide performance evaluation studies and their acceptance limits are described in the draft document, *Environmental Radioactivity Performance Evaluation Studies Program and Radioactive Standards Distribution Program* (EPA 600/4–81–004 and EPA 600/4–80–044), available from NERL-LV.

Table VI-1 Methods for Radionuclide Analysis CFR 141.25 (as of 2/1997)

Contaminant	Methodology	Reference (method or page number)								
		EPA[1]	EPA[2]	EPA[3]	EPA[4]	SM[5]	ASTM[6]	USGS[7]	DOE[8]	Other
Naturally occurring										
Gross alpha[11] and beta	Evaporation	900.0	p 1	00-01	p 1	302, 7110 B	....	R-1120-76	....	
Gross alpha[11]	Co-precipitation	.....	....	00-02	....	7110 C	....	....	....	
Radium 226	Radon emanation, Radiochemical	903.1 903.0	p 16 p 13	Ra-04 Ra-03	p 19	7500-Ra C 304, 305, 7500-Ra B	D 3454-91 D 2460-90	R-1141-76 R-1140-76	Ra-05	N.Y.[9]
Radium 228	Radiochemical	904.0	p 24	Ra-05	p 19	304, 7500-Ra D	....	R-1142-76	....	N.Y.[9] N.J.[10]
Uranium[12]	Radiochemical Fluorometric	908.0 908.1	....	....	....	7500-U B 7500-U C (17th Ed.)	.... D 2907-91	R-1180-76 R-1181-76	.... U-04	
	Alpha spectrometry	.....	....	00-07	p 33	7500-U C (18th or 19th Ed.)	D 3972-90	R-1182-76	U-02	
	Laser Phosphorimetry					....	D 5174-91			
Man-made										
Radioactive cesium	Radiochemical Gamma ray spectrometry	901.0 901.1	p 4	....	.... p 92	7500-Cs B 7120 (19th Ed.)	D-2459-72 D 3649-91	R-1111-76 R-1110-76	.... 4.5.2.3	
Radioactive iodine	Radiochemical Gamma ray spectrometry	902.0 901.1	p 6 p 9	....	.... p 92	7500-IB 7500-IC 7500-ID 7120 (19th Ed.)	D 3649-91 D 4785-88	....	4.5.2.3	
Radioactive Strontium 89, 90	Radiochemical	905.0	p 29	Sr-04	p. 65	303, 7500-Sr B	....	R-1160-76	Sr-01 Sr-02	
Tritium	Liquid scintillation	906.0	p 34	H-02	p. 87	306, 7500-3H B	D 4107-91	R-1171-76	....	
Gamma emitters	Gamma ray Spectrometry	901.1 902.0 901.0	....	....	p 92	7120 (19th Ed.) 7500-Cs B 7500-1B	D 3649-91 D 4785-88	R-1110-76	4.5.2.3	

The procedures shall be done in accordance with the documents listed below. The incorporation by reference of the following documents was approved by the Director of the Federal Register in accordance with 5 U.S.C. 552(a) and 1 CFR part 51. Copies of the documents may be obtained from the sources listed below. Information regarding obtaining these documents can be obtained from the Safe Drinking Water Hotline at 800-426-4791. Documents may be inspected at EPA's Drinking Water Docket, 401 M Street, SW., Washington, DC 20460 (Telephone: 202-260-3027); or at the Office of Federal Register, 800 North Capitol Street, NW., Suite 700, Washington, DC.

1. "Prescribed Procedures for Measurement of Radioactivity in Drinking Water", EPA 600/4-80-032 , August 1980. Available at U.S. Department of Commerce, National Technical Information Service (NTIS), 5285 Port Royal Road, Springfield, VA 22161 (Telephone 800-553-6847), PB 80-224744.

2. "Interim Radiochemical Methodology for Drinking Water", EPA 600/4-75-008(revised), March 1976. Available at NTIS, ibid. PB 253258.

3. "Radiochemistry Procedures Manual", EPA 520/5-84-006, December 1987. Available at NTIS, ibid. PB 84-215581.

4. "Radiochemical Analytical Procedures for Analysis of Environmental Samples", March 1979. Available at NTIS, ibid. EMSL LV 053917.

5. "Standard Methods for the Examination of Water and Wastewater", 13th, 17th, 18th, 19th Editions, 1971, 1989, 1992, 1995. Available at American Public Health Association, 1015 Fifteenth Street N.W., Washington, D.C. 20005 . All methods are in the 17th, 18th and 19th editions except 7500-U C Fluorometric Uranium was discontinued after the 17th Edition, 7120 Gamma Emitters is only in the 19th Edition, and 302, 303, 304, 305 and 306 are only in the 13th Edition.

6. Annual Book of ASTM Standards, Vol. 11.02, 1994. Available at American Society for Testing and Materials, 100 Barr Harbor Drive, West Conshohocken, PA 19428.

7. "Methods for Determination of Radioactive Substances in Water and Fluvial Sediments", Chapter A5 in Book 5 of Techniques of Water-Resources Investigations of the United States Geological Survey, 1977. Available at U.S. Geological Survey (USGS) Information Services, Box 25286, Federal Center, Denver, CO 80225-0425.

8. "EML Procedures Manual", 27th Edition, Volume 1 , 1990. Available at the Environmental Measurements Laboratory, U.S. Department of Energy (DOE), 376 Hudson Street, New York, NY 10014-3621.

9. "Determination of Ra-226 and Ra-228 (Ra-02)", January 1980, Revised June 1982. Available at Radiological Sciences Institute Center for Laboratories and Research, New York State Department of Health, Empire State Plaza, Albany, NY 12201.

10. "Determination of Radium 228 in Drinking Water", August 1980. Available at State of New Jersey, Department of Environmental Protection, Division of Environmental Quality, Bureau of Radiation and Inorganic Analytical Services, 9 Ewing Street, Trenton, NJ 08625.

11. Natural uranium and thorium-230 are approved as gross alpha calibration standards for gross alpha with co-precipitation and evaporation methods; americium-241 is approved with co-precipitation methods.

12. If uranium (U) is determined by mass, a 0.67 pCi/μg of uranium conversion factor must be used. This conservative factor is based on the 1:1 activity ratio of U-234 to U-238 that is characteristic of naturally occurring uranium.

Table VI-2 Sample Handling, Preservation, and Instrumentation

Parameter	Preservative [a]	Container [b]	Maximum holding time [c]	Instrumentation [d]
Gross alpha	Conc. HCl or HNO_3, to pH < 2 [e]	P or G	6 months	A, B, or G
Gross beta	Conc. HCl or HNO_3, to pH < 2 [e]	P or G	6 months	A or G
Strontium-89	Conc. HCl or HNO_3, to pH < 2 [e]	P or G	6 months	A or G
Strontium-90	Conc. HCl or HNO_3, to pH < 2 [e]	P or G	6 months	A or G
Radium-226	Conc. HCl or [e] HNO_3, to pH < 2 [e]	P or G	6 months	A, B, D, or G
Radium-228	Conc. HCl or HNO_3, to pH < 2 [e]	P or G	6 months	A or G
Cesium-134	Conc. HCl to pH < 2 [e]	P or G	6 months	A, C, or G
Iodine-131	None	P or G	8 days	A, C, or G
Tritium	None	G	6 months	E
Uranium	Conc. HCl or HNO_3, to pH < 2 [e]	P or G	6 months	F
Photon emitters	Conc. HCl or HNO_3, to pH < 2 [e]	**P or G**	6 months	C

[a] It is recommended that the preservative be added to the sample at the time of collection unless suspended solids activity is to be measured. It is also recommended that samples be filtered, if suspended or settleable solids are present, prior to adding preservative, at the time of collection. However, if the sample has to be shipped to a laboratory or storage area, acidification of the sample (in its original container) may be delayed for a period not to exceed 5 days. A minimum of 16 hr must elapse between acidification and analysis.

[b] P, plastic, hard or soft; G, glass, hard or soft.

[c] Holding time is defined as the period from time of sampling to time of analysis. In all cases, samples should be analyzed as soon after collection as possible. If a composite sample is prepared, a holding time cannot exceed 12 months.

[d] A, Low background proportional system; B, alpha and beta scintillation system; C, gamma spectrometer [Ge(Hp) or Ge(Li)]; D, scintillation cell system; E, liquid scintillation system (Section C.2.a); F, fluorometer (Section C.1.1); G, low background alpha and beta counting system other than gas-flow proportional.

[e] If HCl is used to acidify samples which are to be analyzed for gross alpha or gross beta activities, the acid salts must be converted to nitrate salts before transfer of the samples to planchets.

7.3. Operating Manuals

Operating manuals and calibration protocols for counting instruments should be available to analysts and technicians.

7.4. Maintenance of Records

Calibration data and maintenance records on all radiation instruments and analytical balances should be maintained in a permanent record.

7.5. Quality Control Requirements

The following are required for each analyte for which the laboratory is certified:

7.5.1. Duplicates: A minimum of one sample for every 10 or fewer compliance samples must be analyzed in duplicate to verify internal laboratory precision for each method. The relative percentage difference between duplicates should be less than or equal to two sigma counting error for the pair of samples. If the difference exceeds the two sigma counting error or 10% of the measured concentration, whichever is greater, prior measurements are suspect; calculations and procedures should be examined and samples should be recounted. All compliance samples, along with the duplicate analyses should be discarded if reanalysis of these samples exceeds the prescribed limits for duplicate analyses or if they cannot be recounted.

7.5.2. Fortified Sample Matrix: Matrix spikes are prepared by adding a known quantity of traceable standard (if available) solution to an actual sample. They are prepared with the sample batch and processed identically to the actual samples. The added concentration should not be less than the background. A matrix spike sample must be prepared and used for each batch of samples, regardless of batch size. Agreement of the measured value ± 20% (95% CL) of the expected value indicates validity of the calibration curve. If the percentage recovery exceeds this limit, the result should be flagged as suspect due to matrix interference. Over time, an effort should be made to spike samples from all the matrixes analyzed.

7.5.3. Counting and Background Check: A counting standard and a background sample should be measured with each set of 20 or fewer samples processed in a day.

7.5.4. Composited Samples: Samples may be composited by the utility or the laboratory provided that sample aliquots are filtered before preservation and preservative is added to the first and subsequent sample aliquots. Generally, it is preferred to have the laboratory composite the samples to insure integrity over time. Analysis of composited samples should be completed within 1 year after the first sample or within 30 days of the last sample if the time between samples was less than 90 days.

7.6. Instrument Performance Charts/Records

Quality control performance charts or records should be maintained for each instrument.

7.7. QA Plan

The laboratory should prepare and follow a written QA plan (see Chapter III).

8. RECORDS AND DATA REPORTING
8.1. Legal Defensibility

Compliance monitoring data should be made legally defensible by keeping thorough and accurate records. The QA plan and/or SOPs should describe the policies and procedures used by the facility for record retention and storage. If samples are expected to become part of a legal action, chain-of-custody procedures should be used (see Appendix A).

8.2. Maintenance of Records

Public Water Systems are required to maintain records of radionuclide analyses of compliance samples for 10 years (40 CFR 141.33). The laboratory should maintain easily accessible records for 5 years or until the next certification data audit is complete, whichever is longer. The client water system should be notified before disposing of records so they may request copies if needed. This includes all raw data, calculations, and quality control data. These data files may be either hard copy, microfiche, or electronic. Electronic data should always be backed up by protected tape or disk or hard copy. If the laboratory changes its computer hardware or software, it should make provisions for transferring old data to the new system so that it remains retrievable within the time frames specified above. Data that are expected to become part of a legal action will probably need to be maintained for a longer period of time. Check with your legal counsel. See Appendix H, Section 3.0, and *Good Automated Laboratory Practices*, EPA 2185, Office of Information Management, Research Triangle Park, NC 27711, 8/10/95.

8.3. Sampling Records

Data should be recorded in ink with any changes lined through such that original entry is visible. Changes should be initialed and dated. The following information should be readily available in a summary or other record(s):

8.3.1.   Date, location (including name of utility and PWSS ID #), site within the system, time of sampling, name, organization and phone number of the sampler, and analyses required;

8.3.2.   Identification of the sample as to whether it is a routine distribution system sample, check sample, raw or finished water sample, repeat or confirmation sample or other special purpose sample;

8.3.3. Date of receipt of the sample;

8.3.4. Sample volume/weight, container type, preservation and holding time and condition on receipt;

8.3.5. pH and disinfectant residual at time of sampling (from plant records);

8.3.6. Transportation and delivery of the sample (person/carrier, conditions).

8.4. Analytical Records

Data should be recorded in ink with any changes lined through such that original entry is visible. Changes should be initialed and dated. The following information should be readily available:

1. Laboratory and persons responsible for performing analysis;
2. Analytical techniques/methods used;
3. Date and time of analysis;
4. Results of sample and quality control analyses;
5. Calibration and standards information;
6. Counting and detection limit data;
7. Results of analyses.

8.5. Computer Programs

Computer programs should be verified initially and periodically by manual calculations and the calculations should be available for inspection. Access to computer programs and electronic data should be limited to appropriate personnel.

9. ACTION RESPONSE TO LABORATORY RESULTS

When a laboratory is responsible, either by contract or state policy, to report sample results which would cause a system to be out of compliance, the proper authority should be promptly notified and a request should be made for resampling from the same sampling point immediately. See Chapter III.

Laboratory_____

Street_____ .

City_____State_____ .

Survey By_____

Affiliation_____

Date_____Telephone No._____

Sample forms for on-site evaluation of laboratories invalved in analysis of public water supplies radiochemistry.

Laboratory_____Evaluator_____Date_____

Location_____

PERSONNEL

Position/Title	Name	Academic Training	Present Specialty	Years Experience (chemistry)	Years Experience (radiochemistry)
Laboratory Director					
Quality Assurance Officer					
Section/Division chief/Director (if applicable)					
Supervisory Analyst					
Chemical Analyst(s)					
Chemical Technician(s)					
Computer Support Technician					
Electronics Support Technician					

Laboratory_____Evaluator_____Date____

Location_____

LABORATORY FACILITIES

Item	Available		Comments
	Yes	No	
Laboratory			
Electrical outlets 120V ac. grounded			
Distilled or deionized water or ASTM type 1, 2, or 3			
Exhaust Hood			
Vacuum source			
Counting Room Separate from wet chemistry, sample and standards preparation area			
Regulated power supply			
Reagents			

Evaluator_____Date_____

Location_____

GENERAL LABORATORY EQUIPMENT AND INSTRUMENTS

Item	No. of Units	Manufacturer	Model	Satisfactory Yes No	
Analytical Balance 0.1 mg sensitivity stable base ASTM type 1 or 2 weights or better					
pH meter ±0.5 units readability ±0.1 units line or battery					
Conductivity meter Readable in ohms or mhos Range of 2 ohms or mhos Line or battery					
Drying oven gravity or convection controlled from room temp to 180°C or higher (±2°C)					
Infrared lamp may be substituted for drying oven					
Desiccator Glass or plastic					
Hot plate temperature control					
Refrigerator					
Magnetic Stirrer variable speed Teflon coated stir bar					
Balance, top loading					
Glassware					
Thermometers					
Muffle furnace to 450°C					
Centrifuge to 3000 rpm to hold 4 x 50 mL					

Laboratory_____Date_____

Location_____Evaluator_____

ALL INSTRUMENTATION

	Yes	No	Comments
Are operating manuals readily available to the operator			
Are calibration protocols available to the operator			
Are calibrations kept in a permanent control chart			
Are permanent service maintenance records kept			

Laboratory_____Date_____

Location_____Evaluator_____

THIN WINDOW GAS-FLOW PROPORTIONAL COUNTER

Instrument number	Manufacturer	Model	Year	Sample Changing		
				Manual	Automatic	Capacity

	Counting Gas	Window Density (g/cm²)	Instrument Background			
			Alpha Operating Voltage cpm		Beta Operating Voltage cpm	

Calibration Standard Type: Alpha Beta Supplier: Alpha Beta	Calibration Frequency[1]				Service Maintenance Frequency[2]				Condition[3]		
	D	W	M	Other	Q	S	A	Other	G	R	N

WINDOWLESS GAS-FLOW PROPORTIONAL COUNTER

Instrument number	Manufacturer	Model	Year	Sample Changing		
				Manual	Automatic	Capacity

	Counting Gas	Window Density (g/cm²)	Instrument Background	
			Operating cpm Voltage	Alpha Operating cpm Voltage Beta

Calibration Standard Type: Alpha Beta Supplier: Alpha Beta	Calibration Frequency[1]				Service Maintenance Frequency[2]				Condition[3]		
	D	W	M	Other	Q	S	A	Other	G	R	N

1. Daily, Weekly, Monthly.
2. Quarterly, Semiannually, Annually.
3. Good, operating but needs Repair, Not operating

Laboratory_____Date_____

Location_____Evaluator_____

LIQUID SCINTILLATION COUNTER

Instrument number	Manufacturer		Model		Year		Sample Changing		
							Manual	Auto	Capacity

Calibration Standard Type: Supplier:	Calibration Frequency[1]				Service Maintenance Frequency[2]				Condition[3]		
	D	W	M	Other	Q	S	A	Other	G	R	N

ALPHA SCINTILLATION COUNTER

Instrument number	Manufacturer		Model		Year		Sample Changing		
							Manual	Auto	Capacity

Calibration Standard Type: Supplier:	Calibration Frequency[1]				Service Maintenance Frequency[2]				Condition[3]		
	D	W	M	Other	Q	S	A	Other	G	R	N

RADON-GAS COUNTING SYSTEM

System number	Gas counting cells/system			Manufacturer of gas counting cell				Counting Instrument		
							Make	Model		Year

	Calibration Frequency[1]				Service Maintenance Frequency[2]				Condition[3]		
	D	W	M	Other	Q	S	A	Other	G	R	N

1. Daily, Weekly, Monthly.
2. Quarterly, Semiannually, Annually.
3. Good, operating but needs Repair, Not operating

Laboratory_____Date_____

Location_____Evaluator_____

GAMMA SPECTROMETER SYSTEM

Detector System						
	Type	Make	**System Number** Model		Year	Size
	Make		**Analyzer System** Model	Year	Channels	
Calibration Standard Type						
Supplier	**Calibration Frequency** D W M Other			**Service Maintenance Frequency** Q S A Other		**Condition** G R N

OTHER APPROVED DETECTOR

Detector System						
	Type	Make	**System Number** Model		Year	Size
	Make		**Analyzer System** Model	Year	Channels	
Calibration Standard Type						
Supplier	**Calibration Frequency** D W M Other			**Service Maintenance Frequency** Q S A Other		**Condition** G R N

1. Daily, Weekly, Monthly.
2. Quarterly, Semiannually, Annually.
3. Good, operating but needs Repair, Not operating

Laboratory_____Date_____

Location_____Evaluator_____

GAMMA SPECTROMETER SYSTEM

Detector System	Type		Make		System Number Model		Year		Size	
	Make			Model	Analyzer System Year			Channels		
Calibration Standard Type										
Supplier	Calibration Frequency D W M Other				Service Maintenance Frequency Q S A Other			Condition G R N			

OTHER APPROVED DETECTOR

Detector System	Type		Make		System Number Model		Year		Size	
	Make			Model	Analyzer System Year			Channels		
Calibration Standard Type										
Supplier	Calibration Frequency D W M Other				Service Maintenance Frequency Q S A Other			Condition G R N			

1. Daily, Weekly, Monthly.
2. Quarterly, Semiannually, Annually.
3. Good, operating but needs Repair, Not operating

Laboratory_____Date_____

Location_____Evaluator_____

METHODOLOGY

Parameter	Sample Load/Mo	Method[1] Used - Cite Edition, Year, and Page						Satifactory	
		EPA	SM1	ASTM	USGS	DOE	Other	Yes	No
Gross Alpha Activity									
Gross Beta Activity									
Strontium-89									
Strontium-90									
Radium-226									
Radium-228									
Cesium-134									
Iodine-131									
Tritium									
Uranium									
Photon Emitters Identify:									
a.									
b.									
c.									
d.									
e.									

1 - Methods used must be referenced in the National Primary Drinking Water Regulations (40 CFR 141.25)

Laboratory_____ Date_____

Location_____ Evaluator_____

QUALITY CONTROL

Item	Performance Evaluation Studies	A[1]	B[2]	Blind PE Studies	A[1]	B[2]
Participation in performance evaluation and Blind PE studies	Gross Alpha			Gross Alpha		
	Gross Beta			Gross Beta		
	Sr-89			Sr-89		
	Sr-90			Sr-90		
Reporting Period:	Ra-226			Ra-226		
	Ra-228			Ra-228		
_____ to	Uranium			Uranium		
	Cs-134			Cs-134		
_____	Cs-137			Cs-137		
	Co-60			Co-60		
	Ba-133			Written QA Plan implemented and available for review		
	Zn-65					
	Tritium					
	I-131					
	Frequency	Yes	No	Comments	Satisfactory Yes	No
Duplicate analyses						
Spikes						
Failed PE studies						
Control charts						
Calibration and Maintenance records						

1 - Scheduled frequency of participation by the laboratory, times per year.
2 - Number of acceptable performance results in the past year, where an acceptable result is a normalized deviation from the known value of ≤3.0 sigma.

Laboratory_____ Date_____

Location_____ Evaluator_____

DATA REPORTING

Item	Comments: systems used, frequency, etc.
Records kept for 10 years Actual laboratory reports	
Tabular Summary	
Information included Date	
Place of sampling	
Time of sampling	
Sampler	
Date of sample receipt	
Date of analysis	
Type of analysis	
Laboratory & person responsible	
Other reported data	
Other reported data	
Method(s) used	
Results	

APPENDIX A
Chain-of-Custody Evaluations

A. Introduction

Written procedures for sample handling should be available and followed whenever samples are collected, transferred, stored, analyzed, or destroyed. For the purposes of litigation, it is necessary to have an accurate written record to trace the possession and handling of samples from collection through reporting. The procedures defined here represent a means to satisfy this requirement.

A sample is in someone's "custody" if:

1. it is in one's actual physical possession;
2. it is in one's view, after being in one's physical possession;
3. it is one's physical possession and then locked up, so that no one can tamper with it;
4. it is kept in a secured area, restricted to authorized personnel only.

B. Sample Collection, Handling, and Identification

1. It is important that a minimum number of persons be involved in sample collection and handling. Guidelines established in standard manuals for sample collection preservation and handling should be used (e.g., EPA NPDES Compliance Sampling Inspection Manual, MCD 51, *Standard Methods for Examination of Water and Wastewater*). Field records should be completed at the time the sample is collected and should be signed or initialed, including the date and time, by the sample collectors). Field records should contain the following information:

    a. unique sample or log number;
    b. date and time;
    c. source of sample (including name, location, and sample type);
    d. preservation used;
    e. analyses required;
    f. name of collector(s);
    g. pertinent field data (pH, DO, Cl residual, etc.);
    h. serial number on seals and transportation cases;
    i. comments.

2. Each sample is identified by affixing a pressure-sensitive gummed label or standardized tag on the container(s). This label should contain the sample number, source of sample, preservative used, and the collector(s)' initials. The analysis required should be identified. Where a label is not available, the sample information should be written on the sample container with an indelible marking pen. An example of a sample identification tag is illustrated in Figure A-1.

3. The closed sample container should then be placed in a transportation case along with the chain-of-custody record form, pertinent field records, and analysis request form. The transportation case should then be sealed and labeled. All records should be filled out legibly in waterproof pen. The

GENERAL CHEMISTRY

U.S. EPA REGION

Official Sample No,

SOURCE

Date and Time

Sampler's Signature   Office
Other Parameters:

PH	Acid
Cond	Alk
TS	SO₄
DS	Cl
SS	F
BOD₂	Cr. + 6
Turb	BOD₅
Color	

MICROBIOLOGY

U.S. EPA REGION

Official Sample No.

SOURCE

Date and Time

Sampler's Signature   Office

Tot. Colif.

Fecal Colif.

Fecal Strep.

Salmonella

PESTICIDES, ORGANICS

U.S. EPA REGION

Official Sample No.

SOURCE

Date and Time

Sampler's Signature   Office

Pesticides

PCB's:

Organics:

EPA

Station No.   Date   Time   Sequence No.

Station Location

Grab
Comp.

BOD   Metals
Solids   Oil and Grease
COD   D.O.
Nutrients   Bact.
               Other

Remarks/Preservative:

Samplers:

Figure A-1   Sample identification tag examples.

use of locked or sealed chests will eliminate the need for close control of individual sample containers. However, there will, undoubtedly, be occasions when the use of a chest will be inconvenient. On these occasions, the sampler should place a seal around the cap of the individual sample container that would indicate tampering if removed.

C.   Transfer of Custody and Shipment

1.   When transferring the possession of the samples, the transferee must sign and record the date and time on the chain-of-custody record. Custody transfers, if made to a sample custodian in the field, should account for each individual sample, although samples may be transferred as a group. Every person who takes custody must fill in the appropriate section of the chain-of-custody record.

2.   The field custodian (or field sampler if a custodian has not been assigned) is responsible for properly packaging and dispatching samples to the appropriate laboratory for analysis. This responsibility includes fining out, dating, and signing the appropriate portion of the chain-of-custody record. A recommended chain-of-custody format is illustrated in Figure A-2.

3.   All packages sent to the laboratory should be accompanied by the chain-of-custody record and other pertinent forms. A copy of these forms should be retained by the field custodian (either carbon or photocopy).

4.   Mailed packages can be registered with return receipt requested. If packages are sent by common carrier, receipts should be retained as part of the permanent chain-of-custody documentation.

5.   Samples to be transported must be packed to prevent breakage. If samples are shipped by mail or by other common carrier, the shipper must comply with any applicable Department of Transportation regulations. (Most water samples are exempt unless quantities of preservatives used are greater than certain levels.) The package must be sealed or locked to prevent tampering. Any evidence of tampering should be readily detected if adequate sealing devices are used.

6.   If the field sampler delivers samples to the laboratory, custody may be relinquished to laboratory personnel. If appropriate personnel are not present to receive the samples, they should be locked in a designated area of the laboratory to prevent tampering. The person delivering the samples should make a log entry stating where and how the samples were delivered and secured. Laboratory personnel may then receive custody by noting in a logbook, the absence of evidence of tampering, unlocking the secured area, and signing the custody sheet.

D.   Laboratory Sample Control Procedures

Sample control procedures are necessary in the laboratory from the time of sample receipt to the time the sample is discarded. The following procedures are recommended for the laboratory:

Survey						Samplers: Signature					
Station Number	Station Location	Date	Time	Sample Type				Seq. No.	No. Of Containers	Analysis Required	
				Water		Air					
				Comp	Grab.						

Relinquished by: Signature	Received by: Signature	Date/Time
Relinquished by: Signature	Received by: Signature	Date/Time
Relinquished by: Signature	Received by: Signature	Date/Time
Relinquished by: Signature	Received by Mobile Laboratory for Field analysis: Signature	Date/Time
Dispatched by:	Date/Time	Received for Laboratory by: Signature	Date/Time

Method of Shipment:

Figure A-2 Chain-of-custody record.

1. A specific person must be designated as custodian and an alternate designated to act as custodian in the custodian's absence. All incoming samples must be received by the custodian who must indicate receipt by signing the accompanying custody/control forms and must retain the signed forms as permanent records.

2. The custodian must maintain a permanent logbook to record, for each sample, the person delivering the sample, the person receiving the sample, date and time received, source of sample, date the sample was taken, sample identification log number, how transmitted to the laboratory, and condition received (sealed, unsealed, broken container, or other pertinent remarks). This log should also show the movement of each sample within the laboratory; i.e., who removed the sample from the custody area, when it was removed, when it was returned, and when it was destroyed. A standardized format should be established for logbook entries.

3. A clean, dry, isolated room, building, and/or refrigerated space that can be securely locked from the outside must be designated as a "custody room."

4. The custodian must ensure that heat-sensitive samples, light-sensitive samples, radioactive samples, or other sample materials having unusual physical characteristics, or requiring special handling, are properly stored and maintained prior to analysis.

5. Distribution of samples to the analyst performing the analysis must be made by the custodian.

6. The laboratory area must be maintained as a secured area, restricted to authorized personnel only.

7. Laboratory personnel are responsible for the care and custody of the sample, once it is received by them and must be prepared to testify that the sample was in their possession and view or secured in the laboratory at all times from the moment it was received from the custodian, until the time that the analyses are completed.

8. Once the sample analyses are completed, the unused portion of the sample, together with all identifying labels, must be returned to the custodian. The returned tagged sample must be retained in the custody room, until permission to destroy the sample is received by the custodian.

9. Samples will be destroyed only upon the order of the responsible laboratory official when it is certain that the information is no longer required or the samples have deteriorated. (For example, standard procedures should include discarding samples after the maximum holding time has elapsed.) The same procedure is true for sample tags. The logbook should show when each sample was discarded or if any sample tag was destroyed.

10. Procedures should be established for internal audits of sample control information. Records should be examined to determine traceability, completeness, and accuracy.

APPENDIX B
Recommended Protocol for Regions Conducting On-Site Laboratory Evaluations

Before conducting the on-site evaluation, the Region should:

- plan all the required activities to be completed during the assessment;
- hold a pre-evaluation conference with appropriate laboratory and field activity representatives to establish a schedule that would have a minimum impact on the laboratory activities;
- request and review appropriate records;
- request that a variety of tests be scheduled during the on-site evaluation;
- arrange for the laboratory staff to be available during the on-site visit.

During the on-site visit, the team shall:

- conduct an opening conference or entrance interview;
- evaluate the procedures and equipment used for those specific analyses for which the laboratory has requested certification, using the criteria in this manual;
- review the records and written standard operating procedures for compliance with the required sampling frequency, sample collection, sample holding times, and if appropriate, resample notification;
- perform a data audit on at least one sample and one PE sample for at least one method, but preferably for each method the laboratory performs;
- insure that the laboratory has a quality assurance (QA) plan in effect by:

  - determining if the laboratory has written procedures (QA plan or equivalent) for conducting its QA program;
  - examining the QA data to determine if the QA program is being implemented.

- complete the on-site checklists and other evaluation forms during the visit (see Chapters IV–VI).
- conduct a closing conference or exit interview in which the auditors review the results of the evaluation with the director of the laboratory, the director of State water supply activities, and appropriate staff members, and the review should:

  - discuss any deviations in the observed procedures and records;
  - recommend changes in equipment and supply needs, staffing requirements, and facility improvements, if necessary;
  - discuss possible assistance the Region can provide the laboratory;
  - discuss a time frame for corrective actions and response.

Evaluation Report for Principal State Laboratories and Laboratories in Nonprimacy States

After an on-site inspection, the evaluation team should prepare a narrative report and action memorandum.

This report should contain all information pertinent to the evaluation and also recommend the certification status for all analyses evaluated. The report should then be forwarded for evaluation to the Certification Program Manager for review. After reviewing and, if necessary, revising the report, it should be forwarded to the Certification Authority for signature.

The Certification Authority should decide the certification status of the laboratory within time constraints on page III-7 and notify the State. The State should be sent the complete report. If the report indicates that the laboratory should not be certified for an analysis, the Certification Authority should give the specific reasons.

The narrative report should be attached to each copy of the completed evaluation form. It should include the general headings and information listed in what follows.

Title Page

The title page should contain the following:
  Title: Report of an on-site evaluation of the (name of laboratory)
  At: (city, state, and zip code)
  On: (date)
  By: (name, title, organization, and address of the certification team)

Certification Status

List either "Certified", "Provisionally Certified", "Administratively/Interim Certified", or "Not Certified" for each contaminant evaluated or if applicable (for VOCs, for instance) for each class of compounds evaluated.

List of Deviations

List each deviation by item number used on the evaluation checklists. Describe the exact deviation and recommended changes.

Remarks

Recommend improvements that, while not affecting certification status, would improve laboratory operation. Other remarks might include reasons for failing the on-site evaluation, special recognition for outstanding performance, and description of unusual tests.

List of Personnel

List name and title of personnel along with the individual tests that each normally performs. In addition, identify the critical laboratory personnel.

Signature

Team members should sign the report.

Distribution

Copies of this report should be distributed to the State requesting the evaluation. For local laboratories in nonprimacy States, reports should be distributed to appropriate Regional personnel.

Annually, each Region should submit to OGWDW a listing of laboratories in the Region having U.S. EPA certification. The listing should include the names and location of each laboratory and its certification status for all regulated contaminants. In addition, Regions should notify OGWDW of all changes in status soon after they occur, so that OGWDW can maintain an updated list of certification status.

APPENDIX C
Definitions and Abbreviations

ASTM: American Society of Testing and Materials.
AWWA: American Water Works Association.
NERL-Ci: U.S. EPA National Exposure Research Laboratory in Cincinnati, Ohio (ORD).
NERL-LV: U.S. EPA National Exposure Research Laboratory in Las Vegas, Nevada (ORD).
NPDWR: National Primary Drinking Water Regulations.
OGWDW: U.S. EPA Office of Ground Water and Drinking Water.
ORD: U.S. EPA Office of Research and Development.
SDWA: The Safe Drinking Water Act as amended (42 U.S.C. 300f et seq.).
Accuracy: A measure of the closeness of an individual measurement or the average of a number of measurements to the true value. Accuracy includes a combination of random error (precision) and systematic error (bias) components that are due to sampling and analytical operations. Refer to Standard Methods, Data Quality Section for a more detailed explanation.
Administrator: The administrator of the U.S. EPA or her/his authorized representative. See 40 CFR 142.2.
Agency: The U.S. EPA. See 40 CFR 142.2.
Auditor: A person who evaluates laboratories to determine if they meet the criteria to be certified. This person should be an experienced professional, who has effective communication skills, experience in quality assurance, the analytical techniques being evaluated, and familiarity with the drinking water regulations and this manual.
Bachelor Degree or Equivalent: A college degree with an equivalent 30 semester hours in a specific discipline. Equivalent is at least 4 years of experience in a specific scientific discipline.
Bias: The systematic or persistent distortion of a measurement process that causes errors in one direction.

Certification Authority (CA): The person or designee who has the authority to certify laboratories conducting drinking water analyses and to certify the officials of the State responsible for the State's certification program in accordance with Section 1412 of the Safe Drinking Water Act. This authority is delegated to the Regional Administrator, but may be redelegated.

Certification Program Manager (CPM): The person responsible for managing the certification program which includes tracking the certification status of the State laboratories, ensuring that the Regional and State certification officers are qualified and reviewing the certification evaluation reports.

Certification Officer (CO): A State or Federal laboratory auditor who has passed the NERL certification officers training course (limited at this time to chemistry and microbiology). This person provides information to the CA or CPM for the purpose of making decisions on the certification status of a laboratory.

Code of Federal Regulations (CFR): A compilation of regulations is revised each time a regulation is promulgated. It is published every year in July.

Confirmation: Verification of the presence of a component through the use of an analytical technique based on a different scientific principle from the original method (e.g., second column, alternate wavelength or detector, etc.)

Data Audit: A qualitative and quantitative evaluation of the documentation and procedures associated with measurements to verify that the resulting data are acceptable.

Data Quality Objectives: Qualitative and quantitative specifications used to design a study that will limit uncertainty to an acceptable level.

Data Reduction: The process of transforming the number of data items by arithmetic or statistical calculations, standard curves, concentration factors, etc., and collation into a more useful form. Data reduction is irreversible and generally results in the loss of detail.

Detection: Any concentration of an analyte, which equals or exceeds the laboratory's detection limit. For VOCs, detection limit is defined as 0.0005 mg/L.

Drinking Water Laboratory: A laboratory that analyses samples as part of compliance monitoring for a public water supply.

Federal Indian Land: Areas, which for regulatory purposes, are treated as independent States. On these lands, the Indian tribe has a federally recognized governing body carrying out government duties and powers.

Holding time: The allowed time from when a sample was taken (or extracted) until it must be analyzed.

Initial Demonstration of Capability (IDC): Before analyzing compliance samples, an analytical team must demonstrate acceptable precision, accuracy, sensitivity, and specificity for the method to be used.

Laboratory Reagent Blank (LRB)(Method blank): An aliquot of reagent water or other blank matrix that is treated exactly as a sample to determine if method analytes or other interferences are present.

Laboratory Fortified Blank (LFB)(Spike): An aliquot of reagent water or other blank matrix to which known quantities of the method analytes are added in the laboratory. The LFB is analyzed exactly like a sample to determine whether the method is in control.

Maximum contaminant level (MCL): It means the maximum permissible level of a contaminant in water that is delivered to any user of a public water system. See 40 CFR Part 141.2.

Maximum contaminant level goal (MCLG): It means the maximum level of a contaminant in drinking water at which no known or anticipated adverse effect on the health of persons would occur and which allows an adequate margin of safety. Maximum contaminant level goals are nonenforceable health goals. See 40 CFR 141.2.

Method Detection Limit (MDL): It is defined as the minimum concentration of a substance that can be measured and reported with 99% confidence that the analyte concentration is > 0. The MDL is determined from analysis of a sample in a given matrix containing this analyte. See 40 CFR 136 App. B.

Monitoring Trigger: The concentration of a regulated contaminant that triggers additional monitoring

National Environmental Laboratory Accreditation Conference (NELAC): A voluntary organization of State, Federal, and other groups to establish mutually acceptable standards for accrediting environmental laboratories.

Performance Evaluation Samples (PEs)(Proficiency test sample): A sample provided to a laboratory for the purpose of demonstrating that the laboratory can successfully analyze the sample within specified acceptance limits specified in the regulations. The qualitative and/or quantitative composition of the reference material is unknown to the laboratory at the time of the analysis. See 40 CFR Part 141.2.

Precision: The measure of mutual agreement among individual measurements.

Primacy: Primary responsibility for administration and enforcement of primary drinking water regulations and related requirements are applicable to public water systems within a State.

Principal State Laboratory System: All facilities, whether part of the State laboratory or contracted to the State, producing data for the State and certified by the EPA, fulfilling the requirements for primacy, as listed in the 40 CFR 142.10(b)(4).

Public Water System: A system for the provision of piped water to the public for human consumption, if such system has at least 15 service connections or regularly serves an average of at least 25 individuals daily at least 60 days out of the year. See 40 CFR Part 141.2.

Quality Assurance: An integrated system of management activities involving planning, quality control, quality assessment, reporting, and quality improvement to ensure that a product or service meets defined standards of quality with a stated level of confidence.

Quality Control: The overall system of technical activities whose purpose is to measure and control the quality of a product or service, so that it meets the needs of the users; operational techniques and activities that are used to fulfill requirements for quality.

QA Plan: A comprehensive plan detailing the aspects of quality assurance needed to adequately fulfill the data needs of a program. This document is required before the laboratory is certified.

Regulatory Level: A concentration of a contaminant that is cited in the Federal Regulations (e.g., MCL, detect, etc.)

Shall: Denotes a mandatory requirement.

Should: Denotes a guideline or recommendation.

Standard Operating Procedure: A written document which details the method of an operation, analysis, or action, whose techniques and procedures are thoroughly prescribed and which is officially approved as the method for performing certain routine or repetitive tasks.

Third Party Auditor: Person or persons, not affiliated with a Region or State, who is designated by the Region or State to audit a laboratory. This person must pass the certification-training course prior to auditing any laboratory, unless the person is a part of an audit team that includes a Regional/State certification officer. The third party auditor must also meet the educational/experience requirements specified in this manual. The certification decision remains with the Region.

Third Party Expert: Any person not designated as a certification officer or auditor, who is requested by the Region to assist in the audit of a laboratory because of the person's expertise in a particular area (e.g., asbestos). This person is not required to take the certification officers' course if the person is part of an audit team that includes a certification officer.

"Unregulated" Contaminants: Contaminants for which monitoring is required but which have no MCL.

APPENDIX D

United States Environmental Protection Agency

MEMORANDUM

SUBJECT: The Use of "Third-Parties" in the Drinking Water Laboratory Certification Program

FROM: Cynthia Dougherty, Director,
Office of round Water and Drinking Water

TO: Water Supply Representatives, Regions I–X
Certification Authorities, Regions I–X
Quality Assurance Officers, Regions I–X
Regional Laboratories, Regions I–X

Purpose

This memorandum updates and clarifies the guidance memorandum from Michael Cook dated December 5, 1989 on "Third-Party Certification for Laboratories in Primacy States."

Action

Under 40 CFR 142.10(b)(3), if a state does not perform all analytical measurements in its own laboratory, it must establish and maintain a program for the certification of laboratories as a condition for receiving and maintaining authority to administer the Safe Drinking Water Act in lieu of EPA (primacy). This memorandum notifies states with primacy that they may contract with other organizations (third parties) to assist the state in fulfilling this requirement. The *authority* for making certification *decisions*, however, must remain with the state.

Discussion

Several states have asked USEPA its position on the use of third parties, i.e., private sector organizations that assist the states with their certification program. OGWDW realizes that dwindling state resources may necessitate assistance from third parties in the State certification programs. Consistent with the regulatory requirement at 40 CFR 142.10(b), providing for the "establishment and maintenance of a state program for the certification of laboratories," the state must retain ultimate authority to decide whether individual laboratories will be certified. This decision may not be abdicated to the third party.

This office will not pass judgment on any specific third-party program. It is the responsibility of each primacy state to assess the qualifications of the third party. In assessing whether to choose a particular third party, the state should consider, as a minimum, the following items that are described in the *Manual for the Certification of Laboratories Analyzing Drinking Water*:

- Ability to provide technical assistance and training.
- Availability of records for review by the state.
- Quality assurance program.
- Freedom from conflicts of interest.
- EPA policy, which provides that the auditor should pass an appropriate course on how to audit in the discipline for which he or she will be auditing.
- Experience of the auditor.

  - The auditor should be an experienced professional with at least a bachelor's degree or equivalent education/experience in the discipline for which he or she audits.
  - The auditor should have recent laboratory experience.

Any state certification program using third-party assistance should meet the requirements in the *Manual for the Certification of Laboratories Analyzing Drinking Water*, just as it would if it were using State employees to perform these functions. The Regions should assist the State and third-party agent to assure that the certification program meets EPA guidelines.

Regions and States should be sensitive to potential conflict-of-interest problems between third parties and evaluated laboratories. For instance, inspectors employed by firms that provide analytical services in the drinking water area should not be put in the position of passing judgment on their competitors.

Further information

If you have questions or need additional information or assistance, please contact the OGWDW Technical Support Center at 513–569–7904.

APPENDIX E

Required Analytical Capability for Principle State Laboratory System

---

INORGANICS
(40 CFR 141.23)
Asbestos
Cyanide
Fluoride
Nitrate
Nitrite
Antimony
Arsenic
Barium
Beryllium
Cadmium
Chromium
Mercury
Selenium
Thallium
(40 CFR 141.89)
Copper
Lead
Conductivity
Calcium
Alkalinity
Orthophosphate
Silica
VOLATILE ORGANICS
(40 CFR 141.24)
THMs
Benzene
Carbon tetrachloride
Chlorobenzene
o-Dichlorobenzene
p-Dichlorobenzene
1,2-Dichloroethane
1,1-Dichloroethylene
cis-1,2-Dichloroethylene
Trans-1,2-Dichloroethylene
Dichloromethane
1,2-Dichloropropane
Ethylbenzene
Styrene
Tetrachloroethene

MICRO-ORGANISMS
(40 CFR 141.24)
Total coliforms
Escherichia coli or fecal coliforms
Heterotrophic bacteria
SOCs
(40 CFR 141.24)
Alachlor
Atrazine
Benzo(a)pyrene
Carbofuran
Chlordane
2,4-D
Di(2-ethylhexyl)adipate
Di(2-ethylhexyl)phthalate
Dibromochloropropane
Dalapon
Dinoseb
Dioxin (2,3,7,8-TCDD)
Diquat
Endothall
Endrin
Ethylenedibromide
Glyphosate
Heptachlor
Heptachlor epoxide
Hexachlorobenzene
Hexachlorocyclopentadiene
Lindane
Methoxychlor
Oxamyl
PCBs (as decachlorobiphenyl)
Pentachlorophenol
Picloram
Simazine
Toxaphene
2,4,5-TP
RADIONUCLIDES
(40 CFR 141.25)
Gross Alpha
Uranium

(Continued)

Toluene	Gross Beta
1,2,4-Trichlorobenzene	Cesium-134
1,1,1-Trichloroethane	Strontium-89
1,1,2-Trichloroethane	Iodine-131
Trichloroethene	Strontium-90
Vinyl chloride	Tritium
Xylenes	Other beta/photon emitters
	Radium-226/228

APPENDIX F

Unregulated, Proposed, and Secondary Contaminants

UNREGULATED VOCs
(40 CFR 141.40)
Bromobenzene
Bromodichloromethane
Bromoform
Bromomethane
Chlorodibromomethane
Chloroethane
Chloroform
Chloromethane
o-Chlorotoluene
p-Chlorotoluene
1,1-Dichloroethane
1,3-Dichloropropane
2,2-Dichloropropane
1,1-Dichloropropene
1,3-Dichloropropene(c/t)
m-Dichlorobenzene
Dibromomethane
1,1,1,2-Tetrachloroethane
1,1,2,2-Tetrachloroethane
1,2,3-Trichloropropane
UNREGULATED IOCs
(40 CFR 141.40)
Corrosivity
Sodium
Sulfate
PHASE 6A

UNREGULATED SOCs
(40 CFR 141.40)
Aldicarb
Aldicarb sulfoxide
Aldicarb sulfone
Aldrin
Butachlor
Carbaryl
Dicamba
Dieldrin
3-Hydroxycarbofuran
Methomyl
Metolachlor
Metribuzin
Propachlor
RADIONUCLIDES
(Proposed 7/18/91)
(40 CFR 141.25)
Radon
Uranium
(40 CFR 141.44)
Lead-210
SECONDARY CONTAMINANTS
Aluminum
Chloride
Color
Copper
Fluoride

(Continued)

(Proposed 7/29/94)	Foaming agents
Free and total chlorine	Iron
Combined chlorine	Manganese
Chlorine dioxide	Odor
Trihalomethanes	PH
Haloacetic acids	Silver
Bromate	Sulfate
Chlorite	Total dissolved solids (TDS)
Total organic carbon	Zinc
Alkalinity	
Bromide	

Contaminants to be Monitored at the Discretion of the State

1,2,4-Trimethylbenzene
1,2,3-Trichlorobenzene
n-Propylbenzene
n-Butlybenzene
Naphthalene
Hexachlorobutadiene
1,3,5-Trimethylbenzene
p-Isopropyltoluene
Isopropylbenzene
tert-Butylbenzene
sec-Butylbenzene
Fluorotrichloromethane
Dichlorodifluoromethane
Bromochloromethane

Appendix G

Analytical Methods for Microbiology: 1. Total Coliform Rule (40 CFR 141.21(f))

(f) Analytical methodology. (1) The standard sample volume required for total coliform analysis, regardless of analytical method used, is 100 mL.

(2) Public water systems need only determine the presence or absence of total coliforms; a determination of total coliform density is not required.

(3) Public water systems must conduct total coliform analyses in accordance with one of the analytical methods in the following table. These methods are contained in the 18th edition of *Standard Methods for the Examination of Water and Wastewater,* 1992, American Public Health Association, 1015

Fifteenth Street, NW, Washington, D.C. 20005, U.S.A.A description of the Colisure test may be obtained from the Millipore Corporation, Technical Services Department, 80 Ashby Road, Bedford, MA 01730, U.S.A. The toll-free phone number is (800) 645–5476.

Organism	Methodology	Citation
Total coliforms[a]	Total coliform fermentation technique[b,c,d]	9221A,B
	Total coliform membrane filter technique	9222A,B,C
	Presence–absence (P–A) coliform test [d,e]	9221D
	ONPG-MUG test[f]	9223
	Colisure test[g]	

[a]The time from sample collection to initiation of analysis may not exceed 30 hr. Systems are encouraged but not required to hold samples below 10°C during transit.

[b]Lactose broth, as commercially available, may be used in lieu of lauryl tryptose broth, if the system conducts at least 25 parallel tests between this medium and lauryl tryptose broth using the water normally tested, and this comparison demonstrates that the false-positive rate and false-negative rate for total coliforms, using lactose broth, is < 10%.

[c]If inverted tubes are used to detect gas production, the media should cover these tubes at least one-half to two-thirds, after the sample is added.

[d]No requirement exists to run the completed phase on 10% of all total coliform-positive confirmed tubes.

[e]Six-times formulation strength may be used if the medium is filter sterilized rather than autoclaved.

[f]The ONPG-MUG test is also known as the Autoanalysis Colilert System.

[g]The Colisure test must be incubated for 28 hr before examining the results. If an examination of the results at 28 hr is not convenient, then results may be examined at any time between 28 and 48 hr.

(4) [Reserved].

(5) Public water systems must conduct fecal coliform analysis in accordance with the following procedure. When the MTF technique or P–A coliform test is used to test for total coliforms, shake the lactose-positive presumptive tube or P–A [bottle] vigorously and transfer the growth with a sterile 3-mm loop or sterile applicator stick into brilliant green lactose bile broth and EC medium, to determine the presence of total and fecal coliforms, respectively. For EPA-approved analytical methods which use a membrane filter, transfer the total coliform-positive culture by oas of the following methods: remove the membrane containing the total coliform colonies from the substrate with a sterile forceps and carefully curl and insert the membrane into a tube of EC medium (the laboratory may first remove a small portion of selected colonies for verification), swab the entire membrane filter surface with a sterile cotton swab and transfer the inoculum to EC medium (do not leave the cotton swab in the EC medium), or inoculate individual total coliform-positive colonies

into EC medium. Gently shake the inoculated tubes of EC medium to insure adequate mixing and incubate in a waterbath at 44.5 ± 0.2°C for 24 ± 2 hr. Gas production of any amount in the inner fermentation tube of the EC medium indicates a positive fecal coliform test. The preparation of EC medium is described in the 18th edition of *Standard Methods for the Examination of Water and Wastewater*, 1992, Method 9221E-p. 9–52, paragraph 1a. Public water systems need only determine the presence or absence of fecal coliforms; a determination of fecal coliform density is not required.

(6) Public water systems must conduct analysis of *Escherichia coli* accordance with one of the following analytical methods:

(i)   EC medium supplemented with 50 $\mu g/mL$ of 4-methylumbelliferyl-β-d-glucuronide (MUG)(final concentration). EC medium is described in the 18th edition of *Standard Methods for the Examination of Water and Wastewater*, 1992, Method 9221E-p. 9–52, paragraph 1a. MUG may be added to EC medium before autoclaving. EC medium supplemented with 50 $\mu g/mL$ of MUG is commercially available. At least 10 mL of EC medium supplemented with MUG must be used. The inner inverted fermentation tube may be omitted. The procedure for transferring a total coliform-positive culture to EC medium supplemented with MUG shall be as specified in paragraph (f)(5) of this section for transferring a total coliform-positive culture to EC medium. Observe fluorescence with an ultraviolet light (366 nm) in the dark after incubating tubes at 44.5 + 0.2°C for 24 + 2 hr. or

(ii)   Nutrient agar supplemented with 100 $\mu g/mL$ MUG (final concentration). Nutrient Agar is described in the 18th edition of *Standard Methods for the Examination of Water and Wastewater*, 1992, p. 9–47 to 9–48. This test is used to determine if a total coliform-positive sample, as determined by the membrane filter technique or any other method in which a membrane filter is used, contains *E. coli*. Transfer the membrane filter containing a total coliform colony(ies) to nutrient agar supplemented with 100 $\mu g/mL$ (final concentration) of MUG. After incubating the agar plate at 35°C for 4 hr, observe the colony(ies) under ultraviolet light (366 nm) in the dark for fluorescence. If fluorescence is visible, *E. coli* is present.

(iii)   Minimal medium ONPG-MUG (MMO-MUG) test, as set forth in the article "National Field Evaluation of a Defined Substrate Method for the Simultaneous Detection of Total Conforms and *Escherichia coli* from Drinking Water: Comparison with Presence–Absence Techniques" (Edberg et al.), Applied and Environmental Microbiology, Volume 55, pp. 1003–1008, April 1989. (Note: The autoanalysis Colilert system is an MMO-MUG test). If the MMO-MUG test is total coliform-positive after a 24-hr incubation, test the medium for fluorescence with a 366-nm ultraviolet light (preferably with a 6 W lamp) in the dark. If fluorescence is observed, the sample is *E. coli*-positive. If fluorescence is questionable (cannot be definitively read) after 24-hr incubation, incubate the culture for an additional 4 hr (but not to exceed 28 hr total) and again test the medium for fluorescence. The MMO-MUG test with hepes buffer in lieu of phosphate buffer is the only approved formulation for the detection of *E. coli*.

(iv) The Colisure test. A description of the Colisure test+ may be obtained from the Millipore Corporation, Technical Services Department, 80 Ashby Road, Bedford, MA 01730, U.S.A.

(7) As an option to paragraph (f)(6)(iii) of this section, a system with a total coliform-positive, MUG-negative, MMO-MUG test may further analyze the culture for the presence of E. coli by transferring a 0.1 mL, 28-hr MMO-MUG culture to EC medium + MUG with a pipet. The formulation and incubation conditions of EC Medium + MUG and observation of the results are described in paragraph (f)(6)(i) of this section.

2. Surface Water Treatment Rule (40 CFR 141.74(a)):

(a) Analytical requirements. Only the analytical method(s) specified in this paragraph, or otherwise approved by EPA, may be used to demonstrate compliance with the requirements of Sections 141.71–141.73. Measurements for pH, temperature, turbidity, and residual disinfectant concentrations must be conducted by a party approved by the State. Measurements for total coliforms, fecal coliforms, and HPC must be conducted by laboratory certified by the State or EPA to do such analysis. Until laboratory certification criteria are developed for the analysis of HPC and fecal coliforms, any laboratory certified for total coliform analysis by EPA is deemed certified for HPC and fecal coliform analysis. The following procedures will be performed in accordance with the publications listed in the following section. This incorporation by reference was approved by the Director of the Federal Register in accordance with 5 U.S.C. 552(a) and 1 CFR part 51. Copies of the methods published in *Standard Methods for the Examination of Water and Wastewater* may be obtained from the American Public Health Association et al., 1015 Fifteenth Street, NW, Washington, D.C. 20005, U.S.A.; copies of the minimal medium ONPG-MUG method, as set forth in the article "National Field Evaluation of a Defined Substrate Method for the Simultaneous Enumeration of Total Coliforms and *Escherichia coli* from Drinking Water: Comparison with the Standard Multiple Tube Fermentation Method" (Edberg et al.), Applied and Environmental Microbiology, Volume 54, pp. 1595–1601, June 1988 (as amended under Erratum, Applied and Environmental Microbiology, Volume 54, p. 3197, December 1988), may be obtained from the American Water Works Association Research Foundation, 6666 West Quincy Avenue, Denver, CO 80235, U.S.A.; and copies of the Indigo method as set forth in the article "Determination of Ozone in Water by the Indigo Method" (Bader and Hoigne), may be obtained from Ozone Science & Engineering, Pergamon Press Ltd., Fairview Park, Elmsford, NY 10523, U.S.A. Copies may be inspected at the U.S. Environmental Protection Agency, Room EB15, 401 M Street, SW,

Washington, D.C. 20460, U.S.A.   or at the Office of the Federal Register, 1100 L Street, NW, Room 8401, Washington, D.C., U.S.A.

(1) Public water systems must conduct analysis of pH and temperature in accordance with one of the methods listed at Section 141.23(k)(1). Public water systems must conduct analysis of total coliforms, fecal coliforms, heterotrophic bacteria, and turbidity in accordance with one of the following analytical methods and by using analytical test procedures contained in Technical Notes on Drinking Water Methods, EPA-600/R-94–173, October 1994, which is available at NTIS PB95–104766.

Organism	Methodology	Citation[a]
Total coliforms[b]	Total coliform fermentation technique[c,d,e]	9221A, B, C
	Total coliform membrane filter technique	9222A, B, C
	ONPG-MUG test[f]	9223
Fecal coliforms[b]	Fecal coliform procedure[g]	9221E
	Fecal coliform membrane filter procedure	9221D
Heterotrophic bacteria[b]	Pour plate method	9215B
Turbidity	Nephelometric method	2130B
	Nephelometric method	180.1[h]
	Great lakes instruments	Method 2[i]

[a]Except where noted, all methods refer to the 18th edition of *Standard Methods for the Examination of Water and Wastewater*, 1992, American Public Health Association, 1015 Fifteenth Street NW, Washington, D.C. 20005, U.S.A.

[b]The time from sample collection to initiation of analysis may not exceed 8 hr. Systems are encouraged but not required to hold samples below 10°C during transit.

[c]Lactose broth, as commercially available, may be used in lieu of lauryl tryptose broth, if the system conducts at least 25 parallel tests between this medium and lauryl tryptose broth using the water normally tested, and this comparison demonstrates that the false-positive rate and false-negative rate for total coliforms, using lactose broth, are < 10%.

[d]Media should cover inverted tubes at least one-half to two-thirds after the sample is added.

[e]No requirement exists to run the completed phase on 10% of all total coliform-positive confirmed tubes.

[f]The ONPG-MUG test is also known as the autoanalysis Colilert system.

[g]A-1 Broth may be held up to 3 months in a tightly closed screwcap tube at 4°C.

[h]"Methods for the Determination of Inorganic Substances in Environmental Samples," EPA-600/R-93–100, August 1993. Available at NTIS, PB94–121811.

[i]GLI Method 2, "Turbidity," November 2, 1992, Great Lakes Instruments, Inc., 8855 North 55th Street, Milwaukee, WI 53223, U.S.A.

Appendix H

Record Audits for Drinking Water Laboratories: **Introduction**

This appendix provides information on the records that drinking water laboratories should maintain. It is intended to assist the certification officer in conducting data audits for drinking water laboratories. Certification officers should use the criteria in this appendix, as well as appropriate earlier chapters, for evaluating all laboratories. Evaluation of adequate record keeping/documentation can be facilitated by using the checklists (chemistry, microbiology, and radiochemistry) at the end of this appendix. Auditors are encouraged to develop their own more comprehensive checklists, using these as a starting point. The data audit information in the microbiology checklists (pp. H-21 to H-24) is also included in Chapter 5. The Chapter 5 checklist supersedes this checklist, which is included here only for appendix completeness.

Implementing the recommended record keeping/documentation specified in this appendix will facilitate an "audit of data quality," the tracking of one or more data points from sample receipt by the laboratory (or from sample collection and preservation, if conducted by the laboratory), through preparation, analysis, interpretation, calculations, recording, and reporting of results. All necessary documentation in laboratory records for each step in the data generation process pertaining to any data point should be readily available for the auditor. By tracking this information, the auditor can better assess the quality of data routinely generated by the laboratory.

If any criterion in this appendix differs from a criterion in a specific EPA-approved method, the method takes precedence, because its use is required by regulations.

Data Audits

The EPA Quality Assurance, Glossary, and Acronyms (February 1991) defines an audit of data quality as:

> "A quantitative and qualitative evaluation of the documentation and procedures associated with environmental measurements to verify that the resulting data are of acceptable quality."

The data audit is typically performed on-site and involves tracing the paper and electronic trail of an individual sample or samples from the time of sample collection to the preparation of the final report. Auditors should be competent scientists who are familiar with methodology, the particular data collection technology, and QC procedures.

When a data audit is to be performed, the first step is to read the pertinent standard operating procedures (SOPs) and generate a checklist of items to be examined. Procedures should be verified for sample collection, receipt, storage, preparation, analysis, reporting, and all associated QC.

A checklist should be prepared prior to the on-site visit. When the on-site audit takes place, the sample or samples to be examined should be chosen at random. One of the samples audited should be the most recent performance evaluation (PE) sample. Sample login records, laboratory notebooks, extraction records, analysis records, and any other documents necessary to confirm that procedures were followed as described should be examined.

The audit report should cite deficiencies, if any, in the procedures and/or documentation and suggested corrective actions.

The data audit process involves reviewing all records that document the sampling, storage, and data generation processes from beginning to end. Individual laboratories may use a variety of systems for recording, reporting, and storing such information—including both manual and automated techniques. In addition, the specific information recorded by laboratories may differ on the basis of SOPs and information storage and retrieval systems. The auditor should be able to reconstruct the final reported data for a field or PE sample. A comparison of raw PE data to raw data for standards should verify that the laboratory has analyzed the PE sample at the correct dilution.

Records to Be Reviewed

This appendix describes documentation needed to document the sampling, storage, and data generation processes. These items constitute the minimum suite of records that should comprise a comprehensive laboratory record keeping system. Not all records apply to all types of drinking water analyses, and the specific records needed at a laboratory will depend on the analyses the laboratory runs. Records of laboratory analyses will vary with the method and instrumentation used. For example, calibration records will be more detailed for instruments used in organic and inorganic chemical analyses than in microbiological analyses. Field sampling and data recording and reporting records apply to all analyses.

The laboratory should record all information on individual data sheets or in a logbook using ink. Original entries should not be obliterated. Corrections should be made with a single line through the entry, initialed and dated. All records maintained electronically should be secured to prevent unauthorized changes. *If the laboratory changes to a new system, all records should either be transferred to the new system or be saved in hard copy format.* All records should be archived in an organized manner that allows easy retrieval. The ease with which laboratory managers are able to retrieve data generation records may be the auditor's first indication of the completeness and effectiveness of the laboratory data generation and record keeping system. Record keeping systems should allow correlation between system and laboratory identification codes.

If a laboratory uses a subcontract laboratory for certain analyses, data provided by subcontract laboratories should be clearly identified as such and any needed documentation at the subcontract laboratory should be available to the auditor upon request.

Drinking water compliance data, pertaining to chemical and radiological analyses, must be stored and retained by the utility for a period of 10 years as required by 40 CFR 141.33. Data pertaining to microbiological analyses must be retained for 5 years. Data for lead and copper (CFR 141.91) must be maintained for 12 years. It is recommended that the laboratory maintain the compliance sample and QC data for at least 5 years or until the next audit, whichever is longer. See Chapters IV–VI, Section 8.2. Data should be stored in a manner that allows efficient and accurate retrieval. The auditor should review the data storage and retrieval system carefully to determine whether proper procedures are followed. If a laboratory is unable to locate data requested by the auditor or requires an inordinate period of time to retrieve requested data, the auditor should consider the data storage system deficient.

To assist auditors in conducting data and record keeping audits, a series of checklists has been developed and are included at the end of this appendix. These checklists include specific items that should be checked during the data audit of an individual field or PE sample.

1. Field Sampling

1.1 Sampling Records

At the time of sample collection, field personnel should record the information listed in the Sample Checklist (page H-13, Field Sampling) on a sample tracking form or electronically for each sample.

A sample tracking record should be maintained for samples. This record should be made in indelible ink or electronic form and should identify the sample collector, the person or persons transporting the samples from the system to the laboratory, as well as the individual(s) who received the samples and the condition of the sample upon receipt. If a commercial shipper was used, shipping records should be available.

Sample collection (and related data recording and record keeping) may not be within the scope of the laboratory's responsibilities. Some field records, therefore, may not be available to or maintained by the laboratory. Either the laboratory or the organization responsible for sample collection should keep the records describing how the sample containers were prepared (i.e., cleaning methods, sterilization technique/date/time/batch number), how preservatives were prepared, and how the samples were collected. In all cases, the sample tracking record and shipping form (if available) should accompany samples to the laboratory, and this documentation, at a minimum, should be maintained in the laboratory. Documentation should be available to verify that SOPs were used for sample collection and that the sample collectors were properly trained.

1.2 Field Analytical Records

Field measurements made for parameters that are dynamic should be made immediately. If a laboratory conducts analyses in the field (i.e., temperature, free residual chlorine, and pH), field personnel should record the items listed in the Sample Checklist on page H-13, Field Measurements. Calibrations should be performed in the field over a range of concentrations that bracket the concentrations measured in the unknown samples. For chemical analyses, the lowest concentration used in a calibration should be at least as low as the pertinent regulatory level. Thermometers should be calibrated annually using a reference thermometer traceable to the National Institute of Standards and Technology (NIST). Residual free chlorine comparison devices, such as color wheels and their application by field personnel, should be checked semiannually. Procedures for field calibrations of instruments should be documented in SOPs for field activities.

1.3   Sample Inspection and Acceptance Policy

When samples arrive at the laboratory, whether by mail or other carrier, the laboratory should check samples and sample records to verify that the required preservation steps and holding times have been adhered to. Laboratory personnel should check sample and shipping containers for integrity (e.g., leaks, cracks, broken seals) and for any obvious signs of improper collection (e.g., head space in VOC samples). The laboratory should verify that proper shipping procedures were followed (e.g., ice should still be present with samples collected for organic analyses). This information should be recorded. In larger laboratories, sample custodians should be designated to perform this function and automated laboratory sample tracking systems may be used to record sample receipt and destination. When a bar code is used, the laboratory should calibrate the bar code reader periodically and document the calibration.

The laboratory should add the bullet items in the Sample Checklist (page H-14, Sample Acceptance) to the sample tracking form for each sample at the time of receipt.

1.4   Sample Storage

The auditor should visit the sample receiving area of the laboratory and inspect the condition of samples to determine that container labels are properly completed and securely attached and that all necessary information is recorded in the laboratory. The auditor should also verify that samples requiring special handling (e.g., light sensitive, temperature sensitive, radioactive) are properly stored and that samples are analyzed within their holding time.

Before analysis, the analyst should check samples for the proper preservation technique, pH, chlorine residual, etc. Errors in sample preservation and/or handling are known to bias analytical results. Laboratories must reject compliance samples that do not comply with the method sampling and preservation requirements and request a replacement sample.

2.   Laboratory Analysis

2.1   Analytical Methods

Each laboratory should list and have on hand, in its Quality Assurance Plan, the SOPs (which include the analytical method) for each drinking water analyte measured. This listing should include the name of the method and a complete reference or source (e.g., 18th Edition of *Standard Methods for Examination of Water and Wastewater*, with method number). The auditor should verify that the methods listed are EPA-approved for drinking water (or otherwise acceptable to EPA), that a copy of the method is available to the analyst, and that the methods listed are the only ones actually being used for the analysis of the drinking water compliance samples.

Laboratories should have written SOPs for all analyses and laboratory procedures. The SOPs delineate the specific procedures used to carry out the methods and how, if at all, the SOP differs from the required method. The purpose of the SOPs is to specify analytical procedures in greater detail than those appears in the published method in order to further ensure that analyses are conducted in a standard manner by all analysts in a laboratory. The auditor should verify that analysts are following the written SOPs and that the SOPs are consistent with EPA-approved analytical methods for drinking water and any deviations are clearly identified. The SOPs should be available in the laboratory at the analyst's workstation.

2.2. Instrument Performance

Instrument and equipment performance is a significant factor in laboratory data quality. Consequently, the laboratory should monitor and document instrument/equipment performance characteristics regularly, in accordance with method specifications and the recommendations of equipment manufacturers. For all instruments/equipment, the bullet items in the checklist (page H-15, Instrument Performance) should be maintained (insofar as they apply to analyses conducted by the laboratory).

Performance records are important for all instruments and equipment used to perform drinking water analyses, including analytical balances, incubators, refrigerators, autoclaves, water baths, and block digestors. For such equipment, routine calibrations against standards traceable to the NIST should be performed.

ASTM type 1 or 2 weights or better should be available to make daily checks on balances. A record of these checks should be available for inspection. The specific checks and their frequency should be as prescribed in the laboratory's QA plan and the laboratory's operations manual, if appropriate. The ASTM weights should be recertified annually.

Wavelengths on spectrophotometers should be verified daily with color standards. A record of these checks should be available for inspection. The specific checks and their frequency should be as prescribed in the laboratory's SOPs.

Proper maintenance of all laboratory instruments/equipment is critical to achieve proper laboratory performance. Instrument manufacturers provide suggested maintenance schedules, and laboratories should adhere to such schedules. Laboratories should also record maintenance and repair activities for all instruments, including both determinative instruments (e.g., specific ion detectors, spectrophotometers, etc.) and nondeterminative equipment (e.g., incubators, ovens, balances, etc.). Maintenance logs should also list the maintenance schedule and the date and name of the individual(s) performing routine instrument maintenance.

2.3. Initial Demonstration of Capability

The initial demonstration of capability (IDC) ensures that the analyst and the method are capable of measuring the analytes of interest within the acceptance criteria necessary to comply with the regulations. The IDC should be performed for each analyst and instruments used before compliance samples are analyzed. The variability introduced by multiple sample preparation technicians must also be taken into account. The typical IDC for chemical measurements consists of demonstrating proficiency in four areas: precision, accuracy (bias), method blank background, and method detection limit. For microbiological methods, the specificity, relative trueness, positive and negative deviation, and repeatability should be estimated if possible. This should include a replicate demonstration of negative results on appropriate negative controls and positive results and confirmations on appropriate positive controls. When reviewing the documentation, the auditor should verify that all results, including errors, are reported in the correct units. Laboratories should also record any actions taken to correct performance that does not fall within the acceptable range established by the applicable method. The method detection limits (MDLs) must be calculated in accordance with 40 CFR 136, Appendix B. The qualitative confirmation capability should be determined, ascertaining the detection limits for alternate techniques. When unacceptable performance occurs on QC samples, compliance samples should be reanalyzed after acceptable performance has been established.

Laboratories should maintain complete records for the IDC, which include the bullet items in the checklist on page H-16, Initial Demonstration of Capability.

2.3.1. Method Precision and Bias

Method precision and bias measurements should be made for chemical analyses each time an analytical method is used and periodically, as specified in Chapter V, for microbiological contaminants. Precision is calculated using repeated measurements. Bias measurements are made by spiking samples or reagent water with known concentrations of analytes or suspensions of micro-organisms and measuring the resulting percent recovery. Laboratories should develop SOPs for this purpose that are consistent with the quality control requirements specified in Chapters IV–VI of this manual, unless otherwise specified in the published method.

For microbiology, the laboratory should analyze duplicate positive samples and take part in proficiency testing or interlaboratory comparisons.

For radiochemistry, precision and accuracy are determined from a laboratory fortified sample matrix (spike) with each batch of samples. Samples should be fortified at 1–10 times the MCL or 5–10 times the background.

Laboratories should maintain complete records of method precision and bias, consistent with the specifications of published methods or, if none are included in the method, consistent with stated laboratory SOPs. Method precision and bias records should include the source, preparation date, and concentration(s) of standards used.

These records should be reviewed by the auditor to verify that the limits of precision and bias achieved are consistent with those established in the published method. If measured precision and bias do not fall within acceptable ranges, efforts should be made to ascertain the reason(s) and depending on the nature of the problem, samples may need to be reanalyzed. The laboratory should also record the reason, identify and implement an appropriate corrective action, and document the results of the corrective action.

2.3.2. Demonstration of Low System Background

For chemistry, most EPA methods require the analysis of a reagent blank with each batch of samples to measure the contamination, which may be introduced at the laboratory. In general, the value of the reagent blank should not exceed the MDL. If the blank exceeds the MDL, samples should be reanalyzed after the contamination has been found and the problem corrected.

For microbiology, the laboratory should analyze negative controls to demonstrate that the cultured samples have not been contaminated.

For radiochemistry, blanks are used to determine if activity is added to the sample from the reagents. Typically, the background is the instrument background that is subtracted from all sample counts.

2.3.3. Analytical Method Detection Limit for Chemical and Radiochemical Analyses

The MDL is the minimum concentration of analyte that can be measured and reported with 99% confidence that the analyte concentration is > 0.

Laboratories should determine their own MDLs for all organic, inorganic, and radiochemistry methods, using the procedure in 40 CFR 136 Appendix B. Each analytical team using a particular analytical method also should determine it's own detection limits. In addition, if several different instruments are used for the same procedure, then the detection limit should be determined for each instrument. The same extracts may be used to determine the detection limits on several instruments with similar sensitivity, if appropriate.

A procedure for measuring MDLs is described in 40 CFR Part 136, Appendix B. This procedure involves analyzing seven replicate water samples that have been spiked to, or are known to, contain the analyte at a concentration at or near the estimated detection limit. Laboratories should maintain documentation for the bullet items in the Checklist on page H-17, Method Detection Limits.

Although 40 CFR 136, Appendix B, provides several possible approaches to select an estimated detection limit for purposes of designing the study, laboratory auditors should recognize that the most reliable method involves an iterative process of measuring achievability of successively lower concentrations, until the actual limit of detection is identified. At a minimum, this approach should be used for purposes of establishing the working MDL when a new method is first used by a laboratory. Successive detection limit determinations to verify the limit in the same matrix or establish the limit in a new matrix may not require measurement of detect ability at numerous concentrations, depending on the experience of the analyst. Any procedure that involves measurement of only one concentration should be reviewed carefully to establish that the analyst was able to document that the concentration measured meets the definition of a method detection limit. The spike concentration should be determined by the signal-to-noise ratio for each analyte. The same concentration for all analytes will not produce acceptable results. The extractions/analyses should be performed over a period of at least 3 days to provide a more reasonable MDL.

The required detection limits for radionuclides are listed in the CFR. Detection limits for radionuclides are determined on the basis of nuclear counting statistics in addition to sample and instrument parameters. Proper counting errors should be calculated for each method and reported with each sample analyses.

2.4. Instrument Calibration and Calibration Records

For all types of analytical instruments, it is important that the analyst establish the relationship between the measured value and the concentration of the analyte in the sample, daily, using accurate reference materials. The calibration should include at least three concentrations in the linear portion of the curve. Additional standards should be used if the curve is nonlinear. The concentrations should be selected to bracket the range of concentrations expected to be observed in the samples to be analyzed. Some organic methods recommend using five concentrations. Some inorganic methods require a blank and three standard concentrations.

Calibration for some methods is very time consuming. For these methods, the standard curve should be initially developed and thereafter, at the beginning of each day on which analyses are performed, this curve should be verified by analysis of at least one standard in the expected concentration range of the samples analyzed that day or within the operating range. All checks must be within the control limits specified in the method or the system must be recalibrated.

For radiochemistry, efficiency (counts per minute per disintegration per minute) curves should be used to determine the efficiency for a given mass or nuclide energy. For alpha and beta counters, the change in efficiency versus mass should be plotted to determine the efficiency at a given mass. For gas flow proportional counters, sample composition, size, alpha/beta energy, and distance from the detector influence the efficiency. For gamma counters, two curves should be prepared, energy vs. channel number and efficiency vs. energy. For gamma counters, the primary variable is the distance from the source to the detector. Each type of sample container will have a unique gamma efficiency curve. The shape of the curve is intrinsic to the detector material and will be the same for each geometry subject to differences due to distance. The efficiency is constant for many orders of magnitude. Standards should be prepared at a level of activity, so that >1600 counts above the background are acquired during the counting time. Under this condition, the uncertainty will be 5% (2 SD) at the 95% confidence level. To verify the curve, one or more of the standards should be counted daily when samples are counted. The RSD should not exceed 3 SD or 10% of the expected value. Multiterm polynomial curves should be generated from the relationship between efficiency and either mass or energy.

Laboratory records should include the concentrations of standards used, a graph of the calibration curve, and if possible, a description of statistical procedures used to establish the curve, the source of standard materials, date of preparation and a description of standard preparation steps or reference to the SOP. To simplify these records, laboratories may refer a SOP that specifies the calibration procedures.

The QC samples from an outside source should be used to verify all initial calibrations. Two or three QC standard concentrations chosen to bracket the working range are recommended. For radiochemistry counting, the laboratory should use a check source and a background both at the beginning and at the end of each batch of samples, especially if the counter is not in constant use. Regardless of the number of samples analyzed, a record of a check standard, analyzed at the end of each batch, should be maintained. In general, if the results of a check sample analysis at the end of a batch exceed the confidence limits (CLs) of the method, a system review and corrective action (such as recalibration) are indicated and the samples should be reanalyzed. Laboratories may develop their own guidelines for determining when reanalysis is necessary (where not specified in the method). Such guidelines should be statistically linked to method performance (e.g., ± 1 SD). Where these guidelines are applied, the auditor should verify that the basis for the guideline is recorded.

Many analysts believe that multiple standard concentrations need not be used to calibrate instruments that provide direct concentration readings, because the manufacturers of the instruments specify the use of a single standard concentration for calibration. For such instruments, multiple concentrations that bracket the required working range are strongly recommended. These standards should be treated as samples, and a record of the instrument readings should be maintained by the analyst. A general rule of thumb is that the measured concentration should be $< 10\%$ of the actual concentration.

Most methods specify calibration procedures, and the auditor should verify that procedures are being followed by the analysts and are consistent with those specified by the method. Where methods do not specify calibration requirements, instrument manufacturers' recommendations or other guidance may be available.

2.5. Routine Monitoring of Analytical Method Performance

Laboratory Quality Assurance Plans should specify a reasonable schedule for routine monitoring of method performance based on requirements specified in the approved EPA methods or based on requirements for other similar measurements where the method does not specify a requirement. Most organic and radiological methods specify requirements for periodic monitoring of method performance.

Methods for microbiological analyses require periodic duplicate analyses as a routine check on method performance. Positive and negative controls should also be analyzed on a regular basis whenever new media is prepared or commercially prepared media is received.

Method performance documentation should include the bullet items listed in the checklist (page H-19, Routine Performance Checks). Further discussion of these records has been provided in the previous sections.

2.5.1.   Laboratory Fortified Blanks

The laboratory should analyze a laboratory fortified blank (LFB) with each batch of samples analyzed. The concentration of the LFB should be 10 times the MDL or estimated detection level or equivalent to a mid range standard. At least some portion of the LFBs should be at the laboratory's reporting level. Precision and recovery for the LFBs must be within the criteria specified in the methods, or the corrective action specified in the method must be taken. The LFBs must be processed through the entire analytical procedure (e.g., extraction, derivation, and detection).

Control charts or limits, generated from mean percent recovery and standard deviation of the LFBs, should be maintained by the laboratory. Until sufficient data are available from the laboratory, usually a minimum of 20–30 test results on a specific analysis, the laboratory should use the control limits specified in the methods. See *Standard Methods for the Examination of Water and Wastewater*, part 1020B, or similar QC reference texts, for further information. The laboratory should continue to calculate control limits for each analyte, as additional results become available. After each 5–10 new recovery measurements, new control limits should be calculated using the most recent 20–30 data points. If any of these control limits are tighter than the method specifications, the laboratory should use the tighter criteria. Otherwise, control limits in the methods are required. Control charts are highly recommended even though some of the older methods for inorganic analytes do not include such requirements.

For microbiology, a pure culture of a known positive reaction should be included with a sample batch periodically and when new media is prepared, to demonstrate that the medium can support growth.

2.5.2.   Laboratory Fortified Matrix

The laboratory should also fortify a minimum of 10% (or one per batch, whichever is greater) of the routine samples (except when the method specifies a different percentage, e.g., furnace methods) to determine if there are any matrix effects. The spike concentration should not be substantially less than the background concentration of the sample selected for spiking. If the sample concentration is below the MDL, then the analyst may choose appropriate spike levels (e.g., a percentage of the MCL or operating calibration range). This laboratory fortified matrix must be processed through the entire analytical procedure. Over time, to the extent practical, samples from all routine sample sources should be spiked. If any of these checks are not within the control limits specified and the method was in control as judged by the LFB, the results of that sample should be labeled suspect due to sample matrix.

2.5.3. Quality Control Samples

At least once each quarter, each laboratory should analyze quality control or reference samples from a source other than that from which their standards are purchased (for each method, analyst, and instrument) as a routine check on performance. If errors exceed limits specified in the methods, corrective action should be taken and documented, and a follow-up quality control standard analyzed as soon as possible to demonstrate the problem has been corrected. The laboratory should maintain records for these routine checks.

2.5.4. Blanks

Most of the EPA-approved analytical methods specify analysis of one or more blank samples with every batch of samples analyzed. Follow the method recommendations and requirements. See the discussion of laboratory reagent blanks in Section 2.3.2. Laboratory reagent blanks (method blanks) measure the level of contamination that may be introduced at the laboratory. The results of analyzing blank samples should be recorded by the analyst with all other data and checked by the auditor.

Field blanks should also be analyzed to measure contamination that may be introduced from preservatives, from storage, or at the sampling site. A field blank may be substituted for the reagent blank. However, if the field blank is contaminated, then a reagent blank must be analyzed to determine the source of the contamination. Trip blanks may also be required to determine if contamination occurred during shipping.

All blanks must be processed through the entire analytical procedure.

Most analytical methods specify an acceptable maximum level for the analyte in a blank sample. Some States require that blank results indicating contamination be submitted with the analytical results.

2.5.5. Other Quality Control Requirements

Some methods require duplicate sample analyses, method of standard additions, column or instrument performance check samples, instrument tuning checks, etc. Whenever additional quality control requirements are specified in the method or method manual, they must be followed.

2.5.6. Performance Evaluation Studies

In order to receive and maintain certification, drinking water laboratories must successfully analyze PE samples, in accordance with the requirements in the NPDWR. In addition to these samples, laboratories certified for radiochemical analysis must also participate in other PE studies every year as prescribed by the program. Laboratories are required to maintain records of their performance in those studies and any corrective actions taken in response to performance problems indicated by the study results. A copy of this documentation should be forwarded to the certification officer. Auditors should review these records to determine that all errors or performance problems are addressed and that proper corrective actions have been taken, where appropriate. Auditors should review the raw PE data to verify that the PE samples were diluted according to the instructions provided. The auditor should be able to verify from the raw data (e.g., area counts for GC analysis) that the PE samples were analyzed with the correct dilution factor.

2.6. Instrument Maintenance

Proper maintenance of all laboratory instruments and equipment is critical to achieving proper laboratory performance. Instrument manufacturers provide suggested maintenance schedules and laboratories should adhere to such schedules. Laboratories should also record maintenance and repair activities for all instruments.

3. Data Handling and Reporting

3.1. Calculations

During the data audit, the auditor should review all procedures for calculating final values from the raw data. Laboratories should maintain written SOPs for making all calculations, and all raw data and supporting information needed to recreate the calculations should be available to the auditor. The auditor should verify representative calculations sufficient to provide a reasonable level of confidence that the appropriate procedures are being used consistently. Using the raw data, the auditor should be able to regenerate the sample data reported to the customer.

In cases where the laboratory does not provide documentation for calculation methods and the method used is not readily apparent, the auditor should attempt to verify calculations using techniques normally used by the Regional or State laboratory. If the results do not agree with those reported by the laboratory, the auditor should attempt to ascertain and evaluate the calculation method used. This may require interviewing one or more analysts who routinely make the calculations to determine the step-wise process used and to determine that all analysts in the laboratory are using the same method. To avoid calculation errors, laboratories should implement a policy of requiring that all calculations be cross-checked by a second analyst. Both analysts should certify the data, by signature with date in ink, and note as acceptable to document that the data has been cross-checked. Where possible, calculations should also be verified by a laboratory supervisor.

3.2. Use of Significant Figures

Laboratories should observe conventions concerning proper use of significant figures in making calculations to avoid the appearance that the data are more precise than the method allows. Conventions for the use of significant digits and proper rounding of numbers are discussed in detail in the EPA publication: *Analytical Quality Control in Water and Wastewater Laboratories* (EPA-600/4-79–019) and in *Standard Methods for Examination of Water and Wastewater* (Section 1050 B in the 18th Edition). As a rule, the significance of an analytical result cannot exceed the significance of the least precise step in the procedure. The numbers resulting from calculations cannot reflect greater precision than the data used to make the calculations. In order to verify that proper conventions have been followed, the auditor should identify the least precise step in the analytical process.

Frequently, the least precise step is the measurement of the sample volume necessary for the analysis, which most often involves the use of a graduated cylinder or pipette. The auditor should verify the actual tolerances of glassware used in the laboratory analytical methods (by examining the glassware) and determine that the appropriate number of significant digits has been carried through all data recording and calculations.

3.3. Reporting Greater-Than and Less-Than Values

Less-than values occur frequently in drinking water analysis. For chemical analytes, less-than values should be recorded by the analyst when the measure obtained is below the value of the lowest standard used to generate the calibration curve. These reported less-than values should be followed with the value of the lowest concentration used in the calibration curve (e.g., <0.2). Regulations may require the laboratory to report data, either quantitatively or qualitatively, to the MDL. If this situation exists, laboratories may report data as "detected" less than the lowest standard (e.g., detected <0.2). If the method involves a one-point calibration, values may be reported down to the MDL.

Samples with concentrations greater than the highest standard should be diluted to fall within the calibration curve. If unusual circumstances prevent this, results should be reported as greater than the highest standard and another sample should be obtained.

For microbiological contaminants, if the organism of concern is not detected, the analyst may report total and fecal coliform and *E. coli* as absent and heterotrophic plate count (HPC) as <1 per unit volume. If the concentration is too great, HPC should be reported as too numerous to count.

3.4. Blank Corrections

Correction of sample data for blank results is an important issue in data reporting. In general, analyte measurements in the blank that exceed allowable levels should be interpreted as an indication of laboratory contamination or other performance problem. Data should never be corrected for blank measurements unless specifically stated in the analytical method being utilized. Laboratories should take steps to isolate and eliminate sources of contamination where blank analyses indicate a consistent and significant problem.

3.5. Error Types

The auditor should ensure that the laboratory Quality Assurance Plan specifies safeguards for protecting against common types of data errors. In general, procedures for spot-checking data and for verification by a second analyst are the most common practices used. Some common types of data errors include the following:

- data entry errors;
- transcription errors;
- rounding errors;
- calculation errors;
- failure to record or retain data, especially dilution factors;
- incomplete data reporting (i.e., missing quality control or method performance data);
- omission of units or incorrect units;
- improper data error corrections (a single line should be drawn through an incorrect value such that both the correct and incorrect values are legible; change should be initialed and dated);
- recording data in pencil (data must be recorded in indelible ink to help ensure data integrity);
- errors in use of significant figures (see Section 3.2);
- errors in logic (e.g., applying a blank correction, where allowed by the method, at the wrong step in the calculation);
- for radiochemistry, other errors may occur if the following factors are not considered: yield determinations, decay and/or ingrowth factors, use of proper counting times, and accurate efficiency curves.

Virtually, all of these types of errors can be prevented by requiring that all data be checked, initialed, and dated by a second analyst (and verified by a laboratory supervisor, if possible) before being recorded for final reporting.

3.6. General Quality Control Records

During the data audit, in addition to identifying and reviewing quality control data specific to individual analyses, the auditor should verify that certain general quality control procedures are conducted and that written records or logs are maintained regularly in the laboratory. For example:

- The date on which chemicals arrive in the laboratory and the date that reagents are first opened should be recorded on the bottle label and initialed by the analyst.
- Reference standard calibration solutions should include labels that list the date of preparation, the concentration, and the name of the analyst who prepared the standard.
- Training records for laboratory analysts should include academic background, specialized training courses completed, and chemical analysis and instrument experience.

- Stock standard solution logs should list the preparation date, the concentration, and the name of the analyst who prepared the solution and the dilution solvent.
- GC injection logs should list each sample injected, the time and date of analysis, and the name of the analyst who performed the analysis.

3.7. Data Security and Backup

Laboratories should maintain data in a secure system to which access is limited. A system for logging files in and out should be maintained, so that laboratory files are known to be complete at all times. An original copy of the data report showing all corrections or changes should be maintained on file. Additional copies for use by analysts may also be maintained and tracked. A laboratory manager should certify, in writing (or by using a unique code identifier for electronic reporting), the authenticity of each data report and maintain such certification records for inspection.

The trend to replace manual operations with computers is expanding rapidly, and the data management practices used to protect the integrity of electronic data are becoming increasingly important. Minimum records to be maintained include a description of the hardware and software used; written SOPs that document procedures for generating, validating, and reviewing computer data; and results of periodic in-house QC inspections of electronic data generation and reporting.

Any data stored electronically should be supported with backup files generated in accordance with Agency guidelines specified in the EPA document, *2185—Good Laboratory Practices for Automated Data*, 1995 Edition (Office of Information Resources Management, RTP, NC 27711). This guidance document suggests appropriate frequencies for generating backup files and suggestions for off-site storage of copies. One important disadvantage of electronic formats is that they are software and hardware dependent.

DRINKING WATER CERTIFICATION
RECORD KEEPING AUDIT

SAMPLE CHECKLIST			
Field Records	**Yes**	**No**	**Comments**
1.1 Field Sampling			
The following items should be recorded in ink at the time of sample collection as part of the sample tracking record			
● Sample identification number			
● Public Water System ID number (where applicable)			
● Date and time of sample collection			
● Sample type (i.e., compliance, confirmation, etc.)			
● Analyses required			
● Sample container type and size, preservatives, holding time			
● Sample volume or weight collected			
● Name, phone number and organization of sampler			
● Preservatives added to the sample, concentration and amount added			
● Sampling location (site ID number, treatment information, where applicable)			
● Sample shipping procedure and holding time			
● Sample container preparation (i.e., cleaning methods, sterilization technique/date/time/batch number)			
● pH and disinfectant residual from plant measurements if available			
1.2 Field Measurements*			
Field analytical records include the following items:			
● Sample identification number			
● Date and time of each field measurement and time of sample collection			
● Pertinent field data (i.e., temperature, free residual chlorine, and pH)			
● Name of analyst			
● A list of calibration steps and concentrations of standards			
● Date of instrument calibration			
● The type, concentration, preparation date and source of any calibration standards used (e.g., reference buffers for pH and reference materials for disinfectant residual)			

*If performed by laboratory personnel.

SAMPLE CHECKLIST			
Laboratory Records	**Yes**	**No**	**Comments**
1.3　Sample Acceptance			
Sample tracking records from field samplers are maintained for all samples			
The following entries are added to the sample tracking form for all samples			
• Name of the person delivering the sample			
• Name of the person receiving the sample			
• Date and time of sample collection and receipt			
• Date and time of sample receipt			
• Condition of samples received (i.e., sealed, unsealed, broken containers, temperature, preservatives noted on label)			
• Sample irregularities are documented at the time of sample receipt (i.e., improper collection, shipping procedure not followed)			
• New sample ID (if assigned by laboratory), which can be cross-referenced to the original sample ID, methods, analysts assigned			
Deficient samples are rejected in accordance with written laboratory policy			
Sample rejection log is maintained			
Maximum holding time not exceeded			
Replacement sample requested for rejected sample			
1.4　Sample Storage			
Samples are properly labeled and securely attached			
Samples exceeding holding times are discarded			
Volatiles are hermetically sealed			
Samples requiring special handling are properly stored (i.e., light sensitive, temperature sensitive, radioactive)			

DRINKING WATER CERTIFICATION
DATA AUDIT

CHEMISTRY LABORATORY ANALYSIS REVIEW CHECKLIST			
Laboratory Records	**Yes**	**No**	**Comments**
2.1 Analytical Methods			
Analytical procedures (SOPs) are listed and referenced in the laboratory's Quality Assurance (QA) Plan for all analytes measured and this listing accurately reflects the analytical methods employed by the laboratory.			
Only EPA-approved methods are used to analyze drinking water compliance samples			
Method procedures are followed exactly by the laboratory personnel or allowed deviations are listed in SOPs			
Written Standard Operating Procedures (SOPs) are consistent with the approved methods and are followed by the laboratory personnel			
2.2 Instrument Performance			
Laboratory monitors and documents instrument performance characteristics regularly, in accordance with the method specifications and the equipment manufacturer's recommendations			
Instrument performance records are maintained and include the following items:			
• Initial demonstration of capability			
• Determination of linear dynamic range			
• Method Detection Limits			
• Initial and routine instrument calibration			
• Performance on standard reference materials and/or QC check samples			
• Instrument sensitivity and stability			
• Tuning checks			
Laboratory equipment such as analytical balances and thermometers are calibrated against standards traceable to NIST			
Equipment stability records such as dry oven, incubator, refrigerator, autoclave, and block digester temperatures are maintained			
2.3 Initial Demonstration of Capability			
Method performance is demonstrated as specified by the published method or if not specified, A minimum of four replicates of a quality control or reference sample are processed through all steps of the analytical procedure*			
Method performance is validated for all analytes measured			
Analytes are measured at levels within the required method performance range			
Initial method precision and bias criteria are met			
The method is validated for each analyst and each instrument the analyst uses			
Variation due to multiple sample preparation technicians is taken into account			
System background is below the MDL			
MDLs are calculated for all analytes			

*If no other requirements are specified by the published method.

CHEMISTRY LABORATORY ANALYSIS REVIEW CHECKLIST			
Laboratory Records	**Yes**	**No**	**Comments**
Qualitative capability is identified for each analyte			
Method validation records are maintained and include the following items:			
• Name and/or number of the analytical method			
• Name of analyst			
• The type of test			
• Date and time of analyses			
• Instrument identification number			
• Number of replicates analyzed			
• Concentrations of the standards used			
• A description of any standard preparation steps (preparation date, name of the analyst who prepared the standard, and the reagents used)			
• Source of the standard material and preparation date			
• All calculations and supporting data			
Calculated error is expressed in the same units as the reported data			
Corrective action is documented and resulting performance is reported where method performance is outside acceptable ranges			
2.3.1 Method Precision and Bias			
Method precision and bias measurements are conducted as specified by the published method or Precision and bias measurements follow the standard operating procedures developed by the laboratory*			
Method precision and bias are determined for all analytes measured			
Method precision and bias are determined at the correct frequency			
Method precision and bias control limits are met for all analytes measured			
Precision records are maintained for each instrument and include the following items:			
• Name and/or number of the method			
• Name of analyst			
• Date and time of analyses			
• The number of replicates analyzed			
• All calculations and supporting data			
A minimum of 10% of routine samples are spiked and analyzed (unless the method specifies a different percentage)			
Spiked sample records are maintained and include the following items:			
• Name and/or number of the method			
• Name of the analyst			
• Date and time of analyses			
• Spike amounts added to samples			

*Where no requirements are specified by the published method.

CHEMISTRY LABORATORY ANALYSIS REVIEW CHECKLIST			
Laboratory Records	**Yes**	**No**	**Comments**
• Concentration of the check standard used			
• Source of the standard material			
• A description of any standard preparation steps (preparation date, name of the analyst who prepared the standard, and reagents used)			
• All calculations and supporting data			
Quality Control charts (% recovery vs time) are maintained or QC limits calculated			
Corrective action is documented and re-analysis results are reported where precision and bias limits fall outside acceptable ranges			
2.3.2 Method Detection Limits			
Method detection limits (MDLs) are measured for all analytical methods			
Each analyst measures their own detection limit for each analytical method and instrument used in the procedure			
MDLs are determined by the procedure described in 40 CFR Part 136 Appendix B			
MDL records are maintained and include the following items:			
• Name and/or number of the method			
• Date of sampling and date of analyses			
• Identification of the analyst			
• The MDL achieved and the method's published MDL (in appropriate units)			
• Analyte level spiked or in the sample			
• Number of replicates analyzed			
• A description of any allowable-variations used in the method			
• A description of the type of water in which the MDL is measured			
• A description of the procedure used to determine the MDL(iterative process used?) and a description of the process used to estimate the MDL			
• Analyte recovery values of reference materials or spiked samples			
• All raw data and calculations necessary to reconstruct MDL determination			
2.4 Initial Instrument Calibration			
Initial calibration is conducted as specified by the published method or Three or more concentrations should be analyzed. The lowest calibration standard measured should be near the reporting level, and the remaining concentrations bracket analyte levels*			
Instruments are calibrated for all analytes measured			
Standard calibration materials are from a different source than the QC standards			
Analytes are measured at concentrations covering the sample concentration range			
Instruments are calibrated at the correct frequency as specified by the method and/or the instrument manufacturer's recommendations			

*Where no requirements are specified by the published method.

CHEMISTRY LABORATORY ANALYSIS REVIEW CHECKLIST			
Laboratory Records	**Yes**	**No**	**Comments**
Calibration records are maintained for each instrument and include the following items:			
• Name and/or number of the analytical method			
• Date and time of calibration measurements			
• Name of the analyst			
• Concentrations of the standards used			
• A description of any standard preparation steps (preparation date, the name of the analyst who prepared the standard, and reagents used)			
• Source of the standard material			
• Labeled graph of the calibration curve			
• A description of any statistical procedures used to establish the curve			
• All calculations and supporting data			
Instrument sensitivity (calibration curve slope) is documented			
Continuing Instrument Calibration			
Calibration check standards are routinely measured as specified by the published method or One or more check standards chosen to bracket analyte concentrations are measured each working day, and a mid-point calibration standard is measured after each batch*			
Instrument response is checked for all analytes measured at the required frequency			
Analytes are measured at concentrations appropriate for the sample concentration range			
Calibration check standards are analyzed at the correct frequency as specified by the method or the Manual for Certification of Laboratories Analyzing Drinking Water			
Calibration curves are verified for all analytes measured			
Routine calibration records are maintained for each instrument and include the following items:			
• Name and/or number of the analytical method			
• Date and time of check standard measurements			
• Name of the analyst			
• Concentration of check standards used			
• Source of standard materials			
• A description of any standard preparation steps			
• All calculations and supporting data			
Calibration procedures are consistent with those specified in the methods and are followed by laboratory analysts			
Corrective action (recalibration) is documented and re-analysis results are recorded where check standard results do not meet method criteria			
Written guidelines for recalibration and re-analysis are available*			

*Where no requirements are specified by the published method.

CHEMISTRY LABORATORY ANALYSIS REVIEW CHECKLIST			
Laboratory Records	**Yes**	**No**	**Comments**
2.5 Routine performance checks			
Quality control (QC) or referenced materials are analyzed quarterly or as specified by the published method			
LFBs are analyzed for all analytes measured at the required concentration			
LFBs are analyzed at the correct frequency			
Method bias (% recovery) criteria is met for all analytes measured			
QC or reference check samples are routinely analyzed for each method, analyst, and instrument			
Performance check records are maintained and include the following items:			
• Name and/or number of the analytical method			
• Name of the analyst			
• The type of test			
• Date of analysis			
• Concentration of the standard used			
• A description of any standard preparation steps (preparation date, the name of the analyst who prepared the standard, and reagents used)			
• Source of the standard material			
• Method precision and bias			
• QC sample results			
• PE sample results			
• Matrix spike results			
• All calculations and supporting data			
2.5.1 Laboratory Performance Evaluation			
The laboratory analyzes, at least annually, water supply performance evaluation samples within EPA limits			
Performance evaluation study records are maintained			
Corrective action is documented where performance problems are indicated by the study results			
2.6 Instrument Maintenance			
Instrument maintenance schedules from manufacturers are followed and maintenance activities are documented			
Repair activities are recorded for all instruments			

DRINKING WATER CERTIFICATION
DATA AUDIT

(The data audit information in this checklist is also included in Chapter 5. Microbiology laboratories do not have to refer to this checklist. The Chapter 5 checklist superseeds this checklist which is included only for Appendix completeness.)

MICROBIOLOGY LABORATORY ANALYSIS REVIEW CHECKLIST			
Laboratory Records	**Yes**	**No**	**Comments**
Laboratory Equipment, Supplies, and Materials			
pH Meter			
Commercial buffer solutions are dated when received and discarded before expiration date			
pH meter is standardized each use period with pH 7.0 and eithr 4.0 buffers or 10.0 buffer			
pH buffer solutions not reused to calibrate pH meter			
Balance			
Balance is calibrated monthly using ASTM type 1 or 2 weights. If non-reference weights are used, non-reference weights are calibrated using type 1 or 2 reference weights			
Correction data available with type 1 or 2 weights			
Annual service contract or internal maintenance protocol and records maintained			
Temperature Monitoring Device			
Glass or electronic thermometers are checked annually and dial thermometers checked quarterly at the temperature used against a reference NIST thermometer or one meeting the requirements of NIST Monograph SP 250-23			
Continuous recording devices used to monitor incubator temperature are recalibrated annually against a NIST thermometer or one meeting the requirements of NIST Monograph SP 250-23			
Incubator			
Temperature is recorded twice daily for days in use, with reading separated by a least four hours			
Autoclaves			
Date, contents, sterilization time, and temperature are recorded for each cycle			
Service contract or maintenance protocol is established			
Heat sensitive tape, spore strips or ampules, or maximum temperature registering thermometer are used during each autoclave cycle			
Automatic timing mechanism is checked for accuracy with a stop watch			
Hot air ovens			
Date, sterilization time, and temperature are recorded for each cycle			
Conductivity meter			
Conductivity meter is calibrated monthly with a 0.01 M KCl solution or lower concentration			

MICROBIOLOGY LABORATORY ANALYSIS REVIEW CHECKLIST			
Laboratory Records	**Yes**	**No**	**Comments**
Refrigerator			
Temperature is recorded for days in use			
Membrane filters and pads			
Lot numbers of membrane filters and the date received are recorded			
Sterility is determined of each lot of membrane filters by placing one membrane filter in non-selective broth medium			
Ultraviolet lamp			
Lamp used for sanitization is tested every quarter			
General Laboratory Practices			
Sample containers			
Sterility of each lot of sample bottles or pre sterilized sample bags is determined by adding non-selective broth, incubating at 35°C for 24 hours and checking for growth			
Reagent water			
Reagent water is tested to assure the minimum requirements are met (see the Manual for the Certification of Laboratories Analyzing Drinking Water, Chapter V, section 4.3.2, for parameter values and test frequencies)			
Dilution/rinse water			
pH of stock phosphate buffer solution is 7.2 ±0.2			
pH of peptone water is 6.8 ±0.2			
Dilution water is checked for sterility			
Glassware washing			
Inhibitory residue test is performed on clean glassware			
Analytical media			
Media preparation records include:			
- Date of preparation			
- Type of medium			
- Lot number			
- Sterilization time and temperature			
- Final pH			
- Technician's initials			

MICROBIOLOGY LABORATORY ANALYSIS REVIEW CHECKLIST			
Laboratory Records	**Yes**	**No**	**Comments**
For liquid media prepared commercially, the following are recorded:			
- Date received			
- Type of medium			
- Lot number			
- pH verification			
Each commercial lot of medium and each batch of laboratory prepared medium is checked before use with positive and negative culture controls, and results recorded			
EC medium + MUG (for E. coli)			
Each lot of commercially prepared medium, or batch of laboratory-prepared medium, is checked with positive and negative culture controls, and results recorded			
Nutrient Agar Medium + MUG (for E. coli)			
Quality of medium lot/batch is evaluated by spot-inoculating control bacteria			
Performance evaluation sample is satisfactorily analyzed annually (if available)			
Analytical Methodology			
A coliform test is conducted quarterly on known coliform-positive and fecal- or E. coli-positive sample			
Duplicate analyses are performed on 5% of samples[2]			
Membrane filter technique (for total coliform in drinking water)			
Sterility check is conducted on each funnel in use at the beginning and end of each filtration series. If control indicates contamination, all data is rejected and another sample obtained			
EC Medium + MUG Test (for E. coli)			
At least 5% of both MUG-positive results and turbid MUG-negative results are verified for E. coli by the use of a multi-test system (API 20E or equivalent); standard biochemical tests (e.g., citrate, indole, and urease tests); serotyping after biochemical identification; or the indole test at 44.5°C and growth in citrate			

---

[2]Standard Methods for the Examination of Water and Wastewater, 18th Ed., 9020

MICROBIOLOGY LABORATORY ANALYSIS REVIEW CHECKLIST			
Laboratory Records	Yes	No	**Comments**
MF procedure (for total coliform in source water)			
Sterility check is conducted on each funnel in use at the beginning and end of each filtration series. If control indicates contamination, all data is rejected and another sample obtained			
If two or more analysts are available, each counts the total coliform colonies on same membrane at least monthly. Colony counts agree within 10%			
Multiple tube fermentation technique (for total coliform in source water)			
Completed test is performed quarterly on coliform-positive tube(s)/bottles			
Fecal coliform membrane filter procedure (for fecal coliform in source water)			
Sterility check is conducted at beginning and end of each filtration series. If control indicates contamination, data rejected and another sample obtained			
If two or more analysts are available, each counts the total coliform colonies on same membrane at least monthly. Colony counts agree within 10%			
Date and start time are recorded for each analysis			

DRINKING WATER CERTIFICATION
DATA AUDIT

RADIOCHEMISTRY LABORATORY ANALYSIS REVIEW CHECKLIST			
Laboratory Records	**Yes**	**No**	**Comments**
Analytical Methods			
Analytical procedures are listed and referenced in the laboratory's Quality Assurance (QA) plan for all analytes measured and this listing accurately reflects the analytical methods employed by the laboratory.			
Only EPA-approved methods are used to measure radioactivity in drinking water			
Published methods are followed exactly by the laboratory personnel			
Written Standard Operating Procedures (SOPs) are consistent with the approved methods and are followed by the laboratory personnel			
Initial Demonstration of Capability			
Initial Instrument Calibration			
Initial instrument calibration is conducted as specified by the published method or A background sample and one or more standard materials are analyzed at a minimum of 3 different concentrations.			
All radionuclides are calibrated against standards traceable to NIST			
Standards are measured at concentrations covering the sample concentration range			
Instruments are calibrated at the correct frequency as specified by the method and/or the instrument manufacturer's recommendations			
Calibration standards are reported with the analytical results[3]			
The following items are recorded for the standard material used for each radionuclide analyzed[4]:			
A description of the solution (i.e., the principal radionuclide, mass or volume, and chemical composition)			
The reference time and date			
The measurement result (activity of the principal and possible daughter radionuclides/liter of solution)			
The measurement method			

[3]Standard Methods for the Examination of Water and Wastewater, 16th Ed., 1985.
[4]Handbook for Analytical Quality Control in Radioanalytical Laboratories, EPA/600/7-77/088, pg. 4–2.

RADIOCHEMISTRY LABORATORY ANALYSIS REVIEW CHECKLIST			
Laboratory Records	**Yes**	**No**	**Comments**
A statement of purity (a list of known impurities, their activities, and how they are measured)			
Decay information			
An estimate of errors (from the measurement themselves and those created by the decay assumption)			
Calibration records are maintained for each instrument and include the following items:			
• The name and/or number of the analytical method			
• Name of the analyst			
• Date and time of analyses			
• Source of the standard material			
• A description of any standard preparation steps (preparation date, name of the analyst who prepared the standard, and reagents used)			
• Counting times			
• Concentrations of the standards used			
• A labeled graph of the calibration curve			
• A description of any statistical procedures used to establish the curve			
• All calculations and supporting data			
Continuing Instrument Calibration			
A background sample and counting check sample are measured every 20 samples or daily if less than 20 samples are analyzed each day[5]			
Radionuclides are calibrated against a standard traceable to NIST			
Standards are measured at a concentration in the sample concentration range			
Counting standards are analyzed at the correct frequency as specified by the method or the Manual for the Certification of Laboratories Analyzing Drinking Water			
Calibration curves are verified for all radionuclides measured			
Check standard records are maintained and include the following items:			
• The name and/or number of the analytical method			
• Name of the analyst			
• Date and time of analyses			

---

[5] Manual for the Certification of Laboratories Analyzing Drinking Water, Chapter VI, section 7.6.2.

RADIOCHEMISTRY LABORATORY ANALYSIS REVIEW CHECKLIST			
Laboratory Records	Yes	No	Comments
• Source of the standard material			
• A description of any standard preparation steps (preparation date, name of the analyst who prepared the standard, and reagents used)			
• Concentration of the counting standard used			
• Counting times			
• All calculations and supporting data			
Precision and Bias			
A minimum of 10 % duplicate samples are analyzed[6]			
Method precision control limits are met for all radionuclides measured			
Precision is determined at the correct frequency			
Precision records are maintained and include the following items:			
• Name and/or number of the method			
• Name of the analyst			
• Date and time of analyses			
• The counter instrument used			
• All calculations and supporting data			
Spiked samples are measured regularly[7]			
Standard materials traceable to NIST are used to spike samples			
Method bias is determined for all radionuclides measured			
Method bias control limits are met for all radionuclides measured			
Spiked sample records are maintained and include the following items:			
• Name and/or number of the method			
• Name of the analyst			
• The date and time of analyses			
• The source of the standards used			
• A description of any standard preparations steps			

[6] Manual for the Certification of Laboratories Analyzing Drinking Water, Chapter VI section 7.6.1

[7] Handbook for Analytical Quality Control in Radioanalytical Laboratories, EPA/600/7-77/088, pg. 4–11.

RADIOCHEMISTRY LABORATORY ANALYSIS REVIEW CHECKLIST			
Laboratory Records	**Yes**	**No**	**Comments**
• Spike amounts added to the samples			
• All calculations and supporting data			
Instrument Performance			
Laboratory monitors and documents instrument performance characteristics regularly, in accordance with the method specifications and the equipment manufacturer's recommendations			
Instrument performance records are maintained and include the following items:			
• Initial and routine instrument calibration			
• Analytical results from field and laboratory blanks			
• Precision and bias			
• Results of inter comparison cross check studies and performance evaluation studies			
• Analytical results from background samples			
Detection Limits[8]			
The detector meets the minimum detectable activity requirements cited in 40 CFR 141.25			
Counting times are listed for each method, instrument, and radionuclide measured. (Sample volumes should be documented for the corresponding counting time)			
Laboratory Performance Evaluation			
The laboratory analyzes, at least annually, water supply performance evaluation samples within EPA limits			
Performance evaluation records are maintained			
The laboratory analyzes, at least semi-annually, EPA inter comparison cross check samples within EPA limits[9]			
Inter comparison cross check records are maintained			
Corrective action is documented where performance problems are indicated by the PE study or inter comparison cross check study results			
Instrument Maintenance			
Instrument maintenance schedules from manufacturers are followed and maintenance records on all radiation instruments and analytical balances are maintained in a permanently bound record[10]			
Repair activities are recorded for all instruments			

---

[8] Standard Methods for the Examination of Water and Wastewater, 16th Ed., 1985

[9] Manual for the Certification of Laboratories Analyzing Drinking Water, Chapter VI section 7.2.

[10] Manual for the Certification of Laboratories Analyzing Drinking Water, Chapter VI section 7.5.

DRINKING WATER CERTIFICATION
DATA AUDIT

DATA HANDLING AND REPORTING CHECKLIST			
Laboratory Records	**Yes**	**No**	**Comments**
Calculations			
Written procedures for all calculations are available for review			
Representative calculations are available and indicate that routine calculations are consistent with the written procedures			
All raw data and supporting information needed to recreate calculations are available for review			
All calculations are cross-checked by a second analyst			
All data and calculations are certified by two analysts by signature with date in ink			
QA plan includes SOP for preventing data errors	·		
Sample and PE calculations verified by the auditor			
Significant Figures			
Significant figure conventions are followed			
The appropriate number of significant figures are carried out through all recorded data and calculations			
The least precise step is identified in the calculations and the number of significant figures is an accurate reflection of the actual tolerances of the instrument or equipment used in this step.			
Greater-Than and Less-Than Values			
Analyte concentrations with greater-than values are quantitatively diluted and reanalyzed when possible			
Less-than values are reported as "less than" (<) followed by the concentration of the lowest standard used in the calibration curve or the MDL			
For microbiological contaminants not detected, fecal coliform and E. Coli are reported as absent and Heterotrophic Plate Count (HPC) as < 1 per unit volume			
For microbiological contaminants, greater than values for HPC are reported as "too numerous to count" (TNTC)			
Blank Corrections			
Blank results are recorded with all other data and meet method specifications			
Blank measurements exceeding allowable analyte levels are reported and the affected data are flagged	·		
Data are never corrected for blank measurements unless specifically stated in the analytical method utilized			
Corrective action is documented to isolate and eliminate sources of contamination where blank analyses indicate a consistent and significant problem			

DATA HANDLING AND REPORTING CHECKLIST			
Laboratory Records	**Yes**	**No**	**Comments**
Data Security and Back-Up			
A secure data system is maintained with limited access			
Procedures are in place to prevent unauthorized access			
Procedures are in place to protect the integrity of the data			
A log is maintained for laboratory file entries and retrievals			
Computer software is documented and adequate for use			
Original copies of analytical data and calculations (showing all corrections or changes) are maintained on file and copies are tracked			
Electronically stored data are supported with back up files			

Appendix G

FEDERAL REGISTER (FDA): GOOD LABORATORY PRACTICE REGULATIONS FOR NONCLINICAL LABORATORY STUDIES

SUBPART A—GENERAL PROVISIONS

58.1. Scope

a. This part prescribes good laboratory practices for conducting nonclinical laboratory studies that support or are intended to support applications for research or marketing permits for products regulated by the Food and Drug Administration, including food and color additives, animal food additives, human and animal drugs, medical devices for human use, biological products, and electronic products. Compliance with this part is intended to assure the quality and integrity of the safety data filed pursuant to Secs. 406, 408, 409, 502, 503, 505, 506, 510, 512–516, 518–520, 721, and 801 of the Federal Food Drug, and Cosmetic Act and Secs. 351 and 354–360F of the Public Health Service Act.

b. References in this part to regulatory sections of the Code of Federal Regulations are to chapter I of title 21, unless otherwise noted.

[43 FR 60013, Dec. 22, 1978, as amended at 52 FR 33779, Sept. 4, 1987; 64 FR 399, Jan. 5, 1991]

58.3. Definitions

As used in this part, the following terms shall have the meanings specified:

a. *Act* means the Federal Food, Drug, and Cosmetic Act, as amended (Secs. 201–902, 52 Stat. 1040 *et seq.*, as amended (21 U.S.C. 321–392)).

b. *Test article* means any food additive, color additive, drug, biological product, electronic product, medical device for human use, or any other article subject to regulation under the act or under Secs. 351 and 354–360F of the Public Health Service Act.

c. *Control article* means any food additive, color additive, drug, biological product, electronic product, medical device for human use, or any article other than a test article, feed, or water that is administered to the test system in the course of a nonclinical laboratory study for the purpose of establishing a basis for comparison with the test article.

d. *Nonclinical laboratory study* means in vivo or in vitro experiments in which test articles are studied prospectively in test systems under laboratory conditions to determine their safety. The term does not include studies utilizing human subjects or clinical studies or field trials in animals. The term does not include basic exploratory studies carried out to determine whether a test article has any potential utility or to determine physical or chemical characteristics of a test article.

e. *Application for research or marketing permit* includes:
    1. A color additive petition, described in part 71.
    2. A food additive petition, described in parts 171 and 571.
    3. Data and information regarding a substance submitted as part of the procedures for establishing that a substance is generally recognized as safe for use, which use results or may reasonably be expected to result, directly or indirectly, in its becoming a component or otherwise affecting the characteristics of any food, described in Secs. 170.35 and 570.35.
    4. Data and information regarding a food additive submitted as part of the procedures regarding food additives permitted to be used on an interim basis pending additional study, described in Sec. 180.1.
    5. An *investigational new drug application (INDA)*, described in part 312 of this chapter.
    6. A *new drug application*, described in part 314.
    7. Data and information regarding an over-the-counter drug for human use, submitted as part of the procedures for classifying such drugs as generally recognized as safe and effective and not misbranded, described in part 330.
    8. Data and information about a substance submitted as part of the procedures for establishing a tolerance for unavoidable contaminants in food and food-packaging materials, described in parts 109 and 509.
    9. Data and information regarding an antibiotic drug submitted as part of the procedures for issuing, amending, or repealing regulations for such drugs, described in Sec. 314.300 of this chapter.
    10. A *Notice of Claimed Investigational Exemption for a New Animal Drug*, described in part 511.

11. A *new animal drug application*, described in part 514.
12. [Reserved]
13. An *application for a biologics license*, described in part 601 of this chapter.
14. An *application for an investigational device exemption*, described in part 812.
15. An *Application for Premarket Approval of a Medical Device*, described in Sec. 515 of the act.
16. A *Product Development Protocol for a Medical Device*, described in Sec. 515 of the act.
17. Data and information regarding a medical device submitted as part of the procedures for classifying such devices, described in part 860.
18. Data and information regarding a medical device submitted as part of the procedures for establishing, amending, or repealing a performance standard for such devices, described in part 861.
19. Data and information regarding an electronic product submitted as part of the procedures for obtaining an exemption from notification of a radiation safety defect or failure of compliance with a radiation safety performance standard, described in subpart D of part 1003.
20. Data and information regarding an electronic product submitted as part of the procedures for establishing, amending, or repealing a standard for such product, described in Sec. 358 of the Public Health Service Act.
21. Data and information regarding an electronic product submitted as part of the procedures for obtaining a variance from any electronic product performance standard as described in Sec. 1010.4.
22. Data and information regarding an electronic product submitted as part of the procedures for granting, amending, or extending an exemption from any electronic product performance standard, as described in Sec. 1010.5.

f. *Sponsor* means:

1. A person who initiates and supports, by provision of financial or other resources, a nonclinical laboratory study;
2. A person who submits a nonclinical study to the Food and Drug Administration in support of an application for a research or marketing permit; or
3. A testing facility, if it both initiates and actually conducts the study.

g. *Testing facility* means a person who actually conducts a nonclinical laboratory study, i.e., actually uses the test article in a test system. *Testing facility* includes any establishment required to register under Sec. 510 of the act that conducts nonclinical laboratory studies and any consulting laboratory described in Sec. 704 of the act that conducts such studies. *Testing facility* encompasses only those operational units that are being or have been used to conduct nonclinical laboratory studies.

h. *Person* includes an individual, partnership, corporation, association, scientific or academic establishment, government agency, or organizational unit thereof, and any other legal entity.

i. *Test system* means any animal, plant, microorganism, or subparts thereof to which the test or control article is administered or added for study. *Test system* also includes appropriate groups or components of the system not treated with the test or control articles.

j. *Specimen* means any material derived from a test system for examination or analysis.

k. *Raw data* means any laboratory worksheets, records, memoranda, notes, or exact copies thereof that are the result of original observations and activities of a nonclinical laboratory study and are necessary for the reconstruction and evaluation of the report of that study. In the event that exact transcripts of raw data have been prepared (e.g., tapes which have been transcribed verbatim, dated, and verified accurate by signature), the exact copy or exact transcript may be substituted for the original source as raw data. *Raw data* may include photographs, microfilm or microfiche copies, computer printouts, magnetic media, including dictated observations, and recorded data from automated instruments.

l. *Quality assurance unit* means any person or organizational element, except the study director, designated by testing facility management to perform the duties relating to quality assurance of nonclinical laboratory studies.

m. *Study director* means the individual responsible for the overall conduct of a nonclinical laboratory study.

n. *Batch* means a specific quantity or lot of a test or control article that has been characterized according to Sec. 58.105(a).

o. *Study initiation date* means the date the protocol is signed by the study director.

p. *Study completion date* means the date the final report is signed by the study director.

[43 FR 60013, Dec. 22, 1978, as amended at 52 FR 33779, Sept. 4, 1987; 54 FR 9039, Mar. 3, 1989; 64 FR 56448, Oct. 20, 1999]

58.10. Applicability to Studies Performed Under Grants and Contracts

When a sponsor conducting a nonclinical laboratory study intended to be submitted to or reviewed by the Food and Drug Administration utilizes the services of a consulting laboratory, contractor, or grantee to perform an analysis or other service, it shall notify the consulting laboratory, contractor, or grantee that the service is part of a nonclinical laboratory study that must be conducted in compliance with the provisions of this part.

58.15. Inspection of a Testing Facility

a. A testing facility shall permit an authorized employee of the Food and Drug Administration, at reasonable times and in a reasonable manner, to inspect the facility and to inspect (and in the case of records also to copy) all records and specimens required to be maintained regarding studies within the scope of this part. This records inspection and copying requirements shall not apply to quality assurance unit records of findings and problems, or to actions recommended and taken.

b. The Food and Drug Administration will not consider a nonclinical laboratory study in support of an application for a research or marketing permit if the testing facility refuses to that a nonclinical laboratory study will not be considered in support of an application for a research or marketing permit does not, however, relieve the applicant for such a permit of any obligation under any applicable statute or regulation to submit the results of the study to the Food and Drug Administration.

SUBPART B—ORGANIZATION AND PERSONNEL

58.29. Personnel

a. Each individual engaged in the conduct of or responsible for the supervision of a nonclinical laboratory study shall have education, training, and experience, or combination thereof to enable that individual to perform the assigned functions.

b. Each testing facility shall maintain a current summary of training and experience and job description for each individual engaged in or supervising the conduct of a nonclinical laboratory study.

c. There shall be a sufficient number of personnel for the timely and proper conduct of the study according to the protocol.

d. Personnel shall take necessary personal sanitation and health precautions designed to avoid contamination of test and control articles and test systems.

e. Personnel engaged in a nonclinical laboratory study shall wear clothing appropriate for the duties they perform. Such clothing shall be changed as often as necessary to prevent microbiological, radiological, or chemical contamination of test systems and test and control articles.

f. Any individual found at any time to have an illness that may adversely affect the quality and integrity of the nonclinical laboratory study shall be excluded from direct contact with test systems, test and control articles and any other operation or function that may adversely affect the study until the condition is corrected. All personnel shall be instructed to report to their immediate supervisors any health or medical conditions that may reasonably be considered to have an adverse effect on a nonclinical laboratory study.

58.31. Testing Facility Management

For each nonclinical laboratory study, testing facility management shall:

a. Designate a study director as described in Sec. 58.33, before the study is initiated.

b. Replace the study director promptly if it becomes necessary to do so during the conduct of a study.

c. Assure that there is a quality assurance unit as described in Sec. 58.35.

d. Assure that test and control articles or mixtures have been appropriately tested for identity, strength, purity, stability, and uniformity, as applicable.

e. Assure that personnel, resources, facilities, equipment, materials, and methodologies are available as scheduled.

f. Assure that personnel clearly understand the functions they are to perform.

g. Assure that any deviations from these regulations reported by the quality assurance unit are communicated to the study director and corrective actions are taken and documented.

[43 FR 60013, Dec. 22, 1978, as amended at 52 FR 33780, Sept. 4, 1987]

58.33. Study Director

For each nonclinical laboratory study, a scientist or other professional of appropriate education, training, and experience, or combination thereof shall be identified as the study director. The study director has overall responsibility for the technical conduct of the study, as well as for the interpretation, analysis, documentation, and reporting of results, and represents the single point of study control. The study director shall assure that:

a. The protocol, including any change, is approved as provided by Sec. 58.120 and is followed.

b. All experimental data including observations of unanticipated responses of the test system are accurately recorded and verified.

c. Unforeseen circumstances that may affect the quality and integrity of the nonclinical laboratory study are noted when they occur, and corrective action is taken and documented.

d. Test systems are as specified in the protocol.

e. All applicable good laboratory practice regulations are followed.

f. All raw data, documentation, protocols, specimens, and final reports are transferred to the archives during or at the close of the study.

[43 FR 60013, Dec. 22, 1978; 44 FR 17657, Mar. 23, 1979]

58.35. Quality Assurance Unit

a. A testing facility shall have a quality assurance unit, which shall be responsible for monitoring each study to assure management that the facilities, equipment, personnel, methods, practices, records, and controls are in conformance with the regulations in this part. For any given study, the quality assurance unit shall be entirely separate from and independent of the personnel engaged in the direction and conduct of that study.

b. The quality assurance unit shall:

1. Maintain a copy of a master schedule sheet of all nonclinical laboratory studies conducted at the testing facility indexed by test article and containing the test system, nature of study, date study was initiated, current status of each study, identity of the sponsor, and name of the study director.

2. Maintain copies of all protocols pertaining to all nonclinical laboratory studies for which the unit is responsible.

3. Inspect each nonclinical laboratory study at intervals adequate to assure the integrity of the study and maintain written and properly signed records of each periodic inspection showing the date of the inspection, the study inspected, the phase or segment of the study inspected, the person performing the inspection, findings and problems, action recommended and taken to resolve existing problems, and any scheduled date for reinspection. Any problems found during the course of an inspection, which are likely to affect study integrity, shall be brought to the attention of the study director and management immediately.

4. Periodically submit to management and the study director written status reports on each study, noting any problems and the corrective actions taken.

5. Determine that no deviations from approved protocols or standard operating procedures were made without proper authorization and documentation.

6. Review the final study report to assure that such report accurately describes the methods and standard operating procedures, and that the reported results accurately reflect the raw data of the nonclinical laboratory study.

7. Prepare and sign a statement to be included with the final study report which shall specify the dates inspections were made and findings reported to management and to the study director.

c. The responsibilities and procedures applicable to the quality assurance unit, the records maintained by the quality assurance unit, and the method of indexing such records shall be in writing and shall be maintained. These items including inspection dates, the study inspected, the phase or segment of the study inspected, and the name of the individual performing the inspection shall be made available for inspection to authorized employees of the Food and Drug Administration.

e. A designated representative of the Food and Drug Administration shall have access to the written procedures established for the inspection and may request testing facility management to certify that inspections are being implemented, performed, documented, and followed-up in accordance with this paragraph.

(Information collection requirements approved by the Office of Management and Budget under control number 0910–0203)

[43 FR 60013, Dec. 22, 1978, as amended at 52 FR 33780, Sept. 4, 1987]

SUBPART C—FACILITIES

58.41. General

Each testing facility shall be of suitable size and construction to facilitate the proper conduct of nonclinical laboratory studies. It shall be designed so that there is a degree of separation that will prevent any function or activity from having an adverse effect on the study.

[52 FR 33780, Sept. 4, 1987]

58.43. Animal Care Facilities

a. A testing facility shall have a sufficient number of animal rooms or areas, as needed, to assure proper: (1) separation of species or test systems, (2) isolation of individual projects, (3) quarantine of animals, and (4) routine or specialized housing of animals.

b. A testing facility shall have a number of animal rooms or areas separate from those described in paragraph (a) of this section to ensure isolation of studies being done with test systems or test and control articles known to be biohazardous, including volatile substances, aerosols, radioactive materials, and infectious agents.

c. Separate areas shall be provided, as appropriate, for the diagnosis, treatment, and control of laboratory animal diseases. These areas shall provide effective isolation for the housing of animals either known or suspected of being diseased, or of being carriers of disease, from other animals.

d. When animals are housed, facilities shall exist for the collection and disposal of all animal waste and refuse or for safe sanitary storage of waste before removal from the testing facility. Disposal facilities shall be so provided and operated as to minimize vermin infestation, odors, disease hazards, and environmental contamination.

[43 FR 60013, Dec. 22, 1978, as amended at 52 FR 33780, Sept. 4, 1987]

58.45. Animal Supply Facilities

There shall be storage areas, as needed, for feed, bedding, supplies, and equipment. Storage areas for feed and bedding shall be separated from areas housing the test systems and shall be protected against infestation or contamination. Perishable supplies shall be preserved by appropriate means.

[43 FR 60013, Dec. 22, 1978, as amended at 52 FR 33780, Sept. 4, 1987]

58.47. Facilities for Handling Test and Control Articles

a. As necessary to prevent contamination or mixups, there shall be separate areas for:

1. Receipt and storage of the test and control articles.
2. Mixing of the test and control articles with a carrier, e.g., feed.
3. Storage of the test and control article mixtures.

b. Storage areas for the test and/or control article and test and control mixtures shall be separate from areas housing the test systems and shall be adequate to preserve the identity, strength, purity, and stability of the articles and mixtures.

58.49. Laboratory Operation Areas

Separate laboratory space shall be provided, as needed, for the performance of the routine and specialized procedures required by nonclinical laboratory studies.

[52 FR 33780, Sept. 4, 1987]

58.51. Specimen and Data Storage Facilities

Space shall be provided for archives, limited to access by authorized personnel only, for the storage and retrieval of all raw data and specimens from completed studies.

SUBPART D—EQUIPMENT

58.61. Equipment Design

Equipment used in the generation, measurement, or assessment of data and equipment used for facility environmental control shall be of appropriate design and adequate capacity to function according to the protocol and shall be suitably located for operation, inspection, cleaning, and maintenance.

[52 FR 33780, Sept. 4, 1987]

58.63. Maintenance and Calibration of Equipment

a. Equipment shall be adequately inspected, cleaned, and maintained. Equipment used for the generation, measurement, or assessment of data shall be adequately tested, calibrated, and/or standardized.

b. The written standard operating procedures required under Sec. 58.81(b)(11) shall set forth in sufficient detail the methods, materials, and schedules to be used in the routine inspection, cleaning, maintenance, testing, calibration, and/or standardization of equipment, and shall specify, when appropriate, remedial action to be taken in the event of failure or malfunction of equipment. The written standard operating procedures shall designate the person responsible for the performance of each operation.

c. Written records shall be maintained of all inspection, maintenance, testing, calibrating, and/or standardizing operations. These records, containing the date of the operation, shall describe whether the maintenance operations were routine and followed the written standard operating procedures. Written records shall be kept of nonroutine repairs performed on equipment as a result of failure and malfunction. Such records shall document the nature of the defect, how and when the defect was discovered, and any remedial action taken in response to the defect.

(Information collection requirements approved by the Office of Management and Budget under control number 0910–0203)

[43 FR 60013, Dec. 22, 1978, as amended at 52 FR 33780, Sept. 4, 1987]

SUBPART E—TESTING FACILITIES OPERATION

58.81. Standard Operating Procedures

a. A testing facility shall have standard operating procedures in writing setting forth nonclinical laboratory study methods that management is satisfied are adequate to insure the quality and integrity of the data generated in the course of a study. All deviations in a study from standard operating procedures shall be authorized by the study director and shall be documented in the raw data. Significant changes in established standard operating procedures shall be properly authorized in writing by management.

b. Standard operating procedures shall be established for, but not limited to, the following:

1. Animal room preparation.
2. Animal care.
3. Receipt, identification, storage, handling, mixing, and method of sampling of the test and control articles.
4. Test system observations.
5. Laboratory tests.

6. Handling of animals found moribund or dead during study.
7. Necropsy of animals or postmortem examination of animals.
8. Collection and identification of specimens.
9. Histopathology.
10. Data handling, storage, and retrieval.
11. Maintenance and calibration of equipment.
12. Transfer, proper placement, and identification of animals.

c. Each laboratory area shall have immediately available laboratory manuals and standard operating procedures relative to the laboratory procedures being performed. Published literature may be used as a supplement to standard operating procedures.
d. A historical file of standard operating procedures, and all revisions thereof, including the dates of such revisions, shall be maintained.

[43 FR 60013, Dec. 22, 1978, as amended at 52 FR 33780, Sept. 4, 1987]

58.83. Reagents and Solutions

All reagents and solutions in the laboratory areas shall be labeled to indicate identity, titer or concentration, storage requirements, and expiration date. Deteriorated or outdated reagents and solutions shall not be used.

58.90. Animal Care

a. There shall be standard operating procedures for the housing, feeding, handling, and care of animals.
b. All newly received animals from outside sources shall be isolated and their health status shall be evaluated in accordance with acceptable veterinary medical practice.
c. At the initiation of a nonclinical laboratory study, animals shall be free of any disease or condition that might interfere with the purpose or conduct of the study. If, during the course of the study, the animals contract such a disease or condition, the diseased animals shall be isolated, if necessary. These animals may be treated for disease or signs of disease provided that such treatment does not interfere with the study. The diagnosis, authorizations of treatment, description of treatment, and each date of treatment shall be documented and shall be retained.
d. Warm-blooded animals, excluding suckling rodents, used in laboratory procedures that require manipulations and observations over an extended period of time or in studies that require the animals to be removed from and returned to their home cages for any reason (e.g., cage cleaning, treatment, etc.) shall receive appropriate identification. All information needed to specifically identify each animal within an animal-housing unit shall appear on the outside of that unit.
e. Animals of different species shall be housed in separate rooms when necessary. Animals of the same species, but used in different studies,

should not ordinarily be housed in the same room when inadvertent exposure to control or test articles or animal mixup could affect the outcome of either study. If such mixed housing is necessary, adequate differentiation by space and identification shall be made.

f. Animal cages, racks, and accessory equipment shall be cleaned and sanitized at appropriate intervals.

g. Feed and water used for the animals shall be analyzed periodically to ensure that contaminants known to be capable of interfering with the study and reasonably expected to be present in such feed or water are not present at levels above those specified in the protocol. Documentation of such analyses shall be maintained as raw data.

h. Bedding used in animal cages or pens shall not interfere with the purpose or conduct of the study and shall be changed as often as necessary to keep the animals dry and clean.

i. If any pest control materials are used, the use shall be documented. Cleaning and pest control materials that interfere with the study shall not be used.

(Information collection requirements approved by the Office of Management and Budget under control number 0910–0203)

[43 FR 60013, Dec. 22, 1978, as amended at 52 FR 33780, Sept. 4, 1987; 54 FR 15924, Apr. 20, 1989; 56 FR 32088, July 15, 1991]

SUBPART F—TEST AND CONTROL ARTICLES

58.105. Test and Control Article Characterization

a. The identity, strength, purity, and composition or other characteristics which will appropriately define the test or control article shall be determined for each batch and shall be documented. Methods of synthesis, fabrication, or derivation of the test and control articles shall be documented by the sponsor or the testing facility. In those cases where marketed products are used as control articles, such products will be characterized by their labeling.

b. The stability of each test or control article shall be determined by the testing facility or by the sponsor either: (1) before study initiation, or (2) concomitantly according to written standard operating procedures, which provide for periodic analysis of each batch.

c. Each storage container for a test or control article shall be labeled by name, chemical abstract number or code number, batch number, expiration date, if any, and, where appropriate, storage conditions necessary to maintain the identity, strength, purity, and composition of the test or control article. Storage containers shall be assigned to a particular test article for the duration of the study.

d. For studies of more than 4 weeks' duration, reserve samples from each batch of test and control articles shall be retained for the period of time provided by Sec. 58.195.

(Information collection requirements approved by the Office of Management and Budget under control number 0910–0203)

[43 FR 60013, Dec. 22, 1978, as amended at 52 FR 33781, Sept. 4, 1987]

58.107. Test and Control Article Handling

Procedures shall be established for a system for the handling of the test and control articles to ensure that:

a. There is proper storage.
b. Distribution is made in a manner designed to preclude the possibility of contamination, deterioration, or damage.
c. Proper identification is maintained throughout the distribution process.
d. The receipt and distribution of each batch is documented. Such documentation shall include the date and quantity of each batch distributed or returned.

58.113. Mixtures of Articles With Carriers

a. For each test or control article that is mixed with a carrier, tests by appropriate analytical methods shall be conducted:

1. To determine the uniformity of the mixture and to determine, periodically, the concentration of the test or control article in the mixture.
2. To determine the stability of the test and control articles in the mixture as required by the conditions of the study either:

    i. Before study initiation, or
    ii. Concomitantly according to written standard operating procedures, which provide for periodic analysis of the test and control articles in the mixture.

b. [Reserved]
c. Where any of the components of the test or control article carrier mixture has an expiration date, that date shall be clearly shown on the container. If more than one component has an expiration date, the earliest date shall be shown.

[43 FR 60013, Dec. 22, 1978, as amended at 45 FR 24865, Apr. 11, 1980; 52 FR 33781, Sept. 4, 1987]

SUBPART G—PROTOCOL FOR AND CONDUCT OF A NONCLINICAL LABORATORY STUDY

58.120. Protocol

a. Each study shall have an approved written protocol that clearly indicates the objectives and all methods for the conduct of the study. The protocol shall contain, as applicable, the following information:

1. A descriptive title and statement of the purpose of the study.
2. Identification of the test and control articles by name, chemical abstract number, or code number.
3. The name of the sponsor and the name and address of the testing facility at which the study is being conducted.
4. The number, body weight range, sex, source of supply, species, strain, substrain, and age of the test system.
5. The procedure for identification of the test system.
6. A description of the experimental design, including the methods for the control of bias.
7. A description and/or identification of the diet used in the study as well as solvents, emulsifiers, and/or other materials used to solubilize or suspend the test or control articles before mixing with the carrier. The description shall include specifications for acceptable levels of contaminants that are reasonably expected to be present in the dietary materials and are known to be capable of interfering with the purpose or conduct of the study if present at levels greater than established by the specifications.
8. Each dosage level, expressed in milligrams per kilogram of body weight or other appropriate units, of the test or control article to be administered and the method and frequency of administration.
9. The type and frequency of tests, analyses, and measurements to be made.
10. The records to be maintained.
11. The date of approval of the protocol by the sponsor and the dated signature of the study director.
12. A statement of the proposed statistical methods to be used.

b. All changes in or revisions of an approved protocol and the reasons therefore shall be documented, signed by the study director, dated, and maintained with the protocol.

(Information collection requirements approved by the Office of Management and Budget under control number 0910–0203)

[43 FR 60013, Dec. 22, 1978, as amended at 52 FR 33781, Sept. 4, 1987]

58.130. Conduct of a Nonclinical Laboratory Study

a. The nonclinical laboratory study shall be conducted in accordance with the protocol.
b. The test systems shall be monitored in conformity with the protocol.
c. Specimens shall be identified by test system, study, nature, and date of collection. This information shall be located on the specimen

container or shall accompany the specimen in a manner that pre-
cludes error in the recording and storage of data.

d. Records of gross findings for a specimen from postmortem observations should be available to a pathologist when examining that specimen histopathologically.

e. All data generated during the conduct of a nonclinical laboratory
study, except those that are generated by automated data collection
systems, shall be recorded directly, promptly, and legibly in ink.
All data entries shall be dated on the date of entry and signed or initialed by the person entering the data. Any change in entries shall be
made so as not to obscure the original entry, shall indicate the
reason for such change, and shall be dated and signed or identified
at the time of the change. In automated data collection systems, the
individual responsible for direct data input shall be identified at the
time of data input. Any change in automated data entries shall be
made so as not to obscure the original entry, shall indicate the reason for change, shall be dated, and the responsible individual shall
be identified.

(Information collection requirements approved by the Office of Management
and Budget under control number 0910–0203)

[43 FR 60013, Dec. 22, 1978, as amended at 52 FR 33781, Sept. 4, 1987]

SUBPARTS H—I [RESERVED]

SUBPART J—RECORDS AND REPORTS

58.185. Reporting of Nonclinical Laboratory Study Results

a. A final report shall be prepared for each nonclinical laboratory study
and shall include, but not necessarily be limited to, the following:

1. Name and address of the facility performing the study and the
dates on which the study was initiated and completed.
2. Objectives and procedures stated in the approved protocol,
including any changes in the original protocol.
3. Statistical methods employed for analyzing the data.
4. The test and control articles identified by name, chemical
abstracts number or code number, strength, purity, and composition or other appropriate characteristics.
5. Stability of the test and control articles under the conditions of
administration.
6. A description of the methods used.

7. A description of the test system used. Where applicable, the final report shall include the number of animals used, sex, body weight range, source of supply, species, strain and sub-strain, age, and procedure used for identification.
8. A description of the dosage, dosage regimen, route of administration, and duration.
9. A description of all circumstances that may have affected the quality or integrity of the data.
10. The name of the study director, the names of other scientists or professionals, and the names of all supervisory personnel, involved in the study.
11. A description of the transformations, calculations, or operations performed on the data, a summary and analysis of the data, and a statement of the conclusions drawn from the analysis.
12. The signed and dated reports of each of the individual scientists or other professionals involved in the study.
13. The locations where all specimens, raw data, and the final report are to be stored.
14. The statement prepared and signed by the quality assurance unit as described in Sec. 58.35(b)(7).

b. The final report shall be signed and dated by the study director.
c. Corrections or additions to a final report shall be in the form of an amendment by the study director. The amendment shall clearly identify that part of the final report that is being added to or corrected and the reasons for the correction or addition, and shall be signed and dated by the person responsible.

[43 FR 60013, Dec. 22, 1978, as amended at 52 FR 33781, Sept. 4, 1987]

58.190. Storage and Retrieval of Records and Data

a. All raw data, documentation, protocols, final reports, and specimens (except those specimens obtained from mutagenicity tests and wet specimens of blood, urine, feces, and biological fluids) generated as a result of a nonclinical laboratory study shall be retained.
b. There shall be archives for orderly storage and expedient retrieval of all raw data, documentation, protocols, specimens, and interim and final reports. Conditions of storage shall minimize deterioration of the documents or specimens in accordance with the requirements for the time period of their retention and the nature of the documents or specimens. A testing facility may contract with commercial archives to provide a repository for all material to be retained. Raw data and specimens may be retained elsewhere provided that the archives have specific reference to those other locations.
c. An individual shall be identified as responsible for the archives.
d. Only authorized personnel shall enter the archives.

e. Material retained or referred to in the archives shall be indexed to permit expedient retrieval.

(Information collection requirements approved by the Office of Management and Budget under control number 0910–0203).

[43 FR 60013, Dec. 22, 1978, as amended at 52 FR 33781, Sept. 4, 1987]

58.195. Retention of Records

a. Record retention requirements set forth in this section do not supersede the record retention requirements of any other regulations in this chapter.

b. Except as provided in paragraph (c) of this section, documentation records, raw data and specimens pertaining to a nonclinical laboratory study and required to be made by this part shall be retained in the archive(s) for whichever of the following periods is shortest:

1. A period of at least 2 years following the date on which an application for a research or marketing permit, in support of which the results of the nonclinical laboratory study were submitted, is approved by the Food and Drug Administration. This requirement does not apply to studies supporting INDAs or applications for investigational device exemptions (IDEs), and records of which shall be governed by the provisions of paragraph (b)(2) of this section.

2. A period of at least 5 years following the date on which the results of the nonclinical laboratory study are submitted to the Food and Drug Administration in support of an application for a research or marketing permit.

3. In other situations (e.g., where the nonclinical laboratory study does not result in the submission of the study in support of an application for a research or marketing permit), a period of at least 2 years following the date on which the study is completed, terminated, or discontinued.

c. Wet specimens (except those specimens obtained from mutagenicity tests and wet specimens of blood, urine, feces, and biological fluids), samples of test or control articles, and specially prepared material, which are relatively fragile and differ markedly in stability and quality during storage, shall be retained only as long as the quality of the preparation affords evaluation. In no case shall retention be required for longer periods than those set forth in paragraphs (a) and (b) of this section.

d. The master schedule sheet, copies of protocols, and records of quality assurance inspections as required by Sec. 58.35(c) shall be maintained by the quality assurance unit as an easily accessible system of records for the period of time specified in paragraphs (a) and (b) of this section.

e. Summaries of training and experience and job descriptions required to be maintained by Sec. 58.29(b) may be retained along with all other testing facility employment records for the length of time specified in paragraphs (a) and (b) of this section.

f. Records and reports of the maintenance and calibration and inspection of equipment, as required by Sec. 58.63(b) and (c), shall be retained for the length of time specified in paragraph (b) of this section.

g. Records required by this part may be retained either as original records or as true copies such as photocopies, microfilm, microfiche, or other accurate reproductions of the original records.

h. If a facility conducting nonclinical testing goes out of business, all raw data, documentation, and other material specified in this section shall be transferred to the archives of the sponsor of the study. The Food and Drug Administration shall be notified in writing of such a transfer.

[43 FR 60013, Dec. 22, 1978, as amended at 52 FR 33781, Sept. 4, 1987; 54 FR 9039, Mar. 3, 1989]

SUBPART K—DISQUALIFICATION OF TESTING FACILITIES

58.200. Purpose

a. The purposes of disqualification are:

1. To permit the exclusion from consideration of completed studies that were conducted by a testing facility which has failed to comply with the requirements of the good laboratory practice regulations until it can be adequately demonstrated that such noncompliance did not occur during, or did not affect the validity or acceptability of data generated by, a particular study; and

2. To exclude from consideration all studies completed after the date of disqualification until the facility can satisfy the Commissioner that it will conduct studies in compliance with such regulations.

b. The determination that a nonclinical laboratory study may not be considered in support of an application for a research or marketing permit does not, however, relieve the applicant for such a permit of any obligation under any other applicable regulation to submit the results of the study to the Food and Drug Administration.

58.202. Grounds for Disqualification

The Commissioner may disqualify a testing facility upon finding all of the following:

a. The testing facility failed to comply with one or more of the regulations set forth in this part (or any other regulations regarding such facilities in this chapter);

b. The noncompliance adversely affected the validity of the nonclinical laboratory studies; and

c. Other lesser regulatory actions (e.g., warnings or rejection of individual studies) have not been or will probably not be adequate to achieve compliance with the good laboratory practice regulations.

58.204. Notice of and Opportunity for Hearing on Proposed Disqualification

a. Whenever the Commissioner has information indicating that grounds exist under Sec. 58.202, which in his opinion justify disqualification of a testing facility, he may issue to the testing facility a written notice proposing that the facility be disqualified.

b. A hearing on the disqualification shall be conducted in accordance with the requirements for a regulatory hearing set forth in part 16 of this chapter.

58.206. Final Order on Disqualification

a. If the Commissioner, after the regulatory hearing, or after the time for requesting a hearing expires without a request being made, upon an evaluation of the administrative record of the disqualification proceeding, makes the findings required in Sec. 58.202, he shall issue a final order disqualifying the facility. Such order shall include a statement of the basis for that determination. Upon issuing a final order, the Commissioner shall notify (with a copy of the order) the testing facility of the action.

b. If the Commissioner, after a regulatory hearing or after the time for requesting a hearing expires without a request being made, upon an evaluation of the administrative record of the disqualification proceeding, does not make the findings required in Sec. 58.202, he shall issue a final order terminating the disqualification proceeding. Such order shall include a statement of the basis for that determination. Upon issuing a final order, the Commissioner shall notify the testing facility and provide a copy of the order.

58.210. Actions upon Disqualification

a. Once a testing facility has been disqualified, each application for a research or marketing permit, whether approved or not, containing or relying upon any nonclinical laboratory study conducted by the disqualified testing facility may be examined to determine whether such study was or would be essential to a decision. If it is determined that a study was or would be essential, the Food and Drug Administration shall also determine whether the study is acceptable, notwithstanding the disqualification of the facility. Any study done by a testing facility before or after disqualification may be presumed to be unacceptable, and the person relying on the study may be required to establish that the study was not affected by the circumstances that led to the disqualification, e.g., by submitting validating information. If the study is then determined to be unacceptable, such data will be eliminated from consideration in support of the application; and such elimination may serve as new information justifying the termination or withdrawal of approval of the application.

b. No nonclinical laboratory study begun by a testing facility after the date of the facility's disqualification shall be considered in support of any application for a research or marketing permit, unless the facility has been reinstated under Sec. 58.219. The determination that a study may not be considered in support of an application for a research or marketing permit does not, however, relieve the applicant for such a permit of any obligation under any other applicable regulation to submit the results of the study to the Food and Drug Administration.

[43 FR 60013, Dec. 22, 1978, as amended at 59 FR 13200, Mar. 21, 1994]

58.213. Public Disclosure of Information Regarding Disqualification

a. Upon issuance of a final order disqualifying a testing facility under Sec. 58.206(a), the Commissioner may notify all or any interested persons. Such notice may be given at the discretion of the Commissioner whenever he believes that such disclosure would further the public interest or would promote compliance with the good laboratory practice regulations set forth in this part. Such notice, if given, shall include a copy of the final order issued under Sec. 58.206(a) and shall state that the disqualification constitutes a determination by the Food and Drug Administration that nonclinical laboratory studies performed by the facility will not be considered by the Food and Drug Administration in support of any application for a research or marketing permit. If such notice is sent to another Federal Government agency, the Food and Drug Administration will recommend that the agency also consider whether or not it should accept nonclinical laboratory studies performed by the testing facility. If such notice is sent to any other person, it shall state that it is given because of the relationship between the testing facility and the person being notified and that the Food and Drug Administration is not advising or recommending that any action be taken by the person notified.

b. A determination that a testing facility has been disqualified and the administrative record regarding such determination are disclosable to the public under part 20 of this chapter.

58.215. Alternative or Additional Actions to Disqualification

a. Disqualification of a testing facility under this subpart is independent of, and neither in lieu of nor a precondition to, other proceedings or actions authorized by the act. The Food and Drug Administration may, at any time, institute against a testing facility and/or against the sponsor of a nonclinical laboratory study that has been submitted to the Food and Drug Administration any appropriate judicial proceedings (civil or criminal) and any other appropriate regulatory action, in addition to or in lieu of, and prior to, simultaneously with, or subsequent to, disqualification. The Food and Drug Administration may also refer the matter to another Federal, State, or local government law enforcement or regulatory agency for such action as that agency deems appropriate.

b. The Food and Drug Administration may refuse to consider any particular nonclinical laboratory study in support of an application for a research or marketing permit, if it finds that the study was not conducted in accordance with the good laboratory practice regulations set forth in this part, without disqualifying the testing facility that conducted the study or undertaking other regulatory action.

58.217. Suspension or Termination of a Testing Facility by a Sponsor

Termination of a testing facility by a sponsor is independent of, and neither in lieu of nor a precondition to, proceedings or actions authorized by this subpart. If a sponsor terminates or suspends a testing facility from further participation in a nonclinical laboratory study that is being conducted as part of any application for a research or marketing permit that has been submitted to any Center of the Food and Drug Administration (whether approved or not), it shall notify that Center in writing within 15 working days of the action; the notice shall include a statement of the reasons for such action. Suspension or termination of a testing facility by a sponsor does not relieve it of any obligation under any other applicable regulation to submit the results of the study to the Food and Drug Administration.

[43 FR 60013, Dec. 22, 1978, as amended at 50 FR 8995, Mar. 6, 1985]

58.219. Reinstatement of a Disqualified Testing Facility

A testing facility that has been disqualified may be reinstated as an acceptable source of nonclinical laboratory studies to be submitted to the Food and Drug Administration if the Commissioner determines, upon an evaluation of the submission of the testing facility, that the facility can adequately assure that it will conduct future nonclinical laboratory studies in compliance with the good laboratory practice regulations set forth in this part and, if any studies are currently being conducted, that the quality and integrity of such studies have not been seriously compromised. A disqualified testing facility that wishes to be so reinstated shall present in writing to the Commissioner reasons why it believes it should be reinstated and a detailed description of the corrective actions it has taken or intends to take to assure that the acts or omissions which led to its disqualification will not recur. The Commissioner may condition reinstatement upon the testing facility being found in compliance with the good laboratory practice regulations upon an inspection. If a testing facility is reinstated, the Commissioner shall so notify the testing facility and all organizations and persons who were notified, under Sec. 58.213. of the disqualification of the testing facility. A determination that a testing facility has been reinstated is disclosable to the public under part 20 of this chapter.

BIBLIOGRAPHY

1. Aboul-Enein HY, Stefan RI, Baiulescu GE. Quality and Reliability in Analytical Chemistry. Boca Raton, FL: CRC Press, 2000.
2. Alvarez RJ, et al. QC in microbiology labs: essential for productivity. ASQC Congress Transactions, Detroit, 1982:737–742.
3. American Public Health Association. Standard Methods for the Examination of Water and Wastewater. American Public Health Association, Washington, DC, 1999.
4. American Society for Quality Control, Statistics Division. Glossary and tables for statistical quality control. American Society for Quality Control, Milwaukee, WI, 1983.
5. Association of Official Analytical Chemists. Official Methods of Analysis, volumes 1 and 2. Association of Official Analytical Chemists, Arlington, VA, 1990.
6. Baiulescu GE. Crit Rev Anal Chem 1987; 17:317.
7. Baiulescu GE, Dumitrescu P, Zugravescu PGh. Sampling. Ellis Horwood, Chichester, 1991.
8. Belk WP, Sunderman FW. Survey of the accuracy of clinical analysis in clinical laboratories. Am J Clin Pathol 1947; 77:858–861.
9. Bell MR. Laboratory Accreditation and Quality System Accreditation—A Merging of the Ways. STP 1057. Philadelphia, PA: American Society for Testing and Materials, 1989:120–143.
10. Belsky JS. Design of HVAC systems for laboratories. Pharm Eng 1991; 11(4):31–35.
11. Boothe R. Who defines quality in service industries? Qual Prog 1990; 23(2): 65–67.
12. Borghese RN. Quality auditing as a tool. BioPharm 1990; 10:14–16.
13. Bossert J. Procurement Quality Control. Milwaukee, WI: Quality Press, 1988.
14. Bossert J. Quality Function Deployment. Milwaukee, WI: Quality Press, 1990.
15. Brown SD, Bear RS Jr. Crit Rev Anal Chem 1993; 24:99.

16. Butler EH. (1986) Speech at Annual Meeting of Society for Quality Assurance (1985). Reprinted in Soc Qual Assur Newsletter.
17. Cadner RY. GMP compliance auditing. Pharm Technol 1984; 13(6):36–38.
18. Chapman KG. Validating to protect the competitive edge. Sci Comput Autom 1990; 3:5.
19. Christian GD. Analytical Chemistry. 5th ed. New York: John Wiley & Sons, 1994.
20. Computer System Validation Committee. Validation concepts for computer systems used in the manufacture of drug products. Pharm Technol 1986; 10(5):24–34.
21. Daniel A. Proficiency testing. Med Dev Diag Indus 1990; 12(5):34–37.
22. Day RG. Quality Function Deployment: Linking a Company with Its Customers. Milwaukee, WI: Quality Press, 1993.
23. Department of Health and Human Services, and National Institute of Health. Biosafety in Microbiological and Biomedical Laboratories. U.S. Government Printing Office, Washington, DC, 1984.
24. Environmental Protection Agency. Manual for the Interim Certification of Laboratories Involved in Analyzing Public Drinking Water Supplies—Criteria and Procedures. U.S. Department of Commerce PB-287 118, Washington, DC, 1978.
25. Environmental Protection Agency. Drinking Water Regulations Under the Safe Drinking Water Act, Fact Sheet. Criteria and Standards Division, Office of Drinking Water, Environmental Protection Agency, Washington, DC, 1988.
26. Environmental Protection Agency. Good Automated Laboratory Practices Implementation Manual—Draft. Environmental Protection Agency, IRM, Research Triangle Park, NC, 1990.
27. Etnyre-Zacher P, Miller SM. An educational program for physicians' office laboratory personnel. Am Clin Lab 1990; 9(1):10–17.
28. Federal Register. Human and veterinary drugs: good manufacturing practices and proposed exemptions for certain OTC products 1978; 43(190): 45014–45089.
29. Federal Register. Nonclinical laboratory studies: good laboratory practices regulations 1978; 43(247):59986–60025.
30. Federal Register. Current good manufacturing practice in manufacturing, processing, packing, or holding human food 1979; 44(112):33238–33248.
31. Federal Register. Proposed rules: Medicare, Medicaid and CLIA Programs; revision of the clinical laboratory regulations for the Medicare, Medicaid and the Clinical Laboratories Improvement Act of 1967 Programs 1988; 53:151.
32. Food and Drug Administration. FDA Investigations Operations Manual. Washington, DC: U.S. Government Printing Office, 2003.
33. Food and Drug Administration Compliance Program Guidance Manual. Good Laboratory Practice (Nonclinical Laboratories). FDA No. 7348.808, 1991.
34. Garfield FM. Quality assurance principles for analytical laboratories. Association of Official Analytical Chemists, Inc., Arlington, VA, 1984.
35. Gladhill RL. Advantages of laboratory accreditation. ASTM STP 1057. American Society for Testing and Materials, Philadelphia, PA, 1989:19–23.

36. Guerra J. Validation of analytical methods by FDA laboratories. Pharm Technol 1986; 10(3):74–84.
37. Health, Education and Welfare. Requirements of laws and regulations enforced by the U.S. Food and Drug Administration. Washington, DC: HEW Publication No. (FDA) 79–1042, 1980.
38. Henry JB. Clinical Diagnosis and Management. Chapters 6, 53, and 55. Philadelphia, PA: W.B. Saunders Co, 1984.
39. Illinois State EPA. Certification and Operation of Environmental Laboratories. Title 35, Subtitle A, Chapter II, Part 183, 1983.
40. ISO. Guide to the Expression of Uncertainty in Measurement. ISO, Geneva, Switzerland, 1993.
41. ISO/IEC 17025. General Requirements for the Competence of Testing and Calibration Laboratories, 1999.
42. Jenemann HR. The Chemist's Balance. DECHEMA, Frankfurt, 1997.
43. Johnson RW. Quality assurance of tissue culture media used in the biotechnology industry. BioPharm 1990; 2:40–44.
44. Lord T. Microbes and cGMPs. Pharm Technol 1989; 13(6):36–38.
45. Marquardt D, et al. Vision 2000: the strategy for ISO 9000 series standards in the 90s. Qual Prog 1991; 24(5):25–31.
46. Miller JM, Wentworth BB, eds. Methods for Quality Control in Diagnostic Microbiology. American Public Health Association, Washington, DC, 1985.
47. Mills CA. The Quality Audit. Milwaukee, WI: Quality Press and McGraw Hill, Inc., 1989.
48. Motiska PJ, Shilliff KA. Ten precepts of quality. Qual Prog 1990; 23(2):27–28.
49. Mueller N. Introducing the concept of uncertainty f measurement in testing in association with the application of the standard ISO/IEC 17025, 2002.
50. National Committee for Clinical Laboratory Standards. Selecting and Evaluating a Referral Laboratory. NCCLS Document GP9-P, Villanova, PA, 1985.
51. National Committee for Clinical Laboratory Standards. Labeling of Laboratory Prepared Materials. NCCLS Document GP4-P, Villanova, PA, 1988.
52. National Committee for Clinical Laboratory Standards. Preparation and Testing of Reagent Water in the Clinical Laboratory. NCCLS Document C3-T2, Villanova, PA, 1988.
53. National Committee for Clinical Laboratory Standards. Quality Assurance for Commercially Prepared Microbiological Culture Media. NCCLS Document MJ22-A, Villanova, PA, 1990.
54. Organisation for Economic Cooperation and Development. Good Laboratory Practice in the Testing of Chemicals. Organisation for Economic Cooperation and Development, 1990.
55. Parenteral Drug Association. A Proposed Training Model for the Microbiological Function in the Pharmaceutical Industry. PDA, Bethesda, MD, 2001.
56. Pharmaceutical Manufacturer's Association's Computer System Validation Committee. Computer system validation—staying current: change control. Pharm Technol 1990; 14(1):20–75.
57. Pulido A, Ruisanchez I, Boque R, Rius FX. Uncertainty of results in routine qualitative analysis. Trends Anal Chem 2003; 22(10):647.

58. Pyell U. Basic Course Experiments to demonstrate intercomparisons. In: Neidhart B, Wegscheider W, eds. Quality in Chemical Measurements. Berlin, Germany: Springer-Verlag, 2001.
59. Ratliff T. The Laboratory Quality Assurance System. New York, NY: Van Nostrand Rein-Hold, 1990.
60. Robinson CB. Auditing a Quality System for the Defense Industry. Milwaukee, WI: Quality Press, 1990.
61. Robinson CB. Auditing a quality system. Qual Prog 1990; 4:49–52.
62. Rosander AC. The quest for quality in services. Milwaukee, WI: Quality Press, 1989.
63. Schock HE, ed. Accreditation Practices for Inspections, Tests, and Laboratories. ASTM STP 1057. Philadelphia, PA: American Society for Testing and Materials, 1989.
64. Singer DC, ed. A Laboratory Quality Handbook of Best Practices and Relevant Regulations. Milwaukee, WI: Quality Press, 2001.
65. Singer DC, Upton RP. Guidelines for Laboratory Quality Auditing. New York (NY): Milwaukee (WI), Marcel Dekker, Quality Press, 1993.
66. Singer DC, et al. Standard operating procedures for the microbiology laboratory. Cosm Technol 1981; 3:40–49.
67. Snyder JW. Quality control in clinical microbiology. API Species 5:2. Analytab Products Inc., Plainview, NY: 1981:13–23.
68. Speck ML. Compendium for the Microbiological Examination of Foods. Washington, DC: American Public Health Association, 1984:13–23.
69. Stefan RI, Bairu SG. Monocrystalline diamond paste based electrodes and their applications for the determination of Fe(II) in vitamins. Anal Chem 2003; 75(20):5394.
70. Taylor JK. Handbook for SRM Users. NBS Publication 260-100, National Institute of Technology and Standards. Washington, DC, 1985.
71. Taylor JK. Quality Assurance of Chemical Measurements. Chelsea, MI: Lewis Publishers, 1989.
72. United States Pharmacopeia XXII. National Formulary XVII. Unites States Pharmacopeial Convention, Rockville, MD, 1990.
73. Wadsworth Center for Laboratories and Research. Proficiency Testing Program. New York State Department of Health, Albany, NY, 1989.
74. WHO Technical Report Series No. 968. Good Manufacturing Practices for Pharmaceutical Products. WHO, Geneva, Switzerland, 2003.
75. Willborn W. Audit Standards—A Comparative Analysis. Milwaukee, WI: Quality Press, 1987.
76. Willborn W. Compendium of Audit Standards. Milwaukee, WI: Quality Press, 1983.

INDEX

9 780367 392468